XXXXXXXXXXXXXXXXXXXXXXXXXXXXXXXXXXXXXXXX
TTTTTTTTTTTTTTTTTTTTTTTTTTTTTTTTTTTTTTTT
RRRRRRRRRRRRRRRRRRRRRRRRRRRRRRRRRRRRRRRR
EEEEEEEEEEEEEEEEEEEEEEEEEEEEEEEEEEEEEEEE
MMMMMMMMMMMMMMMMMMMMMMMMMMMMMMMMMMMMMMMM
EEEEEEEEEEEEEEEEEEEEEEEEEEEEEEEEEEEEEEEE
TTTTTTTTTTTTTTTTTTTTTTTTTTTTTTTTTTTTTTTT
EEEEEEEEEEEEEEEEEEEEEEEEEEEEEEEEEEEEEEEE
XXXXXXXXXXXXXXXXXXXXXXXXXXXXXXXXXXXXXXXX
TTTTTTTTTTTTTTTTTTTTTTTTTTTTTTTTTTTTTTTT
IIIIIIIIIIIIIIIIIIIIIIIIIIIIIIIIIIIIIIII
LLLLLLLLLLLLLLLLLLLLLLLLLLLLLLLLLLLLLLLL
EEEEEEEEEEEEEEEEEEEEEEEEEEEEEEEEEEEEEEEE
SSSSSSSSSSSSSSSSSSSSSSSSSSSSSSSSSSSSSSSS
EEEEEEEEEEEEEEEEEEEEEEEEEEEEEEEEEEEEEEEE
XXXXXXXXXXXXXXXXXXXXXXXXXXXXXXXXXXXXXXXX
TTTTTTTTTTTTTTTTTTTTTTTTTTTTTTTTTTTTTTTT
RRRRRRRRRRRRRRRRRRRRRRRRRRRRRRRRRRRRRRRR
EEEEEEEEEEEEEEEEEEEEEEEEEEEEEEEEEEEEEEEE
MMMMMMMMMMMMMMMMMMMMMMMMMMMMMMMMMMMMMMMM
EEEEEEEEEEEEEEEEEEEEEEEEEEEEEEEEEEEEEEEE
TTTTTTTTTTTTTTTTTTTTTTTTTTTTTTTTTTTTTTTT
EEEEEEEEEEEEEEEEEEEEEEEEEEEEEEEEEEEEEEEE
XXXXXXXXXXXXXXXXXXXXXXXXXXXXXXXXXXXXXXXX
TTTTTTTTTTTTTTTTTTTTTTTTTTTTTTTTTTTTTTTT
IIIIIIIIIIIIIIIIIIIIIIIIIIIIIIIIIIIIIIII
LLLLLLLLLLLLLLLLLLLLLLLLLLLLLLLLLLLLLLLL
EEEEEEEEEEEEEEEEEEEEEEEEEEEEEEEEEEEEEEEE
SSSSSSSSSSSSSSSSSSSSSSSSSSSSSSSSSSSSSSSS
EEEEEEEEEEEEEEEEEEEEEEEEEEEEEEEEEEEEEEEE
XXXXXXXXXXXXXXXXXXXXXXXXXXXXXXXXXXXXXXXX
TTTTTTTTTTTTTTTTTTTTTTTTTTTTTTTTTTTTTTTT
RRRRRRRRRRRRRRRRRRRRRRRRRRRRRRRRRRRRRRRR
EEEEEEEEEEEEEEEEEEEEEEEEEEEEEEEEEEEEEEEE
MMMMMMMMMMMMMMMMMMMMMMMMMMMMMMMMMMMMMMMM
EEEEEEEEEEEEEEEEEEEEEEEEEEEEEEEEEEEEEEEE
TTTTTTTTTTTTTTTTTTTTTTTTTTTTTTTTTTTTTTTT
EEEEEEEEEEEEEEEEEEEEEEEEEEEEEEEEEEEEEEEE
XXXXXXXXXXXXXXXXXXXXXXXXXXXXXXXXXXXXXXXX
TTTTTTTTTTTTTTTTTTTTTTTTTTTTTTTTTTTTTTTT
IIIIIIIIIIIIIIIIIIIIIIIIIIIIIIIIIIIIIIII
LLLLLLLLLLLLLLLLLLLLLLLLLLLLLLLLLLLLLLLL
EEEEEEEEEEEEEEEEEEEEEEEEEEEEEEEEEEEEEEEE
SSSSSSSSSSSSSSSSSSSSSSSSSSSSSSSSSSSSSSSS
EEEEEEEEEEEEEEEEEEEEEEEEEEEEEEEEEEEEEEEE
XXXXXXXXXXXXXXXXXXXXXXXXXXXXXXXXXXXXXXXX
TTTTTTTTTTTTTTTTTTTTTTTTTTTTTTTTTTTTTTTT
RRRRRRRRRRRRRRRRRRRRRRRRRRRRRRRRRRRRRRRR
EEEEEEEEEEEEEEEEEEEEEEEEEEEEEEEEEEEEEEEE
MMMMMMMMMMMMMMMMMMMMMMMMMMMMMMMMMMMMMMMM
EEEEEEEEEEEEEEEEEEEEEEEEEEEEEEEEEEEEEEEE
TTTTTTTTTTTTTTTTTTTTTTTTTTTTTTTTTTTTTTTT
EEEEEEEEEEEEEEEEEEEEEEEEEEEEEEEEEEEEEEEE
XXXXXXXXXXXXXXXXXXXXXXXXXXXXXXXXXXXXXXXX
TTTTTTTTTTTTTTTTTTTTTTTTTTTTTTTTTTTTTTTT
IIIIIIIIIIIIIIIIIIIIIIIIIIIIIIIIIIIIIIII
LLLLLLLLLLLLLLLLLLLLLLLLLLLLLLLLLLLLLLLL
EEEEEEEEEEEEEEEEEEEEEEEEEEEEEEEEEEEEEEEE
SSSSSSSSSSSSSSSSSSSSSSSSSSSSSSSSSSSSSSSS
EEEEEEEEEEEEEEEEEEEEEEEEEEEEEEEEEEEEEEEE
XXXXXXXXXXXXXXXXXXXXXXXXXXXXXXXXXXXXXXXX
TTTTTTTTTTTTTTTTTTTTTTTTTTTTTTTTTTTTTTTT
RRRRRRRRRRRRRRRRRRRRRRRRRRRRRRRRRRRRRRRR
EEEEEEEEEEEEEEEEEEEEEEEEEEEEEEEEEEEEEEEE
MMMMMMMMMMMMMMMMMMMMMMMMMMMMMMMMMMMMMMMM
EEEEEEEEEEEEEEEEEEEEEEEEEEEEEEEEEEEEEEEE
TTTTTTTTTTTTTTTTTTTTTTTTTTTTTTTTTTTTTTTT
EEEEEEEEEEEEEEEEEEEEEEEEEEEEEEEEEEEEEEEE
XXXXXXXXXXXXXXXXXXXXXXXXXXXXXXXXXXXXXXXX
TTTTTTTTTTTTTTTTTTTTTTTTTTTTTTTTTTTTTTTT
IIIIIIIIIIIIIIIIIIIIIIIIIIIIIIIIIIIIIIII
LLLLLLLLLLLLLLLLLLLLLLLLLLLLLLLLLLLLLLLL
EEEEEEEEEEEEEEEEEEEEEEEEEEEEEEEEEEEEEEEE
SSSSSSSSSSSSSSSSSSSSSSSSSSSSSSSSSSSSSSSS
EEEEEEEEEEEEEEEEEEEEEEEEEEEEEEEEEEEEEEEE
XXXXXXXXXXXXXXXXXXXXXXXXXXXXXXXXXXXXXXXX
TTTTTTTTTTTTTTTTTTTTTTTTTTTTTTTTTTTTTTTT
RRRRRRRRRRRRRRRRRRRRRRRRRRRRRRRRRRRRRRRR
EEEEEEEEEEEEEEEEEEEEEEEEEEEEEEEEEEEEEEEE
MMMMMMMMMMMMMMMMMMMMMMMMMMMMMMMMMMMMMMMM
EEEEEEEEEEEEEEEEEEEEEEEEEEEEEEEEEEEEEEEE
TTTTTTTTTTTTTTTTTTTTTTTTTTTTTTTTTTTTTTTT
EEEEEEEEEEEEEEEEEEEEEEEEEEEEEEEEEEEEEEEE
XXXXXXXXXXXXXXXXXXXXXXXXXXXXXXXXXXXXXXXX
TTTTTTTTTTTTTTTTTTTTTTTTTTTTTTTTTTTTTTTT
IIIIIIIIIIIIIIIIIIIIIIIIIIIIIIIIIIIIIIII
LLLLLLLLLLLLLLLLLLLLLLLLLLLLLLLLLLLLLLLL
EEEEEEEEEEEEEEEEEEEEEEEEEEEEEEEEEEEEEEEE
SSSSSSSSSSSSSSSSSSSSSSSSSSSSSSSSSSSSSSSS
EEEEEEEEEEEEEEEEEEEEEEEEEEEEEEEEEEEEEEEE
XXXXXXXXXXXXXXXXXXXXXXXXXXXXXXXXXXXXXXXX
TTTTTTTTTTTTTTTTTTTTTTTTTTTTTTTTTTTTTTTT
RRRRRRRRRRRRRRRRRRRRRRRRRRRRRRRRRRRRRRRR

588

528

274

616

584

588

# EXTREME TEXTILES

EXTREME

Matilda McQuaid

WITH ESSAYS BY
Philip Beesley
Susan Brown
Sean Hanna
Cara McCarty
Patricia Wilson
Amanda Young

WITH CONTRIBUTIONS BY
Alyssa Becker
John W. S. Hearle

# TEXTILES

## DESIGNING FOR HIGH PERFORMANCE

Thames & Hudson

in association with

Smithsonian
*Cooper-Hewitt, National Design Museum*

# CONTENTS

# FOREWORD

"Materiality" lies at the heart of contemporary design, yet textiles are frequently overlooked as high-performance materials, with varied structural capabilities. This exhibition and book reveal the extraordinary breadth of applications in which textiles are used, illustrating the latest advances in textile design and engineering, whether in building construction or for bioimplantable materials used in heart surgery. Each combines the light-weight, flexible qualities of woven fibers with outstanding performance characteristics, representing the most dynamic, vital expressions of contemporary design.

It is fitting for such an exploration to be conducted at Cooper-Hewitt, National Design Museum. As part of the Smithsonian Institution—the world's largest and most visited complex of museums and research centers—Cooper-Hewitt embraces science, history, and art, and *Extreme Textiles* itself serves as a nexus for scientific, design, and engineering inquiry.

What makes the theme so visually compelling, when presented in book or exhibition form, is that many of these objects, designed solely with application and function in mind, are nevertheless endowed with a profound, subtle aesthetic. Whether it is a parafoil used to drop hundreds of pounds of food and supplies to relief areas, a textile offering a protective shield against chemical hazards, or a jacket with a communications network embedded within its weave, all of these innovations represent some of the most exciting and radical expressions of design today.

Matilda McQuaid, Head of Cooper-Hewitt's Textiles Department and curator of *Extreme Textiles*, has been researching these high-performance composites and textiles for over a decade with the aim of presenting an exhibition on this theme. I am delighted that Cooper-Hewitt, the United States' National Design Museum, should be the venue for such a thorough and original research project: the first exhibition to present contemporary design concerns specifically through the "lens" of textile fibers and structures. As with her seminal exhibition on contemporary Japanese textiles presented at New York's Museum of Modern Art in 1998, *Extreme Textiles* marks a watershed in the public understanding of contemporary textile design. The Board of Trustees, staff, and I all thank Matilda for her remarkable achievement in curating this project.

Paul Warwick Thompson
Director

# ACKNOWLEDGMENTS

*Extreme Textiles: Designing for High Performance* has been a remarkable collaboration between numerous individuals and institutions. Most critical to the project was Cooper-Hewitt's Director, Paul Warwick Thompson, who made my thirteen-year exhibition dream a reality. From the beginning, he supported the idea that technical textiles and their applications are stunning achievements in contemporary design, and I am very grateful for his trust and confidence.

*Extreme Textiles* celebrates not only objects, but also the more than one hundred designers, manufacturers, engineers, fabricators, and other collaborators—too numerous to list here—who have had the incredible vision and commitment to pursue and create such exceptional artifacts of our time. This exhibition and publication are dedicated to their pursuits and discoveries.

The Museum is most grateful to Target, which provided lead support to the exhibition, as well as to Maharam, whose early commitment to the exhibition proved crucial. The Coby Foundation, Ltd. provided invaluable support, both to the exhibition and to the symposium, resulting in its being open to all participants free of charge. Stephen McKay, Inc., Furthermore: a program of the J. M. Kaplan Fund, Elise Jaffe + Jeffrey Brown, and Foster-Miller, Inc. also contributed generously. In addition, the Industrial Fabrics Foundation helped to underwrite the meetings of the exhibition advisory committee, a group of professionals in the areas of architecture, design, textile science, and engineering. I would like to thank the following committee members for their guidance and valuable time on this project: George Beylerian, Director, Material Connexion; Mark Blackman, Business Manager, Milliken & Co.; Frank Bradenburg, Seaman Corporation; David Brookstein, Dean, School of Textiles and Materials Technology, Philadelphia University; James Carpenter, James Carpenter Design Associates Inc.; Niels Diffrient; Professor John W. S. Hearle, Senior Consultant, Tension Technology International; Robert Kinney, Director, Individual Protection Directorate, U.S. Army Natick Soldier Center; Toshiko Mori, Toshiko Mori Architect; Marcia Rounsaville, President, Industrial Fabrics Foundation; Susan Szenasy, Editor in Chief, *Metropolis* magazine; Patricia Wilson, Ph.D., Senior Engineer, Foster-Miller Inc.; and Bruce Wright, Editor, *Fabric Architecture*.

Many manufacturers and designers provided in-kind donations, lent objects, or created special installations for the exhibition. I am grateful for their generous participation and would like to especially thank the following individuals: at the National Air and Space Museum: Roger Launius, Valerie Neal, Allan Needall, Amanda Young, and Toni Thomas in the Space

History Division, Peter Jakab and Tom Crouch in the Aeronautics Division, Ed McManus in Conservation, and Kate Igoe in Archives; Foster-Miller Inc.: Patricia Wilson and Justyna Teverovsky; ILC Dover Inc.: David Cadogan; U.S. Army Natick Soldier Center: Robert Kinney, Jean Hampel, Heidi Schreuder-Gibson, and Carole Winterhalter; Goetz Custom Boats: Eric Goetz; NASA Jet Propulsion Laboratory: Jack Jones and Anita Sohus; North Sails Nevada: Bill Pearson; International Fashion Machines: Maggie Orth; Rick Young; Squid:Labs: Dan Goldwater, Saul Griffith, and Colin Bulthaup; BMW: Thomas Girst, Roy Oliemuller, Amy Quarmby; BMW WilliamsF1 Team: Christopher Styring; The Exhibit Company: John Trontell; Van Dusen Racing Boats: Edward S. Van Dusen; Vanguard Sailboats: Steve Clark and Amy Larkin; Vertigo Inc.: James Ivie and David Cronk; B.A.G. Corporation: Jodi Simon; Kennedy & Violich Architecture: Sheila Kennedy; BodyMedia: Chris Kasabach and Vanessa Sica; Testa Architects: Peter Testa and Devyn Weiser; Buro Happold: Ian Liddell, Cristobal Correa, and Amy Grahek.

Toshiko Mori with Jolie Kerns and Sonya Lee of Toshiko Mori Architect created a cohesive, dynamic, and elegant installation while maintaining the individuality and importance of the objects on display.

The publication will always stand as a permanent document for the exhibition, and I am grateful to the essayists—Susan Brown, John W. S. Hearle, Alyssa Becker, Philip Beesley, Sean Hanna, Cara McCarty, Amanda Young, and Patricia Wilson—for providing insightful discussions about technical textiles. John Hearle and Patricia Wilson were especially generous with providing suggestions about companies and particular works for the exhibition. I am especially appreciative of Kevin Lippert and Princeton Architectural Press for publishing the book, Mark Lamster for his early involvement, and Megan Carey for her outstanding editing of the essays. Patrick Seymour, Catarina Tsang, and Susan Brzozowski of Tsang Seymour Design created a beautiful book that, like the exhibition, gives importance to the individual objects. And without the enormous generosity of Bill Pearson at North Sails Nevada, we would not have the extraordinary sailcloth for the cover.

The film and video clips—important and lasting records of the subject of technical textiles—were produced by Susan Morris and Christopher Noey of Stereopticon. Each tells a remarkable textile story of a particular object, adding greater dimension and richness to the exhibition. My gratitude to Sheila Hicks for her enormous help in pushing the project forward, and for Ruth Kaufmann's early support.

There were many on the Cooper-Hewitt staff who helped with so many different aspects of the exhibition and publication, and my special appreciation goes to Barbara Bloemink, Curatorial Director, for her early support of the exhibition; in Development, Caroline Baumann, Bruce Lineker, Lauren Gray, and Trevel Balser for their great efforts in finding financial support for the exhibition; in Communications, Jennifer Northrop and Laurie Olivieri for alerting everyone about the show; in Textile Conservation, Lucy Commoner and Sandra Sardjono for their superb guidance and advice about object installation; in Exhibitions, Jocelyn Groom, Michael O'Shea, and Matthew Weaver for carrying out the exhibition design and taking care of all the minute details of the installation perfectly; in the Textiles Department, Barbara Duggan for cheerfully pinch-hitting whenever needed; in Publications, Chul R. Kim and Karl Ljungquist for expertly editing and coordinating all aspects of the book; in Education, Dorothy Dunn, Mei Mah, Monica Hampton, and Bonnie Harris for enriching the exhibition with educational programming; and, in Registrar, Steven Langehough, Wendy Rogers, and Larry Silver for carefully tracking and coordinating movement of the objects. I would like to thank Ellen Lupton for introducing me to her master's student, George Moore, who has created an engaging Web site that at once documents and enhances the exhibition. In addition, the following interns and volunteers have provided an enormous amount of research and help on this project over the last three years: Prachi Asher, Michelle Everidge, Pamela Soohoo, Marsha Heiman, Martha Hally, Miriam Kim, Mel Schierman, Allyson Drucker, Esther Lee, Marty Rainbow, Elyssa Schram, and Quentin Smith.

Finally, I would like to thank Susan Brown, who has been a wonderful partner in this textile odyssey. She has shared her textile expertise, good taste, and humor in all aspects of the exhibition—from object selection to installation—and I am forever grateful for the collaboration. It was great fun.

Matilda McQuaid

fig. 1
Tire reinforcement fabric
Manufactured by KoSa (now Invista)
Mexico, 2003
High-modulus, low-shrinkage polyester technical filament yarn twisted and woven into a fabric; dip chemical treatment promoting adhesion between the polyester and rubber carcass, final heat treatment

Matilda McQuaid

# STRONGER, FASTER, LIGHTER, SAFER, AND SMARTER

What can be stronger than steel, faster than a world's record, lighter than air, safer than chain mail, and smarter than a doctor? Hint: it is in every part of our physical environment—lying under roadbeds, reinforcing concrete columns, or implanted into humans. A riddle with one answer and many parts, it is also the subject of the exhibition and accompanying book *Extreme Textiles: Designing for High Performance*. Textiles are the answer, and the world of technical textiles—high-performance, purely functional, and precisely engineered fabrics—is the vital component.

Technical textiles represent, in volume, the smallest segment of the enormous textile industry, yet they are some of the most innovative and purest examples of design today. Aesthetic and decorative qualities are not requirements for a technical textile, and if one finds such a textile visually arresting, it is by pure coincidence. Some of these materials and their applications represented here are unique, others are experimental, many are collaborations across a variety of disciplines, and all represent extraordinary amounts of research and dedication by artists, designers, scientists, engineers, and visionaries.

The journey to find these often peculiar but essential cultural artifacts of our day has been a long one. For me, it started fifteen years ago at the Museum of

fig. 2
Tire cutaway showing reinforcement fabric

Modern Art (MoMA), when I was reading a catalogue for the 1956 exhibition *Textiles USA*. The catalogue featured a special category of industrial fabrics (with swatches within the pages), which included materials for convertible car tops, tires, and radar deflection. Arthur Drexler, one of the curators of the exhibition, wrote about these fabrics:

> Many industrial fabrics inadvertently heighten properties familiar to us in other materials. The blond opulence of loosely plaited tire cord, though it is always hidden within layers of rubber, rivals fabrics used for formal gowns.... Industrial fabrics rarely if ever are designed for aesthetic effect, yet they seem beautiful largely because they share the precision, delicacy, pronounced texture, and exact repetition of detail characteristic of twentieth-century machine art.

These beautiful and engineered accomplishments, sometimes mundane and at other times monumental, were on par with the core of MoMA's design collection—the machine art as exemplified by the exalted ball bearing and propeller blade. I wondered how technical textiles had evolved nearly fifty years later.

The textiles and applications presented in *Extreme Textiles* are certainly examples of twentieth- and twenty-first-century machine art, but they are also studies in ingenuity, creativity, and perseverance. The objects in the show do not represent the most common uses of technical textiles; instead, the selection is based on objects for extreme applications, such as the textile integral to the first controlled flight by man, future apparel for explorers visiting Mars, and the garment that can monitor the vital signs of and provide live communication with a soldier on the battlefield. They might be unfamiliar to us now, but they have already had repercussions in areas such as aeronautics and the medical industry.

These textiles are causing a quiet revolution. Quiet because the innovations that have occurred over the last forty to fifty years, with the development of high-performance fibers such as aramids and carbon fibers, have been largely contained within the small markets of aerospace and the military. Not until the 1980s did the rest of the world become more familiar with the existence and potential uses of these fibers and textiles, which resulted in exceptional growth in the field. While more mature commercial development occurred in the 1990s, the new millennium has been, and will continue to be, marked by the global networking of these technologies and the further expansion of the markets and applications for these textiles.

There is not an area of our world unaffected by the advances in technical textiles. Architecture, transportation, industry, medicine, agriculture, civil engineering, sports, and apparel have all benefited from the tremendous progress and the unique collaborations that have taken place in the field. Principles of textile science and technology merge with other specialties such as engineering, chemistry, biotechnology, material/polymer science, and information science to develop solutions unimaginable a century ago. Who would have thought that we would have the technology to design and some day build a forty-story tower out of carbon-fiber composite, or walk on a planet that is fifty-one million miles away, or have clothing that can automatically react and adapt to the surrounding environment? These are achievements that rely on an interface between many disciplines, and require a willingness to experiment time and time again.

Because these objects are extreme and their ultimate success is determined by how they perform under very specific conditions, the organization of the exhibition and book has followed the lead of the technical-textile industry. These performance standards will be the barometers and categories in which to assess the textiles and applications: stronger, faster, lighter, safer, and smarter. Some objects fit neatly into one classification, others into several depending upon their ultimate function. Choices for placement usually respond to the primary motivation for the objects' creation. The essays included in this book elucidate significant events across the major areas of technical textiles, examine the different technologies that have made some of these extraordinary inventions possible, and demystify material and technique in order for us to understand how and why textiles play such a significant role in our lives.

## STRONGER

Incredible strength is one advantage of many of the new textile fibers, which have the capability to reinforce as well as lift hundreds of tons. Susan Brown explains through specific case studies the innovative fibers and the extraordinary structures and techniques that have made it possible to

B.A.G. CORP.
HOME OF THE
SUPER SACK® CONTAINER
B.A.G. CORP.
HOME OF THE
SUPER SACK® CONTAINER

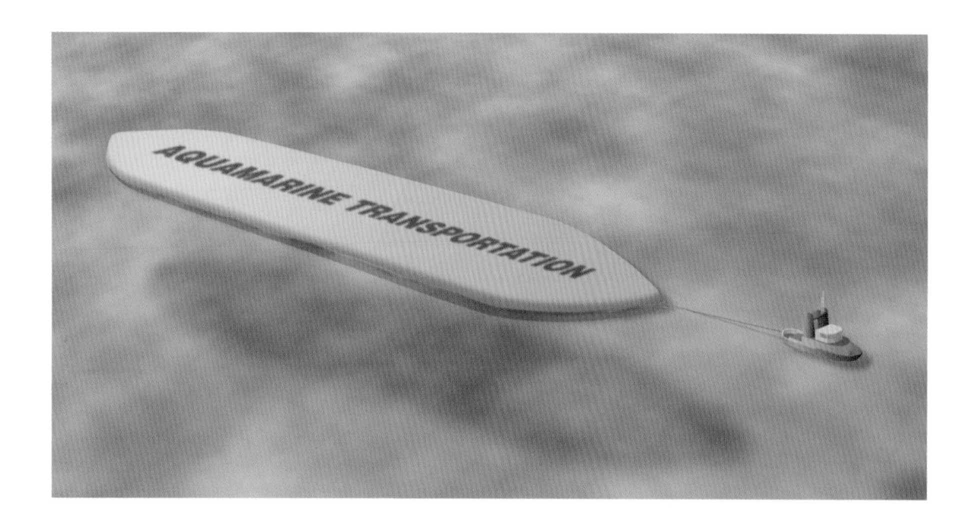


fig. 3 (facing page)
Super Sack® container
Manufactured by B.A.G. Corp.®
U.S.A., 2005
Woven polypropylene
109.2 x 109.2 x 109.2 cm (43 x 43 x 43 in.);
1,996 kg (4,400 lb.) capacity

fig. 4 (above)
Very Large Flexible Barge (VLFB)
Designed by Buro Happold Consulting Engineers for Aquamarine Transportation
England, 2004

achieve such exceptional strength. Basic textile techniques have been around for centuries — weaving, braiding, knitting, and embroidery — but with new fibers, coupled with new types of machinery, or even old looms that have been retooled to accommodate these fibers, the results and final applications are astonishingly different.

An example of a very simple woven structure is tire-cord fabric, which has been used to reinforce tires for over 115 years. Pneumatic tires were originally patented in 1845 in England by R. W. Thompson, but they were first applied to a bicycle in 1888 by John Boyd Dunlop, a Scottish veterinarian, who fitted a rubber hose to his son's tricycle and filled this tire with compressed air. Dunlop patented the pneumatic tire the same year and began limited production; within a decade it had been adopted by the automobile industry. Simultaneously, Dunlop was the first to use a canvas fabric to reinforce the rubber. Over time the canvas was replaced with nylon and rayon, and today primarily steel cord and polyester are used — except when aramids are needed, in specialty vehicles and racing cars. KoSa (now Invista), a leading producer of polyester resin, fiber, and polymer products, has developed a reinforcement fabric made of high-modulus, low-shrinkage polyester industrial filament yarns that is principally used in radial passenger and light-truck tires. Loosely woven and heat-stabilized, it is hidden under layers of rubber, but its significant structural function contributes to successful performance, road handling, and tire durability (figs. 1, 2).

Large, flexible bulk containers, which on first impression seem relatively low-tech and not much more than an oversized tote bag, provide more than reinforcement. These custom-designed containers, such as Super Sack® by B.A.G. Corp.®, are capable of lifting up to twelve tons of liquid or solid. Made of woven recyclable polypropylene, they have been engineered to achieve maximum container capacity while compacting to a fraction of their size when empty (fig. 3).

An even larger container, designed to transport fresh water to the southern and eastern coasts of the Mediterranean, the Gulf states, and southern California, is the Very Large Flexible Barge (VLFB), currently being designed

figs. 5, 6
WilliamsF1 BMW FW26
Manufactured by BMW WilliamsF1 Team; P84 engine manufactured by BMW; chassis manufactured by WilliamsF1; tires manufactured by Michelin
Germany, 2003–04
Chassis: carbon aramid epoxy composite; tires: rubber, steel, aramid fibers; driver's seat: anatomically formed in carbon/epoxy composite material with Alcantara covering
110 cm height x 180 cm width x 460 cm length (43 5/16 in. x 5 ft. 10 7/8 in. x 15 ft. 1 1/8 in.); 605 kg (1333.8 lbs.) including driver (Juan Pablo Montoya) and camera

Just as important as the chassis is the BMW P84 engine, which operates at 19,000 revolutions per minute and 900 brake horse power. It was developed by a team of engineers at BMW F1 Development in close cooperation with specialists from the BMW Research and Innovation Center. New regulations for the 2004 season stipulated that all F1 cars must use a single engine for each vehicle over the entire Grand Prix weekend, increasing the duration of an engine up to 800 kilometers. Every component of the engine was affected by the new specifications, although durability and high performance were not sacrificed. Changes in weight and dimensions were kept to a minimum by developing new heat-treatment procedures that enhanced endurance properties.

by Buro Happold. The concept of towable bags was developed several decades ago, but the bags were used only in small sizes, as larger sizes were unreliable.[1] The new VLFB will have the capacity to transport 250,000 cubic meters of water (over 66,000,000 gallons)—a two-day supply of water for a population of approximately one million people. Its dimensions are 1,148 feet (length) by 236 feet (width) by 46 feet (depth), and, although the material is proprietary, it will likely be made of a polyurethane-coated nylon (fig. 4).

## FASTER

Faster implies a high-performance edge in various types of sporting equipment—cars, sailboats, racing sculls, and bicycles—which have all benefited from the combination of strength, rigidity, and lightness attained in carbon-fiber composites. The WilliamsF1 BMW FW26, the Formula One (F1) car designed and raced in 2004, can reach sixty miles per hour within two and a half seconds, and achieve top engine speed of over two hundred miles per hour. Sailboats are attempting to reach record-breaking times of fifty knots powered only by the wind, and downhill skiers achieve speeds of more than 140 miles per hour. These exceptional performances are due to a combination of physical and mental stamina and material and technological development, and advanced composites provide the successful link to make these events possible.

Advanced composites have been available only since the 1960s, and they were primarily used in aerospace and the military until the early 1980s. All areas of the sports industry realized their enormous potential, with carbon fiber providing the highest stiffness, aramids absorbing the greatest amounts of energy, and both having the ability to replace heavier metal with lighter components. Because lightness ultimately affects speed, textile-reinforced composites are providing major new areas of opportunity for the technical-textile market.

The WilliamsF1 featured here is a blend of endurance and performance, and has achieved these goals through a combination of research in materials and electronics (figs. 5, 6). Although there are many components of the F1 that are either reinforced or made exclusively out of high-performance fibers—for example, brake discs and tires—the largest is the chassis. The car is made of advanced composite materials, such as a carbon-fiber reinforcement within a polymer matrix—mostly taking the form of epoxy resin. Components are molded by laminating layers of the carbon/epoxy material onto a shaped mold (tool) and then curing the resin under heat and pressure. The form of raw materials is the same as that employed in the aerospace industry, i.e., carbon fiber pre-impregnated with epoxy in a "staged" condition (partly cured, not wet, and therefore stable to handle) or what is commonly called "prepreg." Woven carbon fiber is primarily used because it can be draped and tailored into complex shapes, although unidirectional fiber is also employed. Plies of the prepreg are stacked onto the mold and sealed in a vacuum bag, which has the

effect of compacting the laminate prior to curing. This assembly is then put into an autoclave, or a pressurized oven, where nitrogen is applied at around seven bar (seven atmospheres, or one hundred pounds per square inch) to properly consolidate the laminate through the bag. At the same time the temperature in the vessel is raised to approximately 175°C (350°F) in order to cure it. After ninety minutes the part is cooled and is then ejected from the mold as a solid piece.[2]

This same technology, at smaller and larger scales, is used to make everything from high-performance speed-skiing helmets to racing sailboats, as revealed in an interview with master boat builder Eric Goetz. Over the years it has been the elite athlete—whether a race-car driver, skier, or sailor—who has played an important role in the design process. User becomes designer more and more, as racing experience is invaluable in understanding the practical and performance issues of the equipment.

Beat Engel, a downhill racer, started making speed-skiing helmets for himself in the mid-1980s. Over the years he has made helmets for world champions such as Tracie Max Sachs, the 2004 International Ski Federation (FIS)

fig. 7
Tracie Max Sachs, two-time FIS World Cup Champion and Pro World Champion speed-skier, Verbier, Switzerland, Pro-Mondial Final 2004

World Cup champion (fig. 7). Clocking speeds greater than 140 miles per hour, Sachs's performance relies on the highest level of aerodynamics to permit the least resistance as she plummets down the track. Her Speed-monster helmet completely envelops her head and neck so that legs, torso, and head become like one compact bullet (figs. 8, 9). The helmet is a double-shell system with a thin outer layer, used primarily to enhance performance, that breaks away if she should fall, leaving behind an inner helmet for protection. Both shells are made out of woven Kevlar® compressed between two layers of woven and nonwoven glass fiber and applied with polyester resin. It is durable, fast, and light.

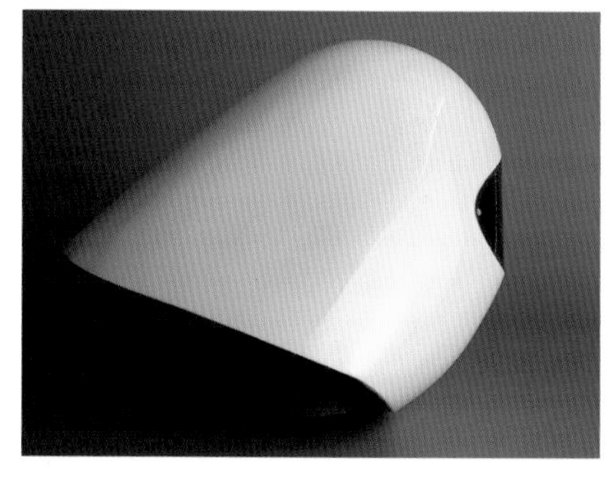

figs. 8, 9
Speed-monster speed-skiing helmet
Designed by Beat Engel, manufactured by Beat Engel Speed Design
Switzerland, 2002
One layer of woven Kevlar® between two layers of woven and nonwoven glass fiber, polyester resin

## LIGHTER

The quality of lightness is always a focus of design for space and aeronautics, as we continue to be fascinated with the ongoing dream of human flight. This dream has inspired some of the most dramatic and curious inventions across all ages.

Beginning with the most rudimentary handcrafted wings made from a variety of materials of their time, humans have attempted to mimic birds to achieve self-powered flight. A group of flying enthusiasts called birdmen have come closest to attaining this vision. Since the 1930s, these men have donned wing suits in order to decelerate free fall and prolong their time aloft for aerial stunts. Most of these early birdmen used a single layer of canvas, stretched from hand to foot like a bat's wing, which allowed little control and virtually no horizontal movement (glide). The breakthrough came in the early 1990s when Patrick de Gayardon invented a wing suit that was neither flat nor rigid, and had wings between his arms and body as well as his legs, with an upper and lower surface that provided an inlet for air—much like a modern

fig. 10
Skyray rigid wing suit
Designed and manufactured by
Alban Geissler; landing and emergency parafoils designed by Daniel Preston and Stane Krajnc, manufactured by Atair Aerospace Inc.
U.S.A., 2003
Rigid composite of Kevlar® and carbon fiber; parafoils of polyamide ripstop nylon; circular braided Spectra® lines; shuttle-loom woven narrow reinforcing tapes

parachute. Since then a number of suits have expanded the idea of skydiving into sky flying, such as Alban Geissler's Skyray, an attachable wing system with a rigid composite made of Kevlar and carbon fiber (fig. 10 ). Daniel Preston and Tom Parker of Atair Aerospace have developed their own wing suit, which consists of a jumpsuit and attached wings made of nonwoven polyethylene laminate and Spectra® fiber (fig. 11). The experience of flying in this suit is different from skydiving, as the wings fill with air as soon as the birdman spreads his limbs. The fabric has no porosity, so the wings remain rigid in flight. The shape of the wing is determined by its three-dimensional inflatable sewn structure and the disposition of the arms and shoulders of the person in the suit.

The birdman still relies on the parachute to land safely on the ground. Parachutes were first used to jump from an airplane in 1912. Atair Aerospace, founded in 2000 by Daniel Preston, grew out of their European counterpart, Atair Aerodynamics, established by Stane Krajnc in 1992. Atair is dedicated to creating state-of-the-art parachute designs as well as flight-navigation systems for all varieties of clients, from the military to major corporations. Their composite parafoil improves upon the most basic building block of

fig. 11
Atair Flexible Wing Suit
Designed by Daniel Preston and Tom Parker, manufactured by Atair Aerospace Inc.
U.S.A., 2004
Wings of composite nonwoven laminate of polyethylene and Spectra® fiber, jumpsuit of ripstop nylon (woven polyamide), Lycra®, and Cordura®.

parachutes by replacing ripstop nylon, whose construction had remained unchanged for more than fifty years, with a flexible nonwoven composite material. This advanced fabric is made by sandwiching an engineered pattern of high-strength fibers, such as ultra-high molecular weight polyethylene (Spectra/Dyneema®) or aramids, between layers of thin polymer foil, and then fusing them under extreme heat (fig. 12). The resulting parafoils have proven to be 300% stronger, 600% less stretchable, and 68% lighter than those constructed in nylon. As the canopy size grows, the strength of this composite material will increase exponentially, and the weight will decrease. This will become an enabling technology for parachutes to be used with extremely heavy cargo weights, where nylon has proven to be a limiting factor.

Orville and Wilbur Wright may not have intentionally mimicked birds, like the birdmen, when the brothers achieved the first fully controlled flight in an aircraft in 1902 (fig. 13). Although this was one year before the landmark day in December when, under power, they sustained heavier-than-air flight, this earlier flight marked the invention of the airplane and officially inaugurated the aerial age.[3]

The textile that they used for covering the wings of the 1902 glider was a type of cotton muslin called "Pride of the West," typically used for ladies' slips. They purchased it off-the-shelf from Rike-Kumler Company, a department store in their hometown of Dayton, Ohio. The brothers used the muslin in its natural state and applied it on the bias. This formed a very tight surface that would distribute landing (or crashing) loads across the wing.[4] They needed a fabric that was flexible and durable in order to achieve their groundbreaking idea for controlling the aircraft, referred to as wing warping, which entails twisting the wing tips of the craft in opposite directions.

fig. 12
Atair Cobalt 95, composite parafoil
Designed by Daniel Preston,
manufactured by Atair Aerospace Inc.
U.S.A., 2003
Top, bottom skin, and ribs of
nonwoven composite laminates of
polyethylene and Spectra® fiber,
lines of circular braided Spectra®
190.5 x 508 cm (6 ft. 3 in. x 16 ft. 8 in.)

fig. 13
Wilbur Wright flying his glider, Kitty Hawk, North Carolina, 1902

fig. 14
Inflatable wings
Developed by ILC Dover Inc. and NASA
U.S.A., 2003
Woven Vectran® wings and
polyurethane bladder
Packed: 50.8 x 12.7 cm (20 x 5 in.);
inflated: 190.5 x 50.8 x 12.7 cm (75 x 20 x 5 in.)

ILC Dover's Unmanned Aerial Vehicle (UAV) is another example of innovation in wing technology (fig. 14). This inflatable wing, made out of a Vectran® restraint, or outer and structural layer, and a polyurethane bladder, can be packed down to a bundle ten times smaller than its deployed wing span of seventy-five inches. It has the potential to fly into any area or situation that would endanger human life—firefighting, military, search and rescue missions—as well as when conditions need to be assessed for risk, such as avalanche/volcanic activity, iceberg patrol, and forest fire survey. Although inflatable wings have been around for several decades, what has evolved during this time are smart materials like electronic textiles for adding functions to the wing. Such electronic textiles are integrated into the UAV, providing a means of controlling direction, communicated remotely. Control can be obtained simply through deformation of the wing geometry.[5] The UAV project has benefited from using technology that ILC Dover implemented in spacesuits and the airbags for the Mars Lander, including the use of high-strength fibers. Fabrics with high strength-to-weight ratios, such as Kevlar and Vectran, have improved the packing efficiency in inflatable wing designs.[6]

There are also more earthbound examples of lightness, which Philip Beesley and Sean Hanna discuss in their essay on textiles and architecture. Exploring areas outside of traditional tensile and membrane structures, Beesley and Hanna find that advanced composites are being used more and more, and on a much larger scale, in architecture. From future projects like Michael Maltzan's house on Leona Drive to Peter Testa's forty-story tower, textile foundations are often at the core of building structures and materials.

### SAFER

Certainly world events have broadened the role of protective applications in recent years, and unique combinations of high-performance fibers and structures are making textiles resistant to cuts, abrasions, bullets, or punctures, and providing protection against extreme cold and heat, chemical or biological hazards, radiation, or high voltages. NASA and the military are playing essential roles in the research and development of textiles in this area, and they are also turning to small, cutting-edge companies such as adventure-gear makers to supply their astronauts and elite soldiers.

Some of these textiles are now very familiar to us—Gore-Tex®, Mylar®, and Kevlar—as they have been integrated into apparel and accessories that may be in our closet today. Cara McCarty discusses and cites examples of this phenomenon, referred to as transfer technology, and acknowledges the important role NASA has played in finding and developing materials that are tested for extreme environments like space, but eventually have great potential on Earth.

Perhaps the ultimate in protective clothing is the spacesuit, a multi-layered body armor and life-support system designed to protect against known and unknown hazards in space. Amanda Young, the official keeper of spacesuits at

fig. 15
A7-L spacesuit (the first type used on the moon), cross section
Fabrics from a typical spacesuit from the 1968/1969 vintage, used during the early Apollo years and early lunar exploration
Multiple layers of Mylar® and Dacron® along with single layers of Nomex® Link Net, Beta cloth, and Teflon® fabric, with the bladder

the Smithsonian Institution's National Air and Space Museum in Washington, D.C., discusses the evolution of the spacesuit, from the first prototypes to the most current developments. Consistent with the process used today, NASA employed the most advanced materials in their prototypes, which led to the white spacesuit that is so familiar to us now. For example, silica Beta cloth, produced by Owens Corning under contract to NASA during the Apollo program, is a nonflammable, Teflon®-coated glass fiber that was used in spacesuits and inside the command module (fig. 15). This was replaced in the mid-1970s with multifibrous Ortho fabric—a combination of Nomex®, Kevlar, and Gore-Tex fibers, and the material of choice for spacesuits throughout assembly of the International Space Station.[7] Chromel-R®, a metallic-fiber fabric, was developed for resistance to abrasions and cuts. The fibers were made of chromium-nickel alloy, which exhibited, at the time, relatively high tensile and tear strength. Although never used in the overall suit (except in an early prototype), it was applied to gloves and boots in the Apollo program.

The gloves that accompany the spacesuit are elaborately customized for each astronaut. Besides fitting properly, they have to be flexible and lightweight while protecting against heat and cold, and must not impede movement or dexterity. Other types of protective gloves may not be customized so much for the specific user as they are for the particular function. SuperFabric® is a new fabric that was first developed for cut and puncture

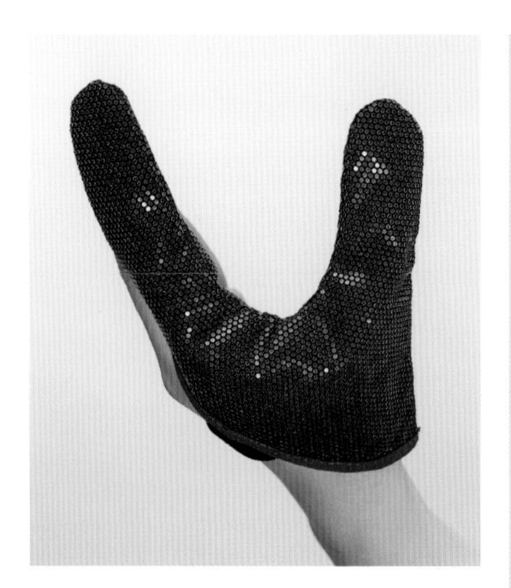

fig. 16
HexArmor™ FingerArmor™
Textile designed by Dr. Young-Hwa Kim, manufactured by HDM Inc.
Textile designed 1996, product designed 2003, manufactured in U.S.A. 2004
SuperFabric® composed of guard plates adhered to nylon base fabric

resistance in the medical profession. It has since been adapted for industrial, military, recreational, and household applications. For instance, FingerArmor™ protects two of the most vulnerable and valuable digits for professional butchers (fig. 16). Miniscule circular guard plates cover all sides and are bound to the nylon base fabric. The base fabric can vary depending upon the use, but it is the guard plates that provide the ultimate protection against cuts. In the Razor-wire gloves, the guard plates are only on the palm side and spaced more widely apart than the FingerArmor (fig. 17). These plates also vary in terms of density, surface texture, and coating, and can fulfill additional performance requirements such as enhanced grip or higher flexibility.

A counterpoint to the SuperFabric gloves is currently being used by the Army for handling razor wire—a hand-cut and sewn suede glove that is covered on the palm side with evenly spaced industrial staples (figs. 18, 19). The "teeth" of the staple face inward, and the interior of the glove has been lined with flannel to protect the hand from being punctured. The positioning of staples takes into account the barbs of the razor wire and performs like chain mail. Although the Army is currently testing SuperFabric to replace the staple-issued gloves, this medieval masterpiece exemplifies the ingenuity that results from necessity and an acute awareness of performance qualities in existing materials.

Motorcycle racing requires unique glove technology that, like the astronaut's glove, provides flexibility, comfort, grip, and resistance to abrasion and moisture. Held, a German company that specializes in gloves, uses kangaroo hide in its Krypton glove along with palm and side hand protection of Kevlar brand fiber ceramic and a lining of Suprotect® shock-absorbing foam (figs. 20, 21). Other areas of the glove are reinforced with these materials to enhance shock resistance and provide the lightest and most protective glove possible.

## SMARTER

Textiles are the natural choice for seamlessly integrating computing and telecommunications technologies to create a more personal and intimate environment. Although clothing has historically been passive, garments of the twenty-first century will become more active participants in our lives, automatically responding to our surroundings or quickly reacting to information that the body is transmitting. These extraordinary examples and uses of electronic textiles are discussed by Patricia Wilson, whose interest in historical metallic embroidery has provided inspiration and guidance in her profession as a material scientist and engineer. She discusses some of the most radical and innovative work being done in this burgeoning field of electronic textiles and, from personal experience, recounts the important collaborations that have taken place between artists, designers, scientists, and engineers.

One of the main incubators for such interdisciplinary study and thought is the MIT Media Lab, which has produced many remarkable designers.

fig. 17
Razor-wire glove
Textile designed by Dr. Young-Hwa Kim, manufactured by HDM Inc.
U.S.A., designed 1996, manufactured 2004
SuperFabric® composed of guard plates adhered to nylon base fabric

figs. 18, 19
Barbed tape-wire handlers' gloves
Developed by U.S. Army Natick
Soldier Center
U.S.A., designed 1967, manufactured 2004
Cowhide, cotton flannel, cotton duck,
with metal staples

fig. 20 (left)
Held Krypton motorcycle racing glove
Manufactured by Intersport Fashions West Inc.
U.S.A., 2004
Kevlar® Ceramic Polymermatrix, kangaroo leather

fig. 21 (right)
Held Ceramic sport touring glove
Manufactured by Intersport Fashions West Inc.
U.S.A., 2004
Kevlar® Ceramic Polymermatrix, Pittards® leather, Kevlar® fiber

fig. 22
Rope and Sound
Designed and engineered by
Squid:Labs LLC
U.S.A., designed 2004, manufactured 2005
Braided polyester rope with integrated
conductive fibers

Three graduates recently formed Squid:Labs, a consulting and research group focused on developing breakthrough technologies in the fields of robotics, materials, and manufacturing. One area they have been investigating is the incorporation of metallic fibers into ropes (fig. 22). These metallic fibers can be used to transmit information and act as antennas for wireless communication, and, potentially more interesting, they can be used as sensors. Squid:Labs has developed an electronic rope made by braiding traditional yarns, such as nylon or polyester, with metallic yarns. There are many variables in the braiding process, including the total number and diameter of yarns, ratio of metallic yarns to polyester/nylon, and the arrangement of metallic yarns. For instance, these yarns could be entirely contained within the rope, but if testing for abrasion, then every few feet, a metallic yarn could migrate to the outside and then back inside the rope. This way, if conductivity is lost in a certain segment of the rope, it is assumed that abrasion has taken place on the external metallic yarn.

There are numerous applications for these intelligent ropes. Mountain climbers could rely on sensors to estimate critical strain in order to know when to retire overly stressed ropes; construction sites could reduce on-site inspection with these sensors, which would indicate when ropes have been compromised because of abrasion; and high-tension power lines, oceanic communication lines, and other electric cables could be enhanced dramatically by adding a thin, intelligent rope around the outside of the cable. All of these examples employ different types of structures that have been used for centuries, but have been transformed into flexible machines or computers that can transmit vital signs of their internal parts.

The variety of applications and design techniques in *Extreme Textiles* attests to the fact that textiles can be anything. They offer the versatility to be hard or soft, stiff or flexible, small or large, structured or arbitrary. They are collectors of energy, vehicles of communication and transport, barriers against physical hazards, and carriers of life-saving cures. They have been created by teams of professionals whose disciplines are diverse, yet who have joined forces with conviction and dedication to chart a course that is reinventing textiles. The future of design lies with these examples of disruptive innovation as textiles continue to push boundaries, eliminate borders between the sciences, and remain a foundation of our physical world.

TTTTTTTTTTTTTTTTTTTTTTTTTTTTTTTTTTTTTTTTTTTTTTTTTTTTTTTTTTTTTTTTTTTTTT

RRRRRRRRRRRRRRRRRRRRRRRRRRRRRRRRRRRRRRRRRRRRRRRRRRRRRRRRRRRRRRRRRRRRRR

OOOOOOOOOOOOOOOOOOOOOOOOOOOOOOOOOOOOOOOOOOOOOOOOOOOOOOOOOOOOOOOOOOOOOO

NNNNNNNNNNNNNNNNNNNNNNNNNNNNNNNNNNNNNNNNNNNNNNNNNNNNNNNNNNNNNNNNNNNNNN

GGGGGGGGGGGGGGGGGGGGGGGGGGGGGGGGGGGGGGGGGGGGGGGGGGGGGGGGGGGGGGGGGGGGGG

EEEEEEEEEEEEEEEEEEEEEEEEEEEEEEEEEEEEEEEEEEEEEEEEEEEEEEEEEEEEEEEEEEEEEE

RRRRRRRRRRRRRRRRRRRRRRRRRRRRRRRRRRRRRRRRRRRRRRRRRRRRRRRRRRRRRRRRRRRRRR

SSSSSSSSSSSSSSSSSSSSSSSSSSSSSSSSSSSSSSSSSSSSSSSSSSSSSSSSSSSSSSSSSSSSSS

TTTTTTTTTTTTTTTTTTTTTTTTTTTTTTTTTTTTTTTTTTTTTTTTTTTTTTTTTTTTTTTTTTTTTT

RRRRRRRRRRRRRRRRRRRRRRRRRRRRRRRRRRRRRRRRRRRRRRRRRRRRRRRRRRRRRRRRRRRRRR

OOOOOOOOOOOOOOOOOOOOOOOOOOOOOOOOOOOOOOOOOOOOOOOOOOOOOOOOOOOOOOOOOOOOOO

NNNNNNNNNNNNNNNNNNNNNNNNNNNNNNNNNNNNNNNNNNNNNNNNNNNNNNNNNNNNNNNNNNNNNN

GGGGGGGGGGGGGGGGGGGGGGGGGGGGGGGGGGGGGGGGGGGGGGGGGGGGGGGGGGGGGGGGGGGGGG

EEEEEEEEEEEEEEEEEEEEEEEEEEEEEEEEEEEEEEEEEEEEEEEEEEEEEEEEEEEEEEEEEEEEEE

RRRRRRRRRRRRRRRRRRRRRRRRRRRRRRRRRRRRRRRRRRRRRRRRRRRRRRRRRRRRRRRRRRRRRR

SSSSSSSSSSSSSSSSSSSSSSSSSSSSSSSSSSSSSSSSSSSSSSSSSSSSSSSSSSSSSSSSSSSSSS

TTTTTTTTTTTTTTTTTTTTTTTTTTTTTTTTTTTTTTTTTTTTTTTTTTTTTTTTTTTTTTTTTTTTTT

RRRRRRRRRRRRRRRRRRRRRRRRRRRRRRRRRRRRRRRRRRRRRRRRRRRRRRRRRRRRRRRRRRRRRR

OOOOOOOOOOOOOOOOOOOOOOOOOOOOOOOOOOOOOOOOOOOOOOOOOOOOOOOOOOOOOOOOOOOOOO

NNNNNNNNNNNNNNNNNNNNNNNNNNNNNNNNNNNNNNNNNNNNNNNNNNNNNNNNNNNNNNNNNNNNNN

GGGGGGGGGGGGGGGGGGGGGGGGGGGGGGGGGGGGGGGGGGGGGGGGGGGGGGGGGGGGGGGGGGGGGG

EEEEEEEEEEEEEEEEEEEEEEEEEEEEEEEEEEEEEEEEEEEEEEEEEEEEEEEEEEEEEEEEEEEEEE

RRRRRRRRRRRRRRRRRRRRRRRRRRRRRRRRRRRRRRRRRRRRRRRRRRRRRRRRRRRRRRRRRRRRRR

SSSSSSSSSSSSSSSSSSSSSSSSSSSSSSSSSSSSSSSSSSSSSSSSSSSSSSSSSSSSSSSSSSSSSS

TTTTTTTTTTTTTTTTTTTTTTTTTTTTTTTTTTTTTTTTTTTTTTTTTTTTTTTTTTTTTTTTTTTTTT

RRRRRRRRRRRRRRRRRRRRRRRRRRRRRRRRRRRRRRRRRRRRRRRRRRRRRRRRRRRRRRRRRRRRRR

OOOOOOOOOOOOOOOOOOOOOOOOOOOOOOOOOOOOOOOOOOOOOOOOOOOOOOOOOOOOOOOOOOOOOO

NNNNNNNNNNNNNNNNNNNNNNNNNNNNNNNNNNNNNNNNNNNNNNNNNNNNNNNNNNNNNNNNNNNNNN

GGGGGGGGGGGGGGGGGGGGGGGGGGGGGGGGGGGGGGGGGGGGGGGGGGGGGGGGGGGGGGGGGGGGGG

EEEEEEEEEEEEEEEEEEEEEEEEEEEEEEEEEEEEEEEEEEEEEEEEEEEEEEEEEEEEEEEEEEEEEE

RRRRRRRRRRRRRRRRRRRRRRRRRRRRRRRRRRRRRRRRRRRRRRRRRRRRRRRRRRRRRRRRRRRRRR

SSSSSSSSSSSSSSSSSSSSSSSSSSSSSSSSSSSSSSSSSSSSSSSSSSSSSSSSSSSSSSSSSSSSSS

TTTTTTTTTTTTTTTTTTTTTTTTTTTTTTTTTTTTTTTTTTTTTTTTTTTTTTTTTTTTTTTTTTTTTT

RRRRRRRRRRRRRRRRRRRRRRRRRRRRRRRRRRRRRRRRRRRRRRRRRRRRRRRRRRRRRRRRRRRRRR

OOOOOOOOOOOOOOOOOOOOOOOOOOOOOOOOOOOOOOOOOOOOOOOOOOOOOOOOOOOOOOOOOOOOOO

NNNNNNNNNNNNNNNNNNNNNNNNNNNNNNNNNNNNNNNNNNNNNNNNNNNNNNNNNNNNNNNNNNNNNN

GGGGGGGGGGGGGGGGGGGGGGGGGGGGGGGGGGGGGGGGGGGGGGGGGGGGGGGGGGGGGGGGGGGGGG

EEEEEEEEEEEEEEEEEEEEEEEEEEEEEEEEEEEEEEEEEEEEEEEEEEEEEEEEEEEEEEEEEEEEEE

RRRRRRRRRRRRRRRRRRRRRRRRRRRRRRRRRRRRRRRRRRRRRRRRRRRRRRRRRRRRRRRRRRRRRR

SSSSSSSSSSSSSSSSSSSSSSSSSSSSSSSSSSSSSSSSSSSSSSSSSSSSSSSSSSSSSSSSSSSSSS

TTTTTTTTTTTTTTTTTTTTTTTTTTTTTTTTTTTTTTTTTTTTTTTTTTTTTTTTTTTTTTTTTTTTTT

RRRRRRRRRRRRRRRRRRRRRRRRRRRRRRRRRRRRRRRRRRRRRRRRRRRRRRRRRRRRRRRRRRRRRR

OOOOOOOOOOOOOOOOOOOOOOOOOOOOOOOOOOOOOOOOOOOOOOOOOOOOOOOOOOOOOOOOOOOOOO

NNNNNNNNNNNNNNNNNNNNNNNNNNNNNNNNNNNNNNNNNNNNNNNNNNNNNNNNNNNNNNNNNNNNNN

GGGGGGGGGGGGGGGGGGGGGGGGGGGGGGGGGGGGGGGGGGGGGGGGGGGGGGGGGGGGGGGGGGGGGG

EEEEEEEEEEEEEEEEEEEEEEEEEEEEEEEEEEEEEEEEEEEEEEEEEEEEEEEEEEEEEEEEEEEEEE

RRRRRRRRRRRRRRRRRRRRRRRRRRRRRRRRRRRRRRRRRRRRRRRRRRRRRRRRRRRRRRRRRRRRRR

SSSSSSSSSSSSSSSSSSSSSSSSSSSSSSSSSSSSSSSSSSSSSSSSSSSSSSSSSSSSSSSSSSSSSS

TTTTTTTTTTTTTTTTTTTTTTTTTTTTTTTTTTTTTTTTTTTTTTTTTTTTTTTTTTTTTTTTTTTTTT

RRRRRRRRRRRRRRRRRRRRRRRRRRRRRRRRRRRRRRRRRRRRRRRRRRRRRRRRRRRRRRRRRRRRRR

OOOOOOOOOOOOOOOOOOOOOOOOOOOOOOOOOOOOOOOOOOOOOOOOOOOOOOOOOOOOOOOOOOOOOO

NNNNNNNNNNNNNNNNNNNNNNNNNNNNNNNNNNNNNNNNNNNNNNNNNNNNNNNNNNNNNNNNNNNNNN

GGGGGGGGGGGGGGGGGGGGGGGGGGGGGGGGGGGGGGGGGGGGGGGGGGGGGGGGGGGGGGGGGGGGGG

EEEEEEEEEEEEEEEEEEEEEEEEEEEEEEEEEEEEEEEEEEEEEEEEEEEEEEEEEEEEEEEEEEEEEE

RRRRRRRRRRRRRRRRRRRRRRRRRRRRRRRRRRRRRRRRRRRRRRRRRRRRRRRRRRRRRRRRRRRRRR

SSSSSSSSSSSSSSSSSSSSSSSSSSSSSSSSSSSSSSSSSSSSSSSSSSSSSSSSSSSSSSSSSSSSSS

TTTTTTTTTTTTTTTTTTTTTTTTTTTTTTTTTTTTTTTTTTTTTTTTTTTTTTTTTTTTTTTTTTTTTT

RRRRRRRRRRRRRRRRRRRRRRRRRRRRRRRRRRRRRRRRRRRRRRRRRRRRRRRRRRRRRRRRRRRRRR

OOOOOOOOOOOOOOOOOOOOOOOOOOOOOOOOOOOOOOOOOOOOOOOOOOOOOOOOOOOOOOOOOOOOOO

NNNNNNNNNNNNNNNNNNNNNNNNNNNNNNNNNNNNNNNNNNNNNNNNNNNNNNNNNNNNNNNNNNNNNN

GGGGGGGGGGGGGGGGGGGGGGGGGGGGGGGGGGGGGGGGGGGGGGGGGGGGGGGGGGGGGGGGGGGGGG

EEEEEEEEEEEEEEEEEEEEEEEEEEEEEEEEEEEEEEEEEEEEEEEEEEEEEEEEEEEEEEEEEEEEEE

RRRRRRRRRRRRRRRRRRRRRRRRRRRRRRRRRRRRRRRRRRRRRRRRRRRRRRRRRRRRRRRRRRRRRR

SSSSSSSSSSSSSSSSSSSSSSSSSSSSSSSSSSSSSSSSSSSSSSSSSSSSSSSSSSSSSSSSSSSSSS

TTTTTTTTTTTTTTTTTTTTTTTTTTTTTTTTTTTTTTTTTTTTTTTTTTTTTTTTTTTTTTTTTTTTTT

RRRRRRRRRRRRRRRRRRRRRRRRRRRRRRRRRRRRRRRRRRRRRRRRRRRRRRRRRRRRRRRRRRRRRR

OOOOOOOOOOOOOOOOOOOOOOOOOOOOOOOOOOOOOOOOOOOOOOOOOOOOOOOOOOOOOOOOOOOOOO

NNNNNNNNNNNNNNNNNNNNNNNNNNNNNNNNNNNNNNNNNNNNNNNNNNNNNNNNNNNNNNNNNNNNNN

GGGGGGGGGGGGGGGGGGGGGGGGGGGGGGGGGGGGGGGGGGGGGGGGGGGGGGGGGGGGGGGGGGGGGG

EEEEEEEEEEEEEEEEEEEEEEEEEEEEEEEEEEEEEEEEEEEEEEEEEEEEEEEEEEEEEEEEEEEEEE

RRRRRRRRRRRRRRRRRRRRRRRRRRRRRRRRRRRRRRRRRRRRRRRRRRRRRRRRRRRRRRRRRRRRRR

SSSSSSSSSSSSSSSSSSSSSSSSSSSSSSSSSSSSSSSSSSSSSSSSSSSSSSSSSSSSSSSSSSSSSS

TTTTTTTTTTTTTTTTTTTTTTTTTTTTTTTTTTTTTTTTTTTTTTTTTTTTTTTTTTTTTTTTTTTTTT

RRRRRRRRRRRRRRRRRRRRRRRRRRRRRRRRRRRRRRRRRRRRRRRRRRRRRRRRRRRRRRRRRRRRRR

OOOOOOOOOOOOOOOOOOOOOOOOOOOOOOOOOOOOOOOOOOOOOOOOOOOOOOOOOOOOOOOOOOOOOO

NNNNNNNNNNNNNNNNNNNNNNNNNNNNNNNNNNNNNNNNNNNNNNNNNNNNNNNNNNNNNNNNNNNNNN

GGGGGGGGGGGGGGGGGGGGGGGGGGGGGGGGGGGGGGGGGGGGGGGGGGGGGGGGGGGGGGGGGGGGGG

EEEEEEEEEEEEEEEEEEEEEEEEEEEEEEEEEEEEEEEEEEEEEEEEEEEEEEEEEEEEEEEEEEEEEE

RRRRRRRRRRRRRRRRRRRRRRRRRRRRRRRRRRRRRRRRRRRRRRRRRRRRRRRRRRRRRRRRRRRRRR

SSSSSSSSSSSSSSSSSSSSSSSSSSSSSSSSSSSSSSSSSSSSSSSSSSSSSSSSSSSSSSSSSSSSSS

TTTTTTTTTTTTTTTTTTTTTTTTTTTTTTTTTTTTTTTTTTTTTTTTTTTTTTTTTTTTTTTTTTTTTT

RRRRRRRRRRRRRRRRRRRRRRRRRRRRRRRRRRRRRRRRRRRRRRRRRRRRRRRRRRRRRRRRRRRRRR

OOOOOOOOOOOOOOOOOOOOOOOOOOOOOOOOOOOOOOOOOOOOOOOOOOOOOOOOOOOOOOOOOOOOOO

NNNNNNNNNNNNNNNNNNNNNNNNNNNNNNNNNNNNNNNNNNNNNNNNNNNNNNNNNNNNNNNNNNNNNN

GGGGGGGGGGGGGGGGGGGGGGGGGGGGGGGGGGGGGGGGGGGGGGGGGGGGGGGGGGGGGGGGGGGGGG

EEEEEEEEEEEEEEEEEEEEEEEEEEEEEEEEEEEEEEEEEEEEEEEEEEEEEEEEEEEEEEEEEEEEEE

RRRRRRRRRRRRRRRRRRRRRRRRRRRRRRRRRRRRRRRRRRRRRRRRRRRRRRRRRRRRRRRRRRRRRR

SSSSSSSSSSSSSSSSSSSSSSSSSSSSSSSSSSSSSSSSSSSSSSSSSSSSSSSSSSSSSSSSSSSSSS

TTTTTTTTTTTTTTTTTTTTTTTTTTTTTTTTTTTTTTTTTTTTTTTTTTTTTTTTTTTTTTTTTTTTTT

RRRRRRRRRRRRRRRRRRRRRRRRRRRRRRRRRRRRRRRRRRRRRRRRRRRRRRRRRRRRRRRRRRRRRR

OOOOOOOOOOOOOOOOOOOOOOOOOOOOOOOOOOOOOOOOOOOOOOOOOOOOOOOOOOOOOOOOOOOOOO

NNNNNNNNNNNNNNNNNNNNNNNNNNNNNNNNNNNNNNNNNNNNNNNNNNNNNNNNNNNNNNNNNNNNNN

GGGGGGGGGGGGGGGGGGGGGGGGGGGGGGGGGGGGGGGGGGGGGGGGGGGGGGGGGGGGGGGGGGGGGG

EEEEEEEEEEEEEEEEEEEEEEEEEEEEEEEEEEEEEEEEEEEEEEEEEEEEEEEEEEEEEEEEEEEEEE

RRRRRRRRRRRRRRRRRRRRRRRRRRRRRRRRRRRRRRRRRRRRRRRRRRRRRRRRRRRRRRRRRRRRRR

SSSSSSSSSSSSSSSSSSSSSSSSSSSSSSSSSSSSSSSSSSSSSSSSSSSSSSSSSSSSSSSSSSSSSS

TTTTTTTTTTTTTTTTTTTTTTTTTTTTTTTTTTTTTTTTTTTTTTTTTTTTTTTTTTTTTTTTTTTTTT

RRRRRRRRRRRRRRRRRRRRRRRRRRRRRRRRRRRRRRRRRRRRRRRRRRRRRRRRRRRRRRRRRRRRRR

OOOOOOOOOOOOOOOOOOOOOOOOOOOOOOOOOOOOOOOOOOOOOOOOOOOOOOOOOOOOOOOOOOOOOO

NNNNNNNNNNNNNNNNNNNNNNNNNNNNNNNNNNNNNNNNNNNNNNNNNNNNNNNNNNNNNNNNNNNNNN

GGGGGGGGGGGGGGGGGGGGGGGGGGGGGGGGGGGGGGGGGGGGGGGGGGGGGGGGGGGGGGGGGGGGGG

EEEEEEEEEEEEEEEEEEEEEEEEEEEEEEEEEEEEEEEEEEEEEEEEEEEEEEEEEEEEEEEEEEEEEE

RRRRRRRRRRRRRRRRRRRRRRRRRRRRRRRRRRRRRRRRRRRRRRRRRRRRRRRRRRRRRRRRRRRRRR

SSSSSSSSSSSSSSSSSSSSSSSSSSSSSSSSSSSSSSSSSSSSSSSSSSSSSSSSSSSSSSSSSSSSSS

TTTTTTTTTTTTTTTTTTTTTTTTTTTTTTTTTTTTTTTTTTTTTTTTTTTTTTTTTTTTTTTTTTTTTT

RRRRRRRRRRRRRRRRRRRRRRRRRRRRRRRRRRRRRRRRRRRRRRRRRRRRRRRRRRRRRRRRRRRRRR

OOOOOOOOOOOOOOOOOOOOOOOOOOOOOOOOOOOOOOOOOOOOOOOOOOOOOOOOOOOOOOOOOOOOOO

NNNNNNNNNNNNNNNNNNNNNNNNNNNNNNNNNNNNNNNNNNNNNNNNNNNNNNNNNNNNNNNNNNNNNN

GGGGGGGGGGGGGGGGGGGGGGGGGGGGGGGGGGGGGGGGGGGGGGGGGGGGGGGGGGGGGGGGGGGGGG

EEEEEEEEEEEEEEEEEEEEEEEEEEEEEEEEEEEEEEEEEEEEEEEEEEEEEEEEEEEEEEEEEEEEEE

RRRRRRRRRRRRRRRRRRRRRRRRRRRRRRRRRRRRRRRRRRRRRRRRRRRRRRRRRRRRRRRRRRRRRR

SSSSSSSSSSSSSSSSSSSSSSSSSSSSSSSSSSSSSSSSSSSSSSSSSSSSSSSSSSSSSSSSSSSSSS

TTTTTTTTTTTTTTTTTTTTTTTTTTTTTTTTTTTTTTTTTTTTTTTTTTTTTTTTTTTTTTTTTTTTT

RRRRRRRRRRRRRRRRRRRRRRRRRRRRRRRRRRRRRRRRRRRRRRRRRRRRRRRRRRRRRRRRRRRRRR

OOOOOOOOOOOOOOOOOOOOOOOOOOOOOOOOOOOOOOOOOOOOOOOOOOOOOOOOOOOOOOOOOOOOOO

NNNNNNNNNNNNNNNNNNNNNNNNNNNNNNNNNNNNNNNNNNNNNNNNNNNNNNNNNNNNNNNNNNNNNN

fig. 1
3DL™ sail
Process designed by J. P. Baudet and Luc Dubois, sail manufactured by North Sails Nevada
U.S.A., process designed 1990–91, manufactured 2004
Molded composite of continuous carbon and aramid fibers laid in a pattern anticipating load paths, laminated between sheets of Mylar®

Susan Brown

# TEXTILES: FIBER, STRUCTURE, AND FUNCTION

With the Mars Exploration Rovers (MER) six miles off the surface of Mars and traveling at twelve thousand miles per hour, the NASA team experiences "six minutes of terror": a rapid-fire series of high-stress events known as entry, descent, and landing. The cruise stage, which provides support for the voyage, is discarded. A parachute opens, slowing the descent of the craft to about 250 miles per hour. The heat shield is jettisoned, and, for a moment, the lander is hanging from a narrow, braided tether—the world on a string. Five seconds before touchdown, braking rockets fire and explosive gas generators inflate the four clusters of airbags attached to the lander. The bags hit the jagged surface at fifty miles per hour and bounce a hundred feet back up in the air, crashing down dozens more times before rolling to a stop on the rocky surface of Mars (fig. 2).[1]

The airbag system was first developed for *Pathfinder* in 1996 as part of a series of low-cost Discovery program missions, and was further refined for the Mars Exploration missions in 2003. While the animations of the projected landings are both amazing and amusing to watch, the bags are highly engineered by any standard, and performance fibers and textiles play an indispensable role in the successful design of the system (fig. 3).

fig. 2
Mars Exploration Rover (MER) lander airbag system
Developed by ILC Dover Inc., Jet Propulsion Laboratory, and NASA
U.S.A., 2002
Animation by Dan Maas, Maas Digital LLC
Animation with CAD overlay of projected airbag bounce

Fibers are considered high-performance if they have exceptional strength, strength-to-weight ratio, chemical or flame resistance, or range of operating temperatures. Advances in fiber strength were made throughout the twentieth century with the introduction of synthetic materials such as nylon in the 1930s and polyester in the 1950s, which still form the bulk of the technical-textiles market. But while polyester provides a 50% increase in strength over cotton, Kevlar delivers a 300% increase in strength and a 1,000% increase in stretch resistance.[2] Performance improvements of this magnitude are the factors leading this second textile revolution, a radical transformation in the way things are made.

The development of such high-performance fibers has caused engineers and designers to reexamine the structural capabilities of traditional textile techniques such as weaving, braiding, knitting, and embroidery. The qualities of textiles are dependent on the interaction between their material properties and their structural geometry, or on the fibers and the way in which those fibers are ordered. Each of the textile techniques represents a very specific architecture of fibers which can be used to create a wide variety of materials for design.

## WEAVING

The textiles that protected the Mars Exploration Rovers on their descent and landing were made using the most fundamental textile technique: the "over-one under-one" interlacing of two perpendicular sets of threads that we learn as children to call weaving, also known as plain weave (fig. 4).

fig. 3 (right)
Mars *Pathfinder* lander airbag prototype
Developed by ILC Dover Inc., Jet Propulsion Laboratory, and NASA
U.S.A., 1996
Plain woven Vectran® (liquid crystal polymer) layers, some with silicone coating for gas retention
Four clusters of bags, each with 4.6 m (15 ft.) length per side; 1.5 m (5 ft.) depth

fig. 4 (below)
Plain weave

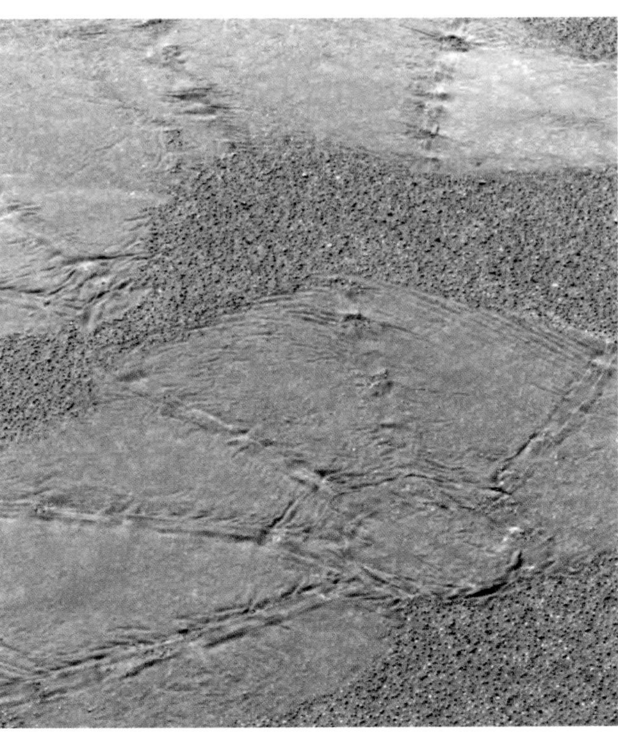

fig. 5
Impressions left by the airbags of the Mars Exploration Rover (MER) *Opportunity* in Martian soil, January 24, 2004

This classic plain weave has the greatest strength and stability of the traditional fabric structures. While no textiles survive from the earliest dates, impressions in clay of basic woven cloth demonstrate its use from at least 7000 BC.[3] Older than metal-working or pottery-making, perhaps even older than agriculture, cloth-weaving has a very primary relationship to the pursuits of humankind.[4]

It is fitting, then, that among the first marks made by man in the soil of Mars was that of a plain woven fabric: an impression made by the impact of the airbags (fig. 5).[5] Each bag has a double bladder and several abrasion-resistant layers made of tightly woven Vectran. Like most synthetic fibers, Vectran liquid crystal polymer is extruded from a liquid state through a spinneret, similar to a shower head, and drawn into filament fibers. The stretching of the fiber during the drawing process orients the polymer chains more fully along the fiber length, creating additional chemical bonds and greater strength. Vectran provides equal strength at one-fifth the weight of steel. Weight is of premium importance for all materials used for space travel, and Warwick Mills, the weaver of the fabric for the bags, achieved a densely woven fabric at a mere 2.4 ounces per square yard, but with a strength of 350 pounds per inch.[6]

The materials are also required to perform at severe temperatures. Because impact occurs two to three seconds after the inflation of the airbags, the fabrics endure their greatest stresses at both extremes of temperature: the explosive gasses that inflate the bags may elevate the temperature inside the bladder layers to over 212°F, but the temperature on the Martian surface is −117°F. Retraction of the airbags to allow the egress of the rovers required that the fabrics remain flexible at these very low temperatures for an extended period of time—about ninety minutes for the deflation and retraction process. Two other fiber types, aramid fibers (Kevlar 29 and Technora T-240) and ultra-high molecular weight polyethylene (UHMWPE) Spectra 1000, were also considered during the development of the Pathfinder airbags. Spectra, a super-drawn fiber, is among the strongest fibers known—fifteen times stronger than steel. However, it performs poorly at extreme temperatures, and so was eliminated early in the development process. Vectran was ultimately selected for the best performance at low temperatures, but Kevlar 129 was used for the tethers inside the bags because of its superior performance at higher temperatures.

The rovers themselves are also textile-based; they are made from super-strong, ultra-lightweight carbon-fiber composites, which are being widely used for aerospace components as well as high-performance sports equipment.[7] As composite reinforcements, textiles offer a high level of customization with regard to type and weight of fiber, use of combinations of fibers, and use of different weaves to maximize the density of fibers in a given direction. Fiber strength is greatest along the length. The strength of composite materials derives from the intentional use of this directional nature. While glass fibers are the most commonly used for composites, for high-performance products the fiber used is often carbon or aramid, or a combination of the two, because of their superior strength and light weight.

One advantage of composite construction is the ability to make a complex form in one piece, called monocoque construction. A woven textile is hand-laid in a mold; the piece is wetted out with resin and cured in an autoclave. The textile can also be impregnated with resin and cured without a wet stage. The same drape or hand that makes twill the preferred weave for most apparel is also desirable for creating the complex forms of boats, paddles, bicycle frames, and other sports equipment. The weft in a twill, rather than crossing under and over each consecutive warp, floats over more than one warp, and with each subsequent weft the grouping is shifted over one warp, creating the marked diagonal effect typical of twills (fig. 8).

Boat builders were among the first to experiment with carbon-reinforced composites. One early innovator, Edward S. ("Ted") Van Dusen, began making carbon-fiber composite racing shells in the 1970s (fig. 7). The critical factor in shell design is the stiffness-to-weight ratio, with greater stiffness meaning that more of the rower's power is translated into forward motion. Van Dusen found that all of the standard construction materials had about the same specific stiffness, or stiffness per unit weight, and began experimenting with glass, boron, and carbon fiber–reinforced composites.[8]

For his Advantage racing shells, Van Dusen uses glass fiber in a complex twill commonly known as satin weave. In a satin, each weft may float over

fig. 6
5/1 satin weave

as many as seven warps (fig. 6). With fewer points of intersection, the fabric has less stability, but more fiber can be packed into the structure. The satin weave is very dense, with nearly five times as many yarns per inch as the plain weave Van Dusen uses to reinforce his riggers. This density minimizes the risk of pinholes forming in the composite, keeping the boat watertight. The glass material is exceptionally fine and light, which allows the textile to be wetted out with a smaller amount of resin, giving a lighter finished product. An aramid honeycomb is sandwiched between two layers of this fabric, and then reinforced with unidirectional carbon-fiber tape. For some pure

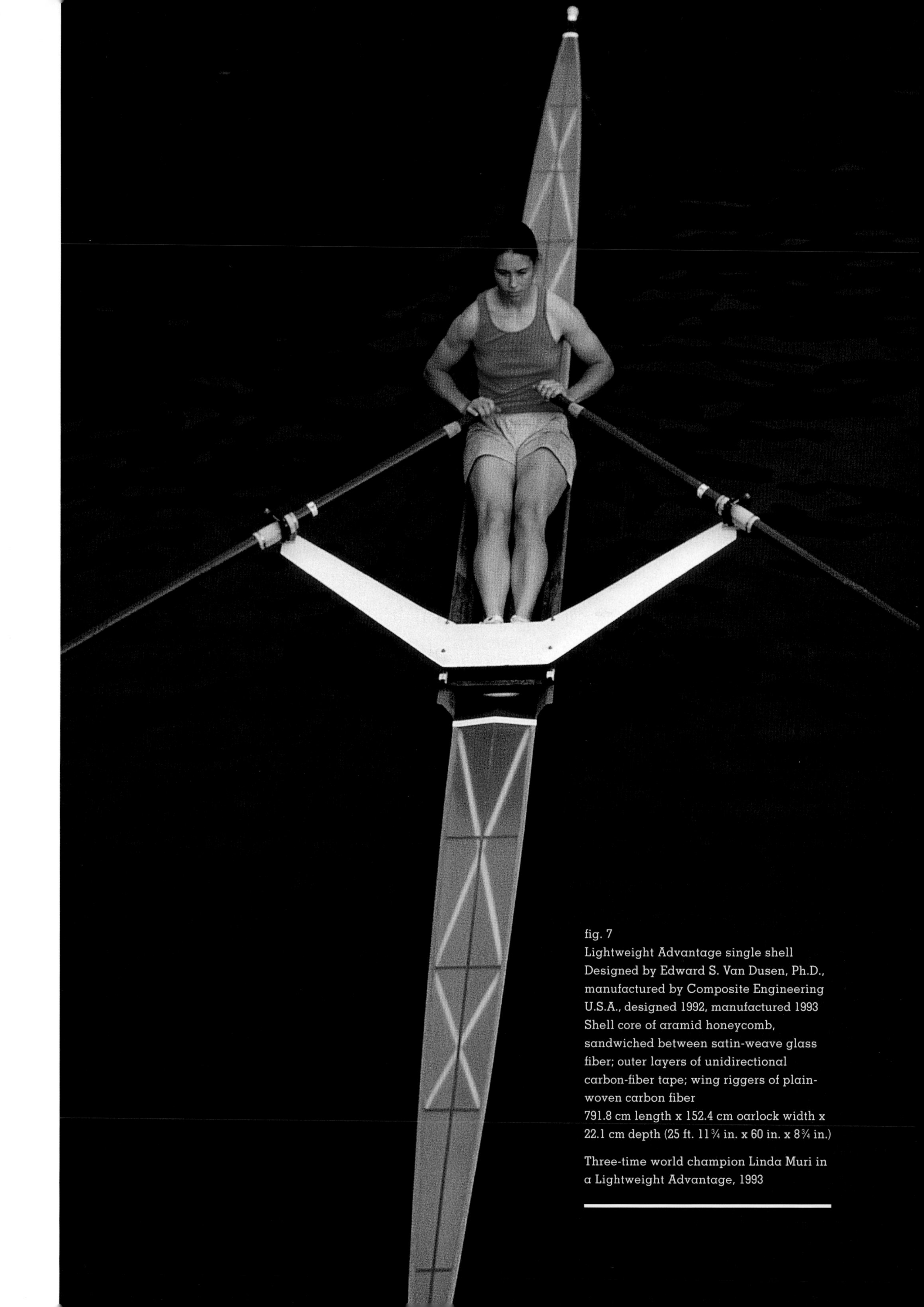

fig. 7
Lightweight Advantage single shell
Designed by Edward S. Van Dusen, Ph.D., manufactured by Composite Engineering U.S.A., designed 1992, manufactured 1993
Shell core of aramid honeycomb, sandwiched between satin-weave glass fiber; outer layers of unidirectional carbon-fiber tape; wing riggers of plain-woven carbon fiber
791.8 cm length x 152.4 cm oarlock width x 22.1 cm depth (25 ft. 11¾ in. x 60 in. x 8¾ in.)

Three-time world champion Linda Muri in a Lightweight Advantage, 1993

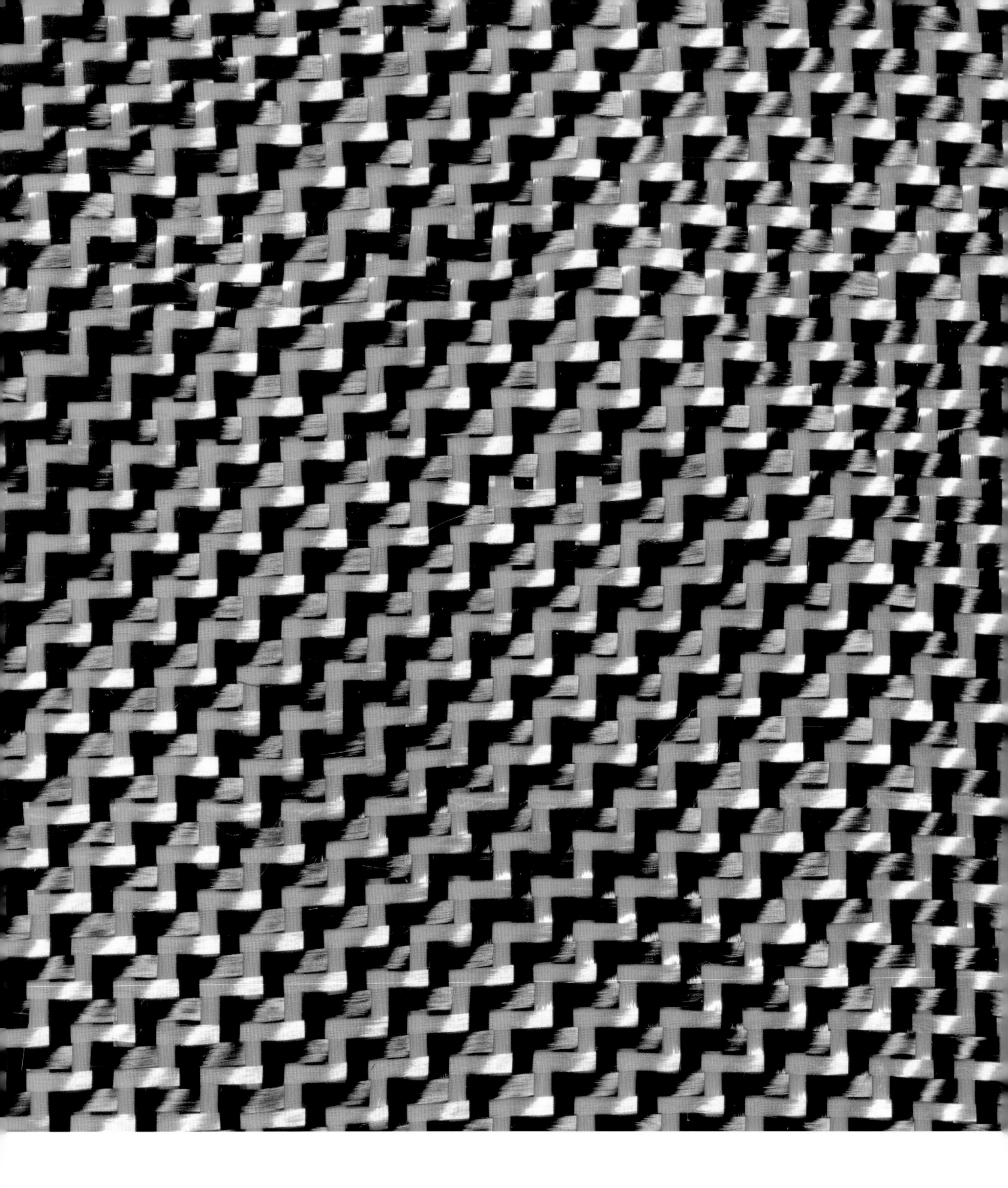

fig. 8
2/1 twill-woven carbon and aramid fiber

figs. 9, 10
Total Eclipse bicycle frame
Designed by Stefan Behrens,
manufactured by Carbonsports GmbH
Germany, 2004
Woven carbon-fiber composite

performance applications like this one, the minor crimping associated with the interlacing of warp and weft is not acceptable, and unidirectional fiber is used to give maximum stiffness in a particular direction.

Van Dusen's early boats weighed as little as twenty-six pounds, as compared with thirty-two to forty pounds for a wooden shell. The competitive advantage was so clear that the legislators of international rowing established the minimum weight for single shells at 30.9 pounds, meaning Van Dusen had to add weight to his shells.[9]

Cycling is another area in which regulation has had a marked impact on equipment design. Triathlon racing is not regulated by the Union Cycliste Internationale, the governing body of international cycling, and therefore some of the more innovative bicycle forms have come from that arena. The Total Eclipse frame designed by Stefan Behrens, an aerospace engineer, is produced by a division of Carbonfibretec, manufacturers of composite aerospace components (figs. 9, 10). The monocoque frame, made from resin-impregnated carbon-fiber twill-woven fabric, gives excellent stiffness to convert pedaling energy into speed, while making possible the suspension seat that reduces rider fatigue. Because composite monocoque frames are made as a single piece, their forms are elegant and aerodynamically curved, taking advantage of the strength inherent in continuous fiber.

fig. 11
Rotor blisk
Designed by Williams International, developed by Foster-Miller Inc. for U.S. Air Force; textile designed and manufactured by Foster-Miller Inc. and Fabric Development
U.S.A., designed 1995, manufactured 1997
Composite form with blades of triaxially braided carbon fiber integrally attached to a polar-woven hub, epoxy matrix
Diameter: 5.2 cm (6 in.); height: 1 cm (⅜ in.)

The process of hand-laying a woven fabric in a mold is extremely time-consuming, and efforts are being made throughout the composites industry to create pre-forms—textiles that can be manufactured in the shape required for the finished product. Braided products such as hoses, ropes, and shoelaces are three-dimensional tubes, while many knitted objects like hosiery and gloves come off the machine in their finished form. These familiar technologies are finding new uses in the technical market.

### BRAIDING

The diversity of braided forms being produced for technical applications is truly astonishing. Braiding is also a millennia-old technique, used for the plaiting of hair, the making of baskets, and the creation of sturdy straps and cloths. Unlike weaving, in which the fibers cross at right angles, the elements of a braid meet at oblique angles. Two features of braids that make them attractive for technical uses are that they take three-dimensional forms easily, and that all of the filaments continue end-to-end, and thus distribute loads or stresses efficiently throughout the structure.

A carbon-fiber mast for a racing yacht, also designed by Ted Van Dusen, is braided over a mandrel to a complex shape, columnar but narrower at the top, extending to a larger, softly triangular shape at the base. Where it is possible to braid such shapes, this saves time over hand-laying a woven textile into a mold, and creates a seamless tube, eliminating the added weight of an overlap seam. The oblique interlacing system, while ideal for conformability to the mandrel, is not an ideal load-bearing structure for a mast.[10] Vertical axis or 0° elements are added to form a biaxial braid, with the vertical

fig. 12
Triaxial fabric
Designed and manufactured by Sakase Adtech Co. Ltd.
Japan, designed 1991, manufactured 2002
Triaxially woven carbon fiber
Cooper-Hewitt, National Design Museum, Gift of The Museum of Modern Art, courtesy of the designer 2002-28-1

elements carrying most of the load of the sail, and the 45° elements providing torsional rigidity and shear strength. Because the mast is hollow, bending stresses could cause the tube to deform or collapse, so a final wrapping of carbon is added at 90°. The overbraiding technique also allows variation in the number of layers or wall thickness, as well as the fiber composition, in this case a combination of carbon and glass. Glass is available in a much finer fiber than carbon, and is used in the oblique elements of the mast to minimize the deflection or crimping of the vertical carbon elements, keeping them as straight and as strong as possible.

A considerably more complex braided structure is a composite rotor blisk for a rocket engine turbopump, designed and produced by Foster-Miller (fig. 11).[11] Extremely lightweight and capable of functioning at 1,500°C (2,700°F), the carbon-fiber blisk is made with triaxially braided blades which are integrally attached to a polar-woven hub (fig. 12). Polar weaving is a method of producing complex forms where very high stresses are anticipated. In polar or cylindrical weaving, the warp runs vertically, while a second set of fibers runs radially and a third circumferentially. As a replacement for a metal part, the

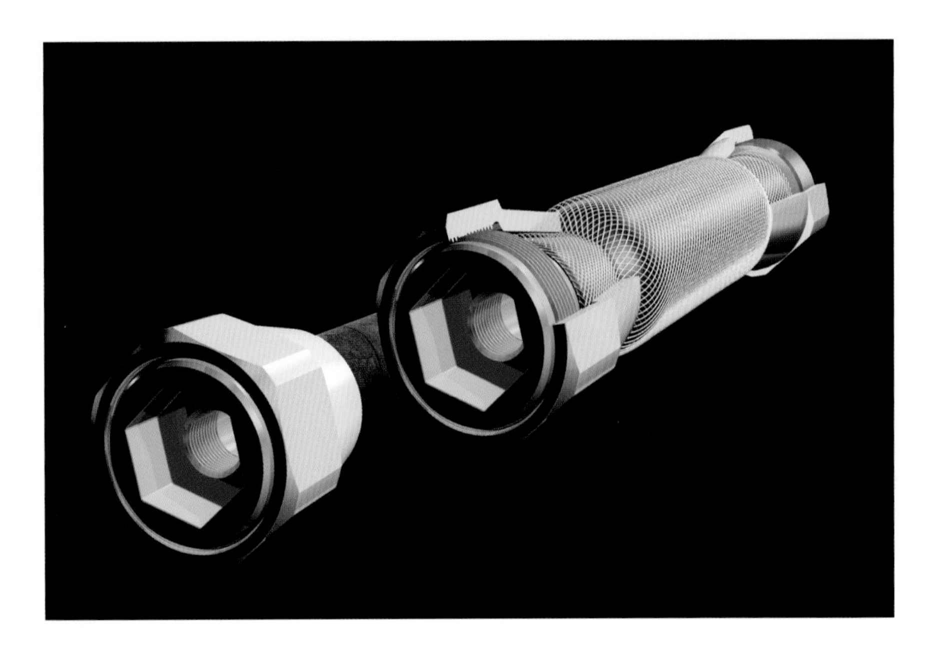

fig. 13
Fluidic Muscle
Textile engineered by Bernd Lorenz, Axel Thallemer, and Dr. Dieter Bergemann; fiber developed by Teijin Twaron®, Arnhem/NL; manufactured by Festo AG & Company KG
Germany, 1996
Spiral weave of aramid fibers, hollow flexible chloroprene core, lightweight metals
Internal diameter of 10, 20, or 40 mm (3/8, 13/16 or 1 9/16 in.)

composite version is quicker to produce due to reduced machining, and is structurally superior because of the continuity between hub and blade.

The structure of Vertigo's AirBeam™ is in some ways similar to the mast, being a seamlessly braided three-dimensional form (fig. 17, p. 118). But because it is inflatable and designed to be easily transported, flexibility is a must for repeated inflation and deflation as well as for packing and shipping. As with the Mars lander airbags, Vectran fiber is used to give high strength with good flex-fatigue resistance. The seamlessness of the braided form is also critical, as the creation of strong, leakproof seams has been a chronic problem for inflatable structures that must remain under high pressure for extended periods of time.

One of the most compelling new uses for a textile is Festo's Fluidic Muscle, which behaves like an industrial-strength human muscle (fig. 13). The muscle is a hydraulic or pneumatic actuator that operates on a membrane contraction system. The shaft of the actuator resembles a braided hose, with aramid fibers laid at oblique angles to one another and encased in a rubbery sheath. By building up internal pressure with air or fluid, the angle of the interlacing fibers is altered, and the hose contracts. This deformation generates a tensile force in the axial direction. The muscle can exert ten times the force of conventional actuators, but weighs far less—about one-eighth the weight of a metal cylinder of the same inner diameter.[12] Because there are no moving mechanical parts, the muscle is free from the jerking associated with the breakaway moment of static friction in a conventional actuator. The smooth operation makes it ideal for precision robotics; combined with its light weight, it shows tremendous potential for use in prostheses (fig. 14).

## KNITTING

Knit fabrics are most commonly used for their stretch characteristics. Knitting is a looping technique, and the knit stitch can be easily distended

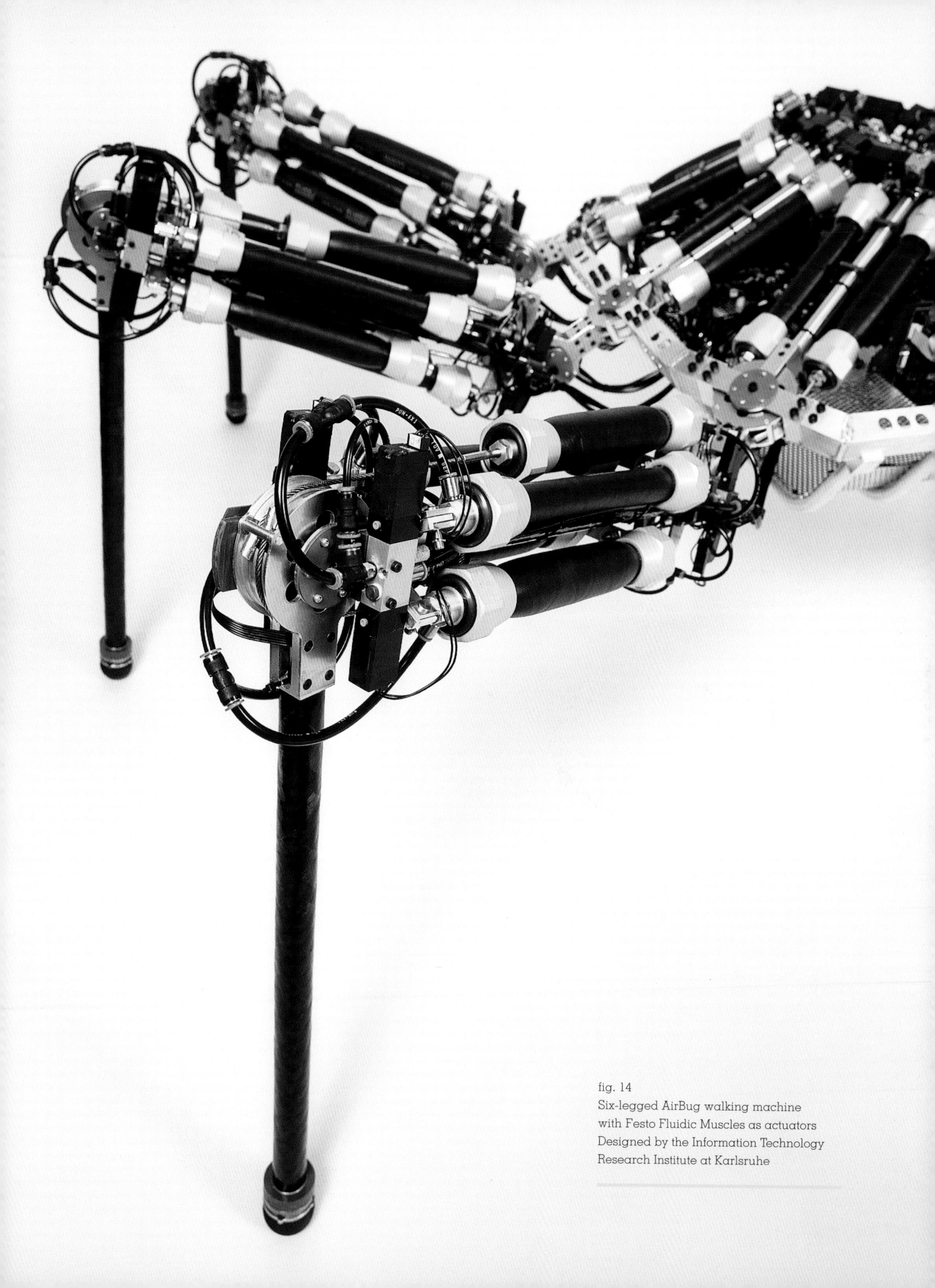

fig. 14
Six-legged AirBug walking machine
with Festo Fluidic Muscles as actuators
Designed by the Information Technology
Research Institute at Karlsruhe

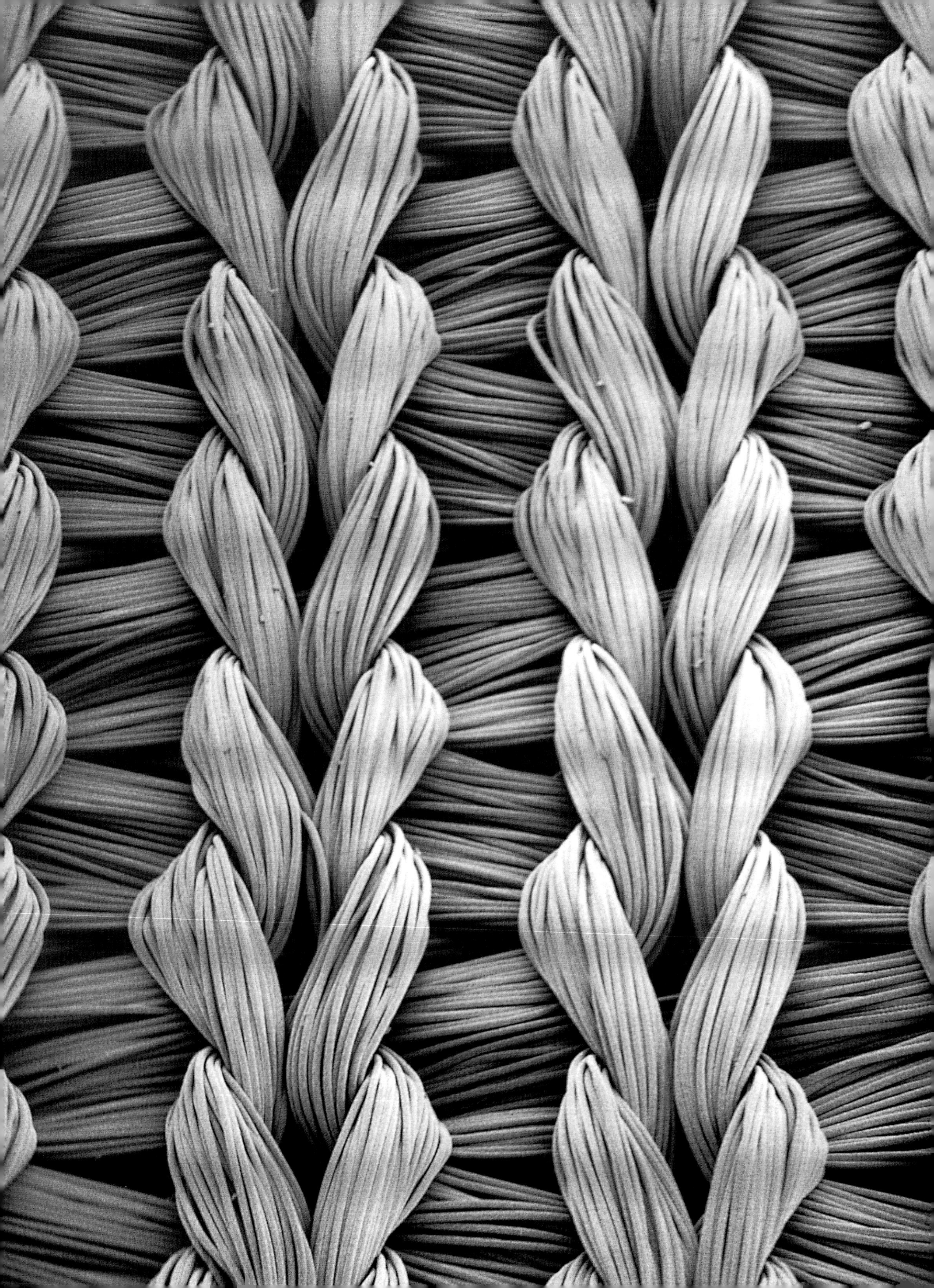

fig. 15 (facing page)
Warp knit

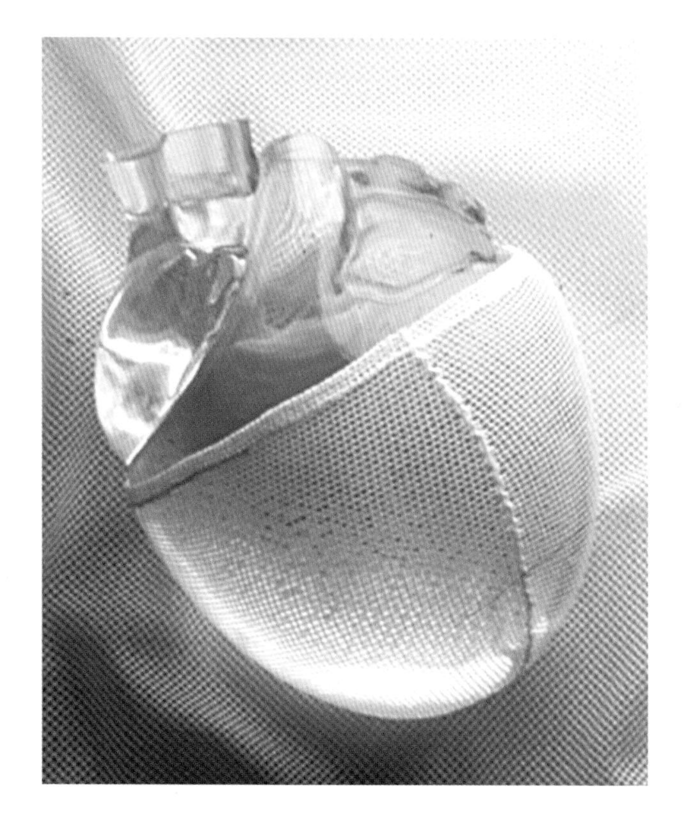

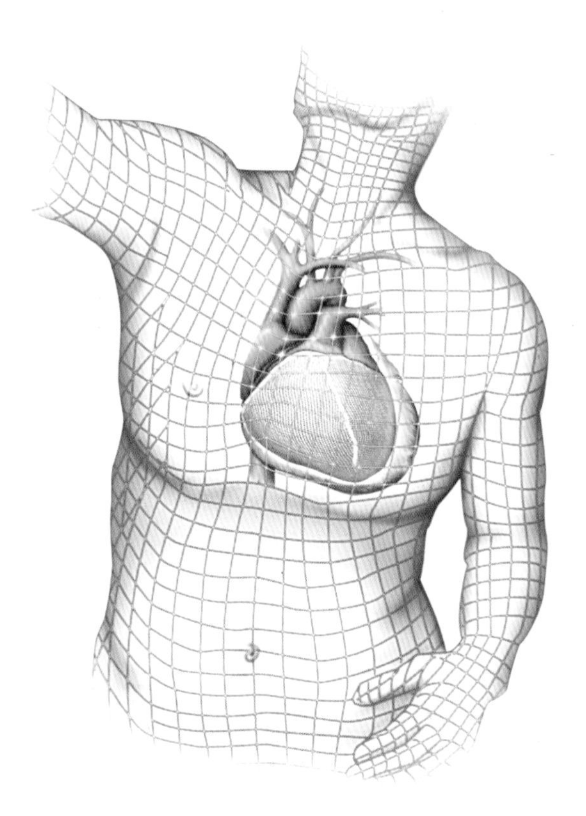

figs. 16, 17 (above)
CorCap™ cardiac support device
Manufactured by Acorn Cardiovascular Inc.™
U.S.A., designed 1998, manufactured 2004
Warp-knit PET polyester multifilament yarn
Diameter: 15.9 cm (6 ¼ in.)

in either direction—known as bidirectional distortion (fig. 15). Acorn Cardiovascular has engineered a warp-knit mesh bag used to treat enlargement of the heart (figs. 16, 17).[13] Degenerative heart failure is a condition in which the compromised heart, damaged by heart attack or coronary disease, loses its ability to pump enough blood to meet the body's needs. The heart compensates by working harder, but over time this damages the heart muscle, causing it to become enlarged and even less efficient. The support device must contain the heart to prevent enlargement, while allowing it to beat normally. Untwisted yarns disperse stresses, preventing the yarns from cutting into the flesh (fig. 18). The knit mesh has more give in the vertical direction, and provides support for the heart wall without interfering with its normal beat.

The difference between knit and woven textiles can be clearly illustrated by the usage of replacement arteries in surgeries. Woven grafts benefit from the dimensional stability of plain weave, and show good resistance to the high pressure of the beat near the heart. Knit grafts, on the other hand, give better flexibility and handling for the physician, potentially reducing the

fig. 18
CorCap™ cardiac support device (detail)

duration of the surgery (fig. 19). Both can be designed with a velour surface to promote tissue in-growth. The small branches are hand-sewn—a surprising reminder of textile handcraft in a high-tech, clean-room environment (fig. 20). In a research collaboration between the University of Rhode Island and Beth Israel Deaconess Medical Center, Dr. Martin Bide, a professor in textiles and dyestuffs, has used his expertise in the area of textile dyeing to help solve the problem of attaching antibiotics to bioimplantable materials, like the vascular grafts, to prevent rejection. In what amounts to a bad dye job, the antibiotics leak out gradually, giving long-term resistance to infection.[14]

## EMBROIDERY

Knitted and woven surgical devices have been successfully implanted for decades, but some needs are not fully addressed by these techniques. New solutions focus on structurally biocompatible implants, combining engineering principles with those of the life sciences. Recently, the advantages of embroidery have been explored by Julian Ellis of Ellis Developments for the creation of surgical implants.

Embroidery has been used for thousands of years to adorn textiles and clothing. While weaving is defined by the interlacing of threads at right angles,

fig. 19 (right)
Hemashield Platinum™ woven double velour vascular graft
Manufactured by Boston Scientific Company Inc.
U.S.A., 2004
Woven, crimped polyester with bovine collagen
Length: 50 cm (19 11/16 in.); diameter: 24mm (15/16 in.); smaller tube diameters: 10, 8, 8, and 10 mm (3/8, 5/16, 5/16, and 3/8 in.)

fig. 20 (below)
Detail showing hand-sewn attachment

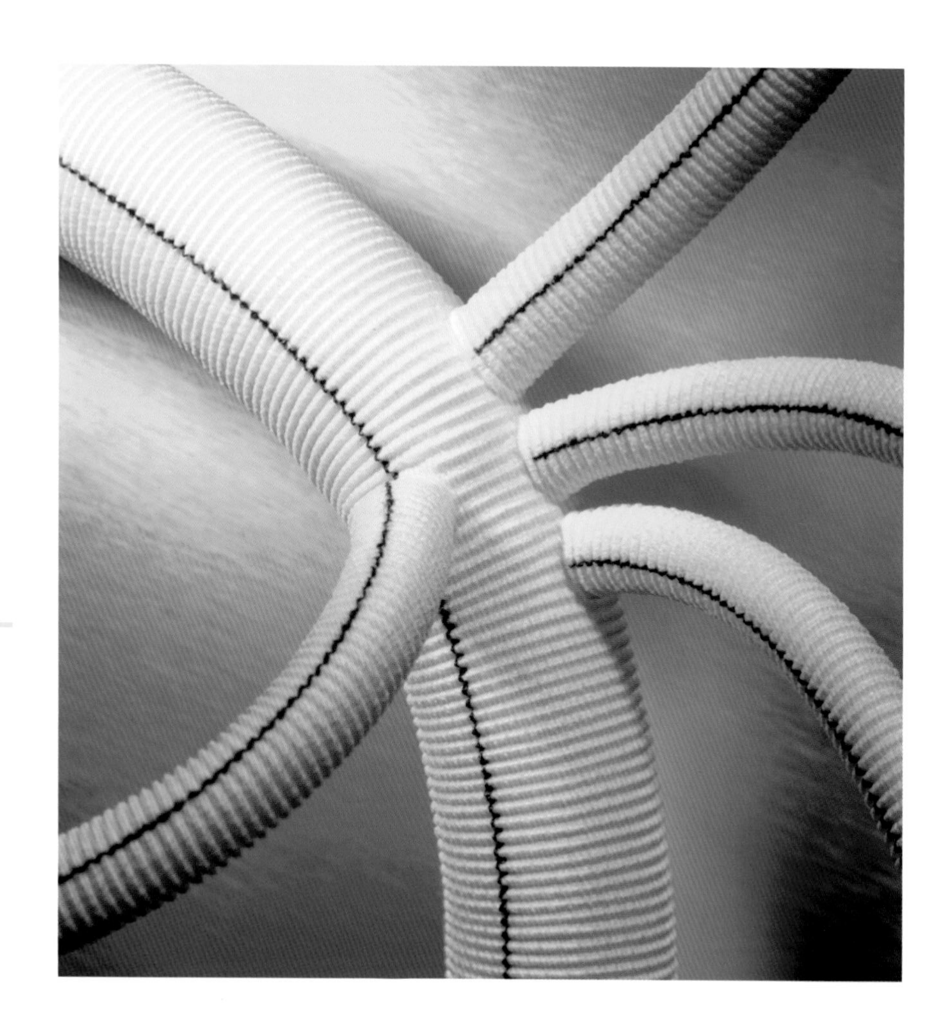

fig. 21
Sample of machine-embroidered lace
U.S.A., ca. 1910
Linen embroidery (ground cloth dissolved)
Cooper-Hewitt, National Design Museum,
Gift of the George Walter Vincent Smith
Art Museum, 1971-26-2

braiding by their interaction at oblique angles, and knitting is created by the formation of loops, embroidery is a surface technique allowing the placement of threads in any position or direction on a base cloth. This freedom makes it the most drawing-like of the textile techniques. Moreover, if appropriate design features have been incorporated into the embroidery design, the base cloth can be dissolved away, leaving an open structure. This technique has been used since the nineteenth century to create machine-made lace (fig. 21).

Modern embroidery uses sophisticated software to quickly produce new designs in order to keep pace with fashion, as well as for personalized items such as monogrammed linens or embroidered name badges. Within the medical context, rapid customization takes on new implications. Modifications to the software, used in conjunction with advanced medical-imaging systems, allow customized implants to be created with equal facility (fig. 22).[15] The oriented fibers can be used to mimic natural fibrous arrays like ligaments, and to match the mechanical properties of the implant to the demands of the host tissue. Embroidery also allows the primary function of such implants—the transference of load—to be achieved by a thread or group of threads, which can be structurally integrated with other features, such as eyelets for the insertion of screws or open mesh areas to promote tissue in-growth. Nature loathes fasteners, as they introduce a point of weakness. Integrated eyelets provide a way of effectively dispersing the strain of the attachment point without compromising the textile (fig. 23).

Bespoke fiber placement is unique to embroidery in the sphere of traditional textile manufacture, but it is being executed on a much larger scale on highly unconventional equipment at North Sails Nevada, with their 3DL™ manufacturing process (fig. 24). Laminate sails have been used in racing for over twenty years, and early laminate sails used traditional sailmaking techniques. Fiber was laid between sheets of Mylar in grid patterns, and these laminates were cut and sewn like ordinary cloth. The grids did not permit spatial variation in fiber density or orientation, and the seams compromised the strength of the sails, negating some of the advantages of high-performance fibers. By contrast, 3DL sails are made as a single piece on an enormous adjustable mold to the precise aerofoil shape ideal for each boat.[16] Just as the stitch heads of an embroidery machine travel freely over the surface of cloth, fiber-laying gantries travel over the surface of the sail molds laying down carbon and aramid fibers on a Mylar scrim (figs. 25, 26). The placement of the fibers reflects the anticipated wind forces and variations in the stress field, and optimizes local strength and stiffness. Unlike the formal, symmetrical microstructure of woven fabrics, this process embraces asymmetry, making the lightest possible sail by putting fiber only where it is needed. This allows the sails to carry astronomical loads: the corner loads on many racing sails are in excess of ten thousand pounds. The specificity of the design and technique is representative of the most radical advances in the field of textiles today.

### NONWOVENS

An underlying theme in the area of technical textiles is, indeed, a second manufacturing revolution, which would move beyond the Industrial Revolution idea of mass production, and its associated mass waste, toward ideas of individualized production or rapid customization. This impetus seems to come from very different sources: from advances in technology that allow accelerated manufacture, and the drive for personal, wearable technology and communications systems, to the idea of biomimicry, which considers that nature produces the world's most superior and most highly specialized materials on an as-needed basis.[17] Many young designers are working with printing and rapid prototyping technologies to bring manufacturing into the home, so that information is bought and sold, but what is produced is only what is actually needed.

fig. 22
Bioimplantable device for reconstructive shoulder surgery
Designed by Prof. Simon Frostick and Dr. Alan McLeod, textile designed by Peter Butcher, developed by Ellis Developments Ltd., manufactured by Pearsalls Ltd.
England, designed 2003, manufactured 2004
Machine-embroidered polyester (base cloth dissolved)
Longest diameter: 14.6 cm (5¾ in.)
Cooper-Hewitt, National Design Museum, 2004-15-1/4

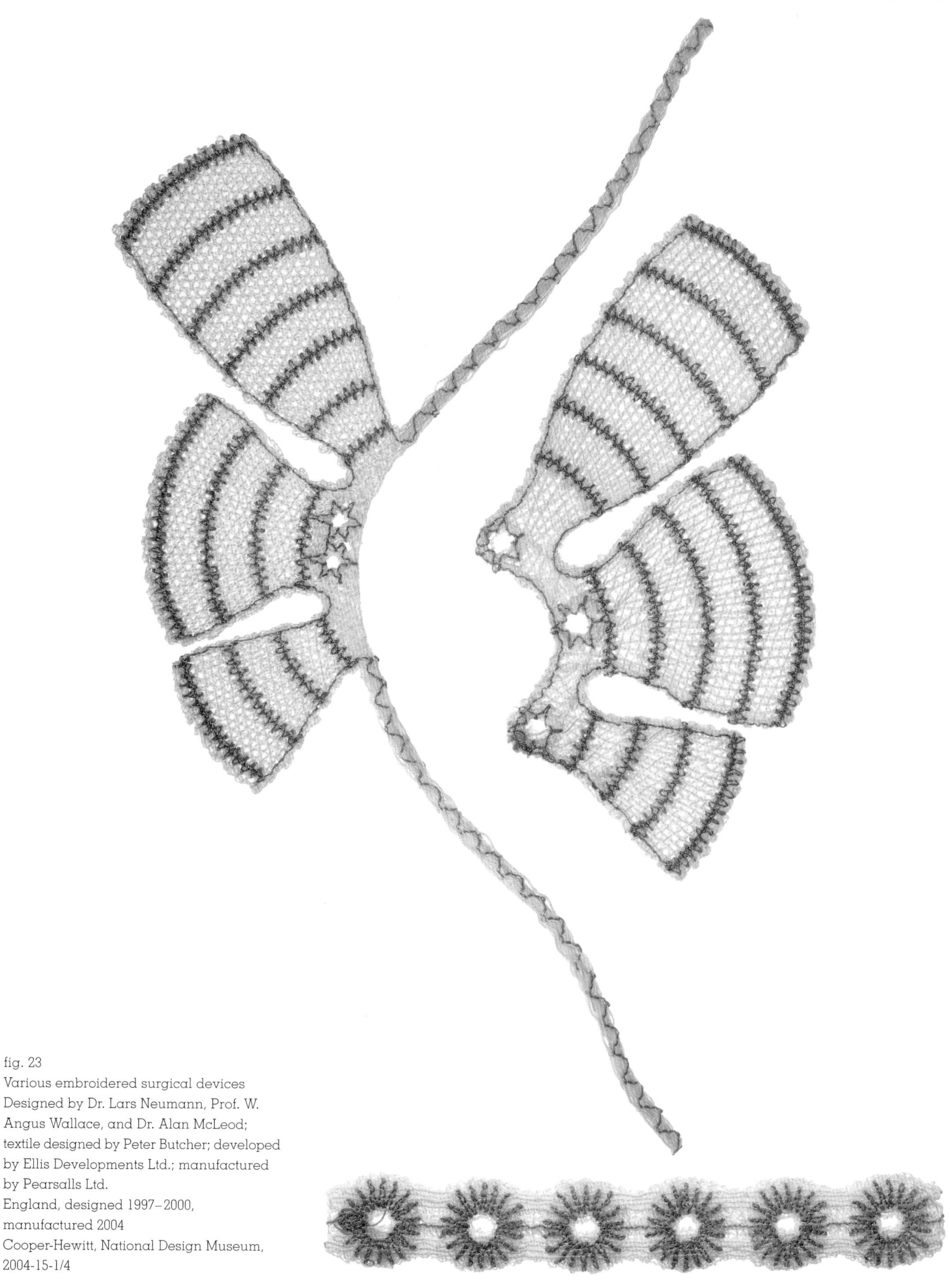

fig. 23
Various embroidered surgical devices
Designed by Dr. Lars Neumann, Prof. W. Angus Wallace, and Dr. Alan McLeod; textile designed by Peter Butcher; developed by Ellis Developments Ltd.; manufactured by Pearsalls Ltd.
England, designed 1997–2000, manufactured 2004
Cooper-Hewitt, National Design Museum, 2004-15-1/4

fig. 24
Farr 50 class boats, some with 3DL™ sails
Starting line at Southern Ocean Racing Conference, Miami Beach, 2001

figs. 25, 26
Fiber head on gantry laying down carbon fiber on the Mylar® film (bottom); aramid fiber sail on the mold table being inspected (top).

With the exception of the sails, all of the objects discussed here are made on traditional textile machinery. But as technology moves toward fibers that are more specialized and difficult to process, new textile-manufacturing techniques will be needed. Nonwovens—fibrous constructions similar to felt, known to most of us as wipes, diapers, or Tyvek®—are the fastest growing area in the textile industry (fig. 28). In nonwoven techniques, precursor polymers are transformed through the fiber stage directly to textiles in a single manufacturing process.

A major focus of interest is nanotechnology—the manipulation of materials on the atomic level. Textiles made from nanofibers, each on the order of 1/180,000 of the breadth of a human hair, offer very small fiber diameter and pore size, high absorption, and a large number of chemical functional groups along their molecular chains. This combination of properties has far-reaching implications in filtration, health care, energy storage, and bioengineering. The very large surface area provides infinite attachment points for molecules—for example, drugs for a bandage or even a clothing-based drug-delivery system, or reactive molecules capable of sensing chemical hazards in the environment. The large surface area could also allow for the entrapment of molecules in all varieties of filtration applications: air, chemical, and blood. Nanoscale fibers made from carbon, called carbon nanotubes, have bonds stronger than those in diamonds. A fiber that is 60% nanotubes is twenty times tougher than

fig. 27
Electrospun fiber mat
Developed by U.S. Army Natick Soldier Center, fabricated by Dr. Heidi Schreuder-Gibson
U.S.A., 2004
Fine layer of electrospun nanofibers of polyethylene oxide (spun onto a clear mask)

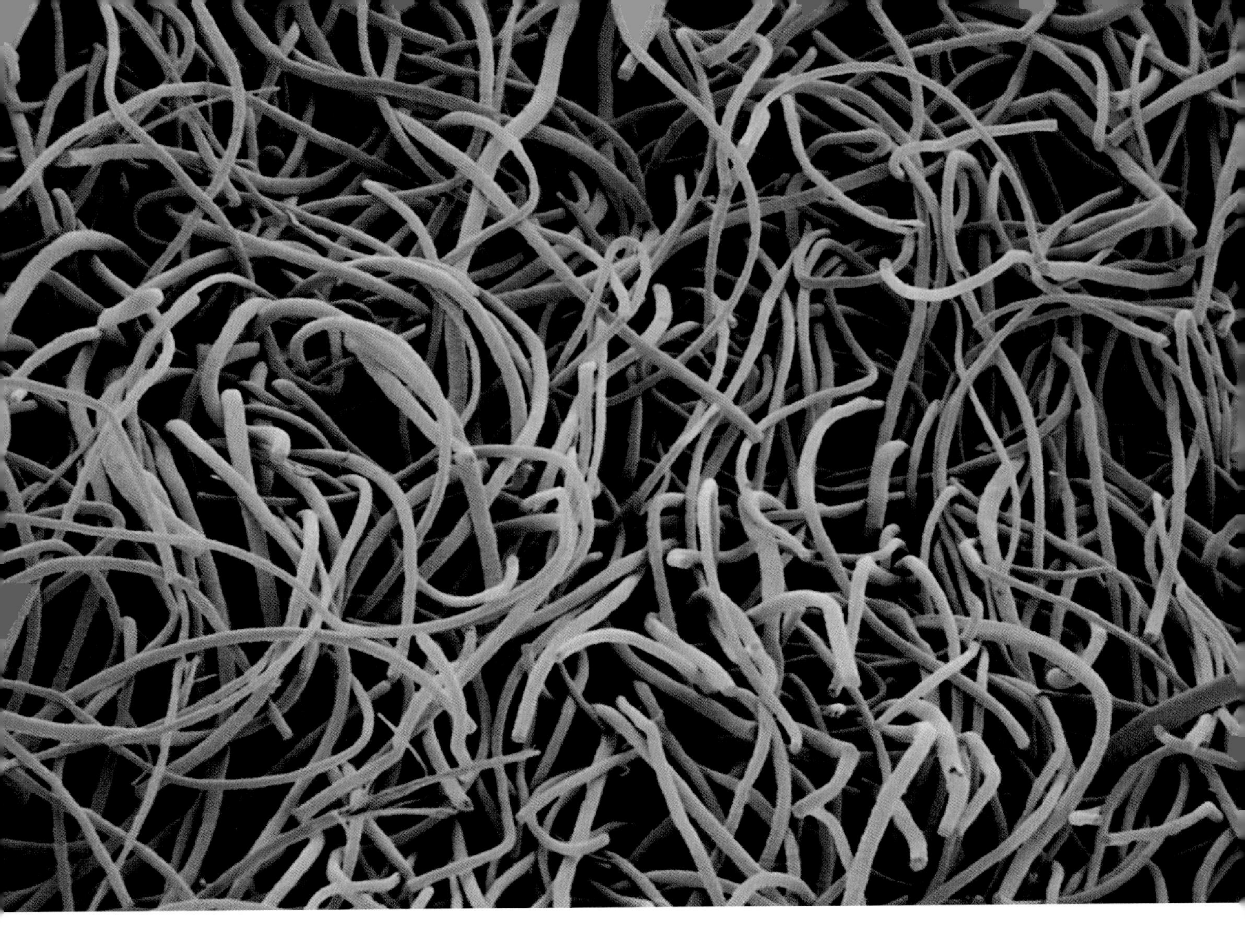

fig. 28
Wool felt

steel. Since the carbon tubes also have electrical properties, they could be pivotal in the creation of responsive materials and molecular-scale electronic devices.

Nanofiber membranes are generally produced by electrospinning, in which a liquid polymer solution is drawn toward a highly charged metal plate, pulling it into nanoscale fibers (fig. 27). While standard textile polymers can be used, some researchers are also working with biopolymers, such as collagen and elastin, both of which occur naturally in the human body. The superfine network of fibers provides an ideal scaffold for tissue engineering, for the replacement of damaged organs or tissues. To return to the example of the vascular graft, a collagen tube made by electrospinning technology could be six times smaller than the smallest available graft, and could grow with the patient.[18]

The idea of manipulating materials at the nanoscale, of integrating functionality at the atomic level, blurs the line between what materials are and what they do. Textile technologies are undergoing a profound change. But the unexpected, creative, and successful consideration of textiles by engineers and designers in widely diverse fields assures that textiles will remain the materials that shape our world.

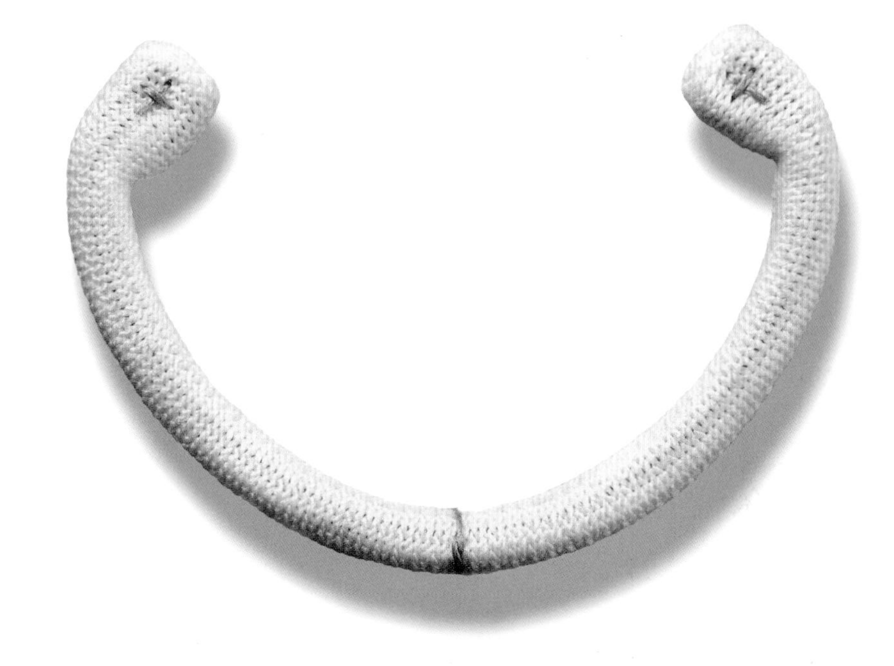

fig. 29 (top right)
C-G Future™ Band valve repair device
Designed by Dr. Aubrey C. Galloway and Dr. Stephen B. Colvin, engineered by Tim Ryan, manufactured by Medtronic Heart Valves
U.S.A., 2004
Metallic core of MP-35N covered by thin layer of silicon, polyester
2.5 x 3.5 cm (1 x 1⅜ in.)

This mitral valve repair device utilizes a metal alloy core to facilitate semi-rigid remodeling of a faulty valve while retaining enough flexibility to allow the valve to close during the contraction of the heart. The core is covered by a loosely woven layer of polyester, which allows for tissue in-growth for incorporation of the device into the heart, while the eyelets ensure secure anchoring.

fig. 30 (bottom right)
Arterial filter
Manufactured by Gish Biomedical Inc., textile designed and manufactured by Sefar Filtration Inc.
U.S.A., designed 2001, manufactured 2004
Pleated twill woven polyester monofilament, 25 micron pore size, 37 micron thread diameter, 450 threads per inch

fig. 31
Self-lubricating gears and lubrication cogwheels
Manufactured by Vereinigte Filzfabriken AG
Germany, 2004
Wool needle felt

fig. 32
Knitted metal catalytic converters and industrial filtration devices
Manufactured by Buck Enterprises LLC
Germany, 2004
Machine-knit stainless steel, compressed

fig. 33
Copsil™ press mat
Designed by Mel Douglas, manufactured by Marathon Belting Ltd.
England, designed 1994, manufactured 2005
Herringbone twill weave of brass wire warp and silicone-encased copper wire weft

Used as a press mat for industrial lamination processes, Copsil™ employs a silicon-elastomer in the weft to prevent deformation of the weave under compression. The use of a herringbone twill weave produces a flat surface with high brass content, maximizing thermal efficiency, while the reversal of the herringbone balances the spring of the elastomer to prevent mat shifting.

fig. 34
Lifting belt sling
Designed by Ruedi Hess, manufactured by Mammut tec AG, engineering partner ETH, Zurich
Germany, designed 2002, manufactured 2004
Parallel strands of PES high-strength polyester monofilament in a channeled sleeve of cut- and abrasion-resistant woven Kevlar®

This twenty-foot-long sling of polyester monofilament has a lifting capacity of fifty tons, or 111,000 pounds of dynamic load. The actual breaking strength is 770,000 pounds. The soft, flexible fabric protects the object being lifted, while the abrasion-resistant Kevlar® sleeve prevents damage to the sling.

# ROPES

John W. S. Hearle

The ancients knew how to make the best use of natural materials to create strong ropes. A relief found by the British adventurer Austen Henry Layard when he excavated Nineveh, in modern-day Iraq, shows a colossus with scores of men pulling it along on thick three-strand ropes. In 480 BC, Xerxes crossed the Hellespont on a bridge of boats lashed to six cables, two of flax and four of papyrus. And in the 1940s, soldiers exploring a cave on the banks of the Nile found a rope wrapped round a huge block of stone similar to those used in the pyramids. The rope, which was dated to about 500 BC, was similar in construction to a modern rope: seven papyrus fibers were twisted into yarns, forty yarns were twisted into strands, and three strands were twisted into the rope.

Ropes, made from natural fibers, changed little in their design for another two thousand years, but from the mid-nineteenth century, steel-wire ropes replaced fiber ropes for heavy engineering uses. Advances in fiber ropes were made in the mid-twentieth century with the introduction of nylon ropes, followed by polyester. For the same strength, these ropes were about half the weight of steel ropes, but about twice the diameter. Ropes made from the second generation of synthetic fibers—aramid, high-modulus polyethylene, Vectran, and polyphenylene benzobisoxazole (PBO)—give diameters similar to steel but in one-tenth its weight (fig. 35).

Whereas natural fibers have to be highly twisted together to prevent the fibers sliding over one another, this is unnecessary with the continuous filament yarns. New low-twist constructions, with just enough structure to give coherence to the ropes, have been developed. Where ropes need to stretch and to absorb high-impact energies, nylon and polyester are the design choices; where resistance to extension is needed, the newer high-modulus fibers are preferred. In familiar uses of ropes, the change in recent years has been in designing for purpose. Until fifty years ago, mountaineers used common hemp ropes; now there is a range of ropes optimized for the different uses in climbing (fig. 36). There is a similar pattern in yachting ropes: one choice for the Olympic racer, and another for the weekend sailor.

In the most demanding applications, fiber ropes are now competing with steel. Since 1997, Petrobras has used polyester ropes to moor about twenty oil rigs in deep water off the coast of Brazil. The first installation in North America was in March 2004 for BP's Mad Dog floating production system in 4,500 feet of water in the Gulf of Mexico. The great advantage for BP and its Mad Dog partners was the low weight of polyester compared to steel. Among textile fibers, polyester has the right balance of properties: rugged durability and enough extensibility to prevent large tensions developing as the rigs rise and fall. For Mad Dog, Marlow Superline polyester rope, with a diameter of ten inches and break load of two thousand tons, was used—the strongest fiber

fig. 35
Neutron-8™ ropes
Manufactured by Samson Rope Technologies U.S.A., designed 2003, manufactured 2004
8 x 3 Strand™ of Dyneema® fiber with Samthane™ coating

fig. 36
Mountain-climbing ropes
Manufactured by Edelrid
Germany, 2004

figs. 37, 38
Marlow Superline polyester rope; installation of BP's Mad Dog spar in the Gulf of Mexico
Rope manufactured by Marlow Ropes Ltd., UK; fiber and yarn manufactured by Honeywell International, U.S.A.
England, 2003
Braided polyester
Diameter: 27 cm (10⅝ in.)

rope ever made (figs. 37, 38). The eleven mooring lines taking over ten miles of rope needed more than one thousand tons of polyester yarns, which, for the fiber alone, would cost over three million dollars.

In marine applications, fiber ropes are well established. The challenge for the twenty-first century is to replace steel in terrestrial applications such as bridge cables, elevator hoists, cranes, and so on. The technical advantages are clear; the limitation is the conservatism of design engineers who do not want to replace a material that has been used for 150 years with a new material untested in use.[1]

# HIGH-PERFORMANCE FIBERS

Alyssa Becker

Can a fiber withstand the shock of a bullet blast? Is it flexible and absorbent enough to wear comfortably, or stiff and strong enough to reinforce a concrete wall? Does it conduct heat quickly, melt into a sticky mass, or simply decompose? Will it ignite with an open flame? What happens to it after months of sunlight exposure, contact with seawater, or battery acid? Can it fold without splitting?

High-performance fibers are noted typically for being extremely stiff and strong. They are described as high-modulus (resisting stretch when pulled) and high-tenacity (breaking only under great force). Neither modulus nor tenacity, however, fully describes a fiber's behavior; other performance criteria include extreme operating temperatures and flame or chemical resistance. Some of the selected fibers described below are stronger and stiffer than others. However, each material has its own advantages, and can be used alone or in combination to maximize performance.

## GLASS

Glass was first spun as a technical fiber in the late 1930s. Glass fibers are strong because their main ingredient, silicon dioxide, forms a three-dimensional network of chemical bonds.[1] The inclusion of other oxides modifies its strength and chemical stability. Lighter than metal, relatively inert, and conducting little heat or sound, glass fibers reinforce boat hulls and heat-sensitive composites such as circuit boards, and make effective insulation and high-temperature filtration materials.[2] Purified glass fibers transmit light in optical cables.[3] Virtually incombustible, glass is also used for high-heat applications in the aerospace and missile industries. It is less suited for heat-resistant apparel: denser than other high-performance fibers and brittle, glass fibers abrade easily and irritate the skin.[4]

## CERAMIC

Ceramic fibers such as Nextel™ (3M) or Sylramic (Dow Corning) are spun from organic or mineral precursor materials, which are then heated or pyrolyzed. They withstand higher temperatures than glass and more corrosive environments than carbon.[5] Ceramics conduct heat quickly: they protect pressurized rocket gas lines from the flames of rocket plumes, and insulate race-car engines to keep temperatures cool in the driver's seat.[6] Woven into protective outerwear, they dissipate heat from molten-metal splashes. Ceramic fibers also make noncorroding reinforcements for cables and metal structures. While manufacturing refinements have improved their dimensional stability at high temperatures, ceramic fibers remain relatively brittle and heavy, limiting their use in protective apparel.[7]

figs. 39–41
Photomicrographs of bicomponent fibers
16-segment pie (top)
Hollow-core 16-segment pie (middle)
"Island in the Stream" (bottom)

The numbers in these tables represent typical values of some important fiber properties; the actual behavior of fibers may differ as variants are produced for diverse end uses. These numbers were compiled from many different sources and are meant for illustration purposes only.

COMPARISON OF YARN STRENGTH

M5
PBO
LCP
HMPE
P-Aramid
Carbon
Ceramic
Glass
Polyester
Nylon
Steel
0 1 2 3 4 5 6 7
Yarn strength based on area of fiber (GPa)
Yarn strength based on weight of fiber (N/tex)

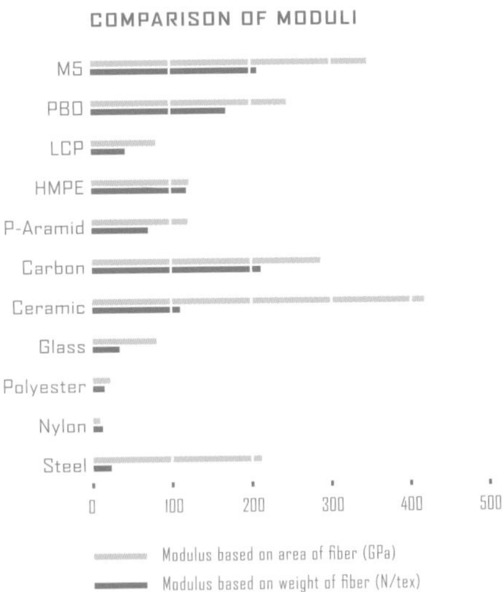


## CARBON

Thomas Edison first used carbon fiber when he employed charred cotton thread to conduct electricity in a lightbulb (he patented it in 1879). Only in the past fifty years, however, has carbon developed as a high-strength, high-modulus fiber.[8] Oxidized then carbonized from polyacrylonitrile (PAN) or pitch precursor fibers, carbon's tenacity and modulus vary depending on its starting materials and process of manufacture.[9]

Less dense than ceramic or glass, lightweight carbon-fiber composites save fuel when used in aerospace and automotive vehicles. They also make for strong, efficient sports equipment. Noncorroding, carbon reinforcements strengthen deep seawater concrete structures such as petroleum production risers.[10] Fine diameter carbon fibers are woven into sails to minimize stretch.[11] In outer apparel, carbon fibers protect workers against open flames (up to 1000°C/1,800°F) and even burning napalm: they will not ignite, and shrink very little in high temperatures.[12]

## ARAMIDS

Aramids, such as Kevlar (DuPont) and Twaron® (Teijin), are famous for their use in bulletproof vests and other forms of ballistic protection, as well as for cut resistance and flame retardance. Initially developed in the 1960s, aramids are strong because their long molecular chains are fully extended and packed closely together, resulting in high-tenacity, high-modulus fibers.[13]

Corrosion- and chemical-resistant, aramids are used in aerial and mooring ropes and construction cables, and provide mechanical protection in optical fiber cables.[14] Like carbon, aramid-composite materials make light aircraft components and sporting goods, but aramids have the added advantages of impact resistance and energy absorption.

## LIQUID CRYSTAL POLYMER (LCP)

Although spun from different polymers and processes, LCPs resemble aramids in their strength, impact resistance, and energy absorption, as well as their sensitivity to UV light. Compared to aramids, Vectran (Celanese), the only commercially available LCP, is more resistant to abrasion, has better flexibility, and retains its strength longer when exposed to high temperatures. Vectran also surpasses aramids and HMPE in dimensional stability and cut resistance: it is used in wind sails for America's Cup races, inflatable structures, ropes, cables and restraint-lines, and cut-resistant clothing.[15] Because it can be sterilized by gamma rays, Vectran is used for medical devices such as implants and surgical-device control cables.[16]

## HIGH-MODULUS POLYETHYLENE (HMPE)

HMPE, known by the trade names Dyneema (Toyobo/DSM) or Spectra (Honeywell), is made from ultra-high molecular-weight polyethylene by a special gel-spinning process. It is the least dense of all the high-performance

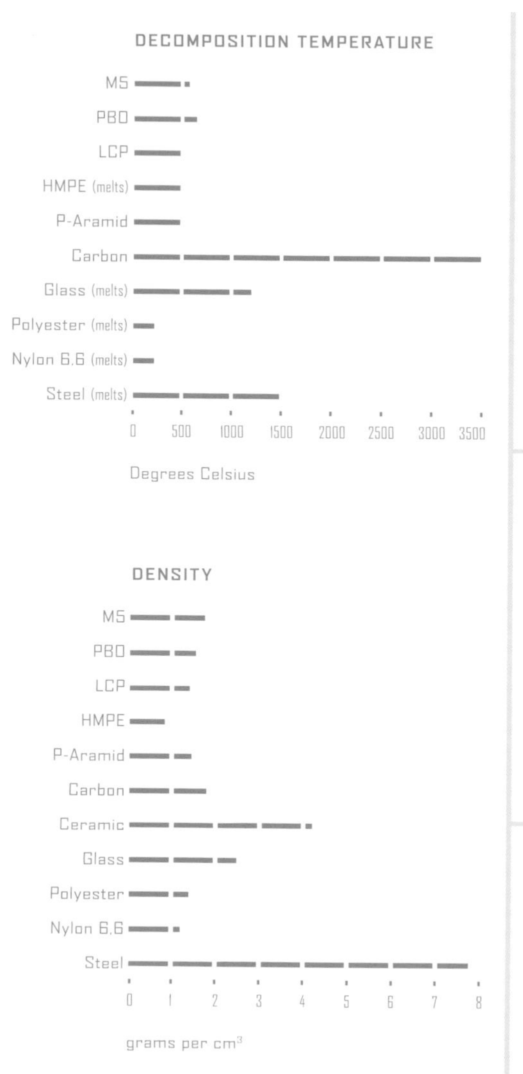


fibers, and the most abrasion-resistant. It is also more resistant than aramids, PBO, and LCP to UV radiation and chemicals.[17] It makes for moorings and fish lines that float and withstand the sun, as well as lightweight, cut-resistant gloves and protective apparel such as fencing suits and soft ballistic armor. In composites, it lends impact resistance and energy absorption to glass- or carbon-reinforced products. HMPE conducts almost no electricity, making it transparent to radar.[18] HMPE does not withstand gamma-ray sterilization and has a relatively low melting temperature of 150°C (300°F)—two qualities that preclude its use where high temperature resistance is a must.

### POLYPHENYLENE BENZOBISOXAZOLE (PBO)

PBO fibers surpass aramids in flame resistance, dimensional stability, and chemical and abrasion resistance, but are sensitive to photodegradation and hydrolysis in warm, moist conditions.[19] Their stiff molecules form highly rigid structures, which grant an extremely high tenacity and modulus. Apparel containing Zylon® (Toyobo), the only PBO fiber in commercial production, provides ballistic protection because of its high energy absorption and dissipation of impact. Zylon is also used in the knee pads of motorcycle apparel, for heat-resistant work wear, and in felt used for glass formation.[20]

### PIPD

PIPD, M5 fiber (Magellan Systems International), expected to come into commercial production in 2005, matches or exceeds aramids and PBO in many of its properties. However, because the molecules have strong lateral bonding, as well as great strength along the oriented chains, M5 has much better shear and compression resistance. In composites it shows good adhesion to resins. Its dimensional stability under heat, resistance to UV radiation and fire, and transparency to radar expands its possible uses. Potential applications include soft and hard ballistic protection, fire protection, ropes and tethers, and structural composites.[21]

### HYBRIDS

A blend of polymers in a fabric, yarn, or fiber structure can achieve a material better suited for its end use. Comfortable fire-retardant, antistatic clothing may be woven primarily from aramid fibers but feature the regular insertion of a carbon filament to dissipate static charge. Yarns for cut-resistant applications maintain good tactile properties with a wrapping of cotton around HMPE and fiberglass cores. On a finer level, a single fiber can be extruded from two or more different polymers in various configurations to exhibit the properties of both.

RRRRRRRRRRRRRRRRRRRRRRRRRRRRRRRRRRRRRRRRRRRRRRRRRRRRRRRRRRRRRRRRRRRRRRRRRRRRRRRRRRRRRRRRRRRRRRRRRRRRRRRRRRRRRRRRRRRRRRRRRRRRRRRRRRRRRRRRRRRRRRRRRRRRRRRRRRRRRRRRRRRRRRRRRRRRRRRR
FFFFFFFFFFFFFFFFFFFFFFFFFFFFFFFFFFFFFFFFFFFFFFFFFFFFFFFFFFFFFFFFFFFFFFFFFFFFFFFFFFFFFFFFFFFFFFFFFFFFFFFFFFFFFFFFFFFFFFFFFFFFFFFFFFFFFFFFFFFFFFFFFFFFFFFFFFFFFFFFFFFFFFFFFFFFFFFFFFFF
AAAAAAAAAAAAAAAAAAAAAAAAAAAAAAAAAAAAAAAAAAAAAAAAAAAAAAAAAAAAAAAAAAAAAAAAAAAAAAAAAAAAAAAAAAAAAAAAAAAAAAAAAAAAAAAAAAAAAAAAAAAAAAAAAAAAAAAAAAAAAAAAAAAAAAAAAAAAAAAAAAAAAAAAAAAAAAAAAAAA
SSSSSSSSSSSSSSSSSSSSSSSSSSSSSSSSSSSSSSSSSSSSSSSSSSSSSSSSSSSSSSSSSSSSSSSSSSSSSSSSSSSSSSSSSSSSSSSSSSSSSSSSSSSSSSSSSSSSSSSSSSSSSSSSSSSSSSSSSSSSSSSSSSSSSSSSSSSSSSSSSSSSSSSSSSSSSSSSSSSS
TTTTTTTTTTTTTTTTTTTTTTTTTTTTTTTTTTTTTTTTTTTTTTTTTTTTTTTTTTTTTTTTTTTTTTTTTTTTTTTTTTTTTTTTTTTTTTTTTTTTTTTTTTTTTTTTTTTTTTTTTTTTTTTTTTTTTTTTTTTTTTTTTTTTTTTTTTTTTTTTTTTTTTTTTTTTTTTTTTTT
EEEEEEEEEEEEEEEEEEEEEEEEEEEEEEEEEEEEEEEEEEEEEEEEEEEEEEEEEEEEEEEEEEEEEEEEEEEEEEEEEEEEEEEEEEEEEEEEEEEEEEEEEEEEEEEEEEEEEEEEEEEEEEEEEEEEEEEEEEEEEEEEEEEEEEEEEEEEEEEEEEEEEEEEEEEEEEEEEEEEE
RRRRRRRRRRRRRRRRRRRRRRRRRRRRRRRRRRRRRRRRRRRRRRRRRRRRRRRRRRRRRRRRRRRRRRRRRRRRRRRRRRRRRRRRRRRRRRRRRRRRRRRRRRRRRRRRRRRRRRRRRRRRRRRRRRRRRRRRRRRRRRRRRRRRRRRRRRRRRRRRRRRRRRRRRRRRRRRRR
FFFFFFFFFFFFFFFFFFFFFFFFFFFFFFFFFFFFFFFFFFFFFFFFFFFFFFFFFFFFFFFFFFFFFFFFFFFFFFFFFFFFFFFFFFFFFFFFFFFFFFFFFFFFFFFFFFFFFFFFFFFFFFFFFFFFFFFFFFFFFFFFFFFFFFFFFFFFFFFFFFFFFFFFFFFFFFFFFFFF
AAAAAAAAAAAAAAAAAAAAAAAAAAAAAAAAAAAAAAAAAAAAAAAAAAAAAAAAAAAAAAAAAAAAAAAAAAAAAAAAAAAAAAAAAAAAAAAAAAAAAAAAAAAAAAAAAAAAAAAAAAAAAAAAAAAAAAAAAAAAAAAAAAAAAAAAAAAAAAAAAAAAAAAAAAAAAAAAAAAA
SSSSSSSSSSSSSSSSSSSSSSSSSSSSSSSSSSSSSSSSSSSSSSSSSSSSSSSSSSSSSSSSSSSSSSSSSSSSSSSSSSSSSSSSSSSSSSSSSSSSSSSSSSSSSSSSSSSSSSSSSSSSSSSSSSSSSSSSSSSSSSSSSSSSSSSSSSSSSSSSSSSSSSSSSSSSSSSSSS
TTTTTTTTTTTTTTTTTTTTTTTTTTTTTTTTTTTTTTTTTTTTTTTTTTTTTTTTTTTTTTTTTTTTTTTTTTTTTTTTTTTTTTTTTTTTTTTTTTTTTTTTTTTTTTTTTTTTTTTTTTTTTTTTTTTTTTTTTTTTTTTTTTTTTTTTTTTTTTTTTTTTTTTTTTTTTTTTTTTT
EEEEEEEEEEEEEEEEEEEEEEEEEEEEEEEEEEEEEEEEEEEEEEEEEEEEEEEEEEEEEEEEEEEEEEEEEEEEEEEEEEEEEEEEEEEEEEEEEEEEEEEEEEEEEEEEEEEEEEEEEEEEEEEEEEEEEEEEEEEEEEEEEEEEEEEEEEEEEEEEEEEEEEEEEEEEEEEEEEEE
RRRRRRRRRRRRRRRRRRRRRRRRRRRRRRRRRRRRRRRRRRRRRRRRRRRRRRRRRRRRRRRRRRRRRRRRRRRRRRRRRRRRRRRRRRRRRRRRRRRRRRRRRRRRRRRRRRRRRRRRRRRRRRRRRRRRRRRRRRRRRRRRRRRRRRRRRRRRRRRRRRRRRRRRRRRRRRRR
FFFFFFFFFFFFFFFFFFFFFFFFFFFFFFFFFFFFFFFFFFFFFFFFFFFFFFFFFFFFFFFFFFFFFFFFFFFFFFFFFFFFFFFFFFFFFFFFFFFFFFFFFFFFFFFFFFFFFFFFFFFFFFFFFFFFFFFFFFFFFFFFFFFFFFFFFFFFFFFFFFFFFFFFFFFFFFFFFFFF
AAAAAAAAAAAAAAAAAAAAAAAAAAAAAAAAAAAAAAAAAAAAAAAAAAAAAAAAAAAAAAAAAAAAAAAAAAAAAAAAAAAAAAAAAAAAAAAAAAAAAAAAAAAAAAAAAAAAAAAAAAAAAAAAAAAAAAAAAAAAAAAAAAAAAAAAAAAAAAAAAAAAAAAAAAAAAAAAAAAA
SSSSSSSSSSSSSSSSSSSSSSSSSSSSSSSSSSSSSSSSSSSSSSSSSSSSSSSSSSSSSSSSSSSSSSSSSSSSSSSSSSSSSSSSSSSSSSSSSSSSSSSSSSSSSSSSSSSSSSSSSSSSSSSSSSSSSSSSSSSSSSSSSSSSSSSSSSSSSSSSSSSSSSSSSSSSSSSSSS
TTTTTTTTTTTTTTTTTTTTTTTTTTTTTTTTTTTTTTTTTTTTTTTTTTTTTTTTTTTTTTTTTTTTTTTTTTTTTTTTTTTTTTTTTTTTTTTTTTTTTTTTTTTTTTTTTTTTTTTTTTTTTTTTTTTTTTTTTTTTTTTTTTTTTTTTTTTTTTTTTTTTTTTTTTTTTTTTTTTT
EEEEEEEEEEEEEEEEEEEEEEEEEEEEEEEEEEEEEEEEEEEEEEEEEEEEEEEEEEEEEEEEEEEEEEEEEEEEEEEEEEEEEEEEEEEEEEEEEEEEEEEEEEEEEEEEEEEEEEEEEEEEEEEEEEEEEEEEEEEEEEEEEEEEEEEEEEEEEEEEEEEEEEEEEEEEEEEEEEEE
RRRRRRRRRRRRRRRRRRRRRRRRRRRRRRRRRRRRRRRRRRRRRRRRRRRRRRRRRRRRRRRRRRRRRRRRRRRRRRRRRRRRRRRRRRRRRRRRRRRRRRRRRRRRRRRRRRRRRRRRRRRRRRRRRRRRRRRRRRRRRRRRRRRRRRRRRRRRRRRRRRRRRRRRRRRRRRRR
FFFFFFFFFFFFFFFFFFFFFFFFFFFFFFFFFFFFFFFFFFFFFFFFFFFFFFFFFFFFFFFFFFFFFFFFFFFFFFFFFFFFFFFFFFFFFFFFFFFFFFFFFFFFFFFFFFFFFFFFFFFFFFFFFFFFFFFFFFFFFFFFFFFFFFFFFFFFFFFFFFFFFFFFFFFFFFFFFFFF
AAAAAAAAAAAAAAAAAAAAAAAAAAAAAAAAAAAAAAAAAAAAAAAAAAAAAAAAAAAAAAAAAAAAAAAAAAAAAAAAAAAAAAAAAAAAAAAAAAAAAAAAAAAAAAAAAAAAAAAAAAAAAAAAAAAAAAAAAAAAAAAAAAAAAAAAAAAAAAAAAAAAAAAAAAAAAAAAAAAA
SSSSSSSSSSSSSSSSSSSSSSSSSSSSSSSSSSSSSSSSSSSSSSSSSSSSSSSSSSSSSSSSSSSSSSSSSSSSSSSSSSSSSSSSSSSSSSSSSSSSSSSSSSSSSSSSSSSSSSSSSSSSSSSSSSSSSSSSSSSSSSSSSSSSSSSSSSSSSSSSSSSSSSSSSSSSSSSSSS
TTTTTTTTTTTTTTTTTTTTTTTTTTTTTTTTTTTTTTTTTTTTTTTTTTTTTTTTTTTTTTTTTTTTTTTTTTTTTTTTTTTTTTTTTTTTTTTTTTTTTTTTTTTTTTTTTTTTTTTTTTTTTTTTTTTTTTTTTTTTTTTTTTTTTTTTTTTTTTTTTTTTTTTTTTTTTTTTTTTT
EEEEEEEEEEEEEEEEEEEEEEEEEEEEEEEEEEEEEEEEEEEEEEEEEEEEEEEEEEEEEEEEEEEEEEEEEEEEEEEEEEEEEEEEEEEEEEEEEEEEEEEEEEEEEEEEEEEEEEEEEEEEEEEEEEEEEEEEEEEEEEEEEEEEEEEEEEEEEEEEEEEEEEEEEEEEEEEEEEEE
RRRRRRRRRRRRRRRRRRRRRRRRRRRRRRRRRRRRRRRRRRRRRRRRRRRRRRRRRRRRRRRRRRRRRRRRRRRRRRRRRRRRRRRRRRRRRRRRRRRRRRRRRRRRRRRRRRRRRRRRRRRRRRRRRRRRRRRRRRRRRRRRRRRRRRRRRRRRRRRRRRRRRRRRRRRRRRRR
FFFFFFFFFFFFFFFFFFFFFFFFFFFFFFFFFFFFFFFFFFFFFFFFFFFFFFFFFFFFFFFFFFFFFFFFFFFFFFFFFFFFFFFFFFFFFFFFFFFFFFFFFFFFFFFFFFFFFFFFFFFFFFFFFFFFFFFFFFFFFFFFFFFFFFFFFFFFFFFFFFFFFFFFFFFFFFFFFFFF
AAAAAAAAAAAAAAAAAAAAAAAAAAAAAAAAAAAAAAAAAAAAAAAAAAAAAAAAAAAAAAAAAAAAAAAAAAAAAAAAAAAAAAAAAAAAAAAAAAAAAAAAAAAAAAAAAAAAAAAAAAAAAAAAAAAAAAAAAAAAAAAAAAAAAAAAAAAAAAAAAAAAAAAAAAAAAAAAAAAA
SSSSSSSSSSSSSSSSSSSSSSSSSSSSSSSSSSSSSSSSSSSSSSSSSSSSSSSSSSSSSSSSSSSSSSSSSSSSSSSSSSSSSSSSSSSSSSSSSSSSSSSSSSSSSSSSSSSSSSSSSSSSSSSSSSSSSSSSSSSSSSSSSSSSSSSSSSSSSSSSSSSSSSSSSSSSSSSSSS
TTTTTTTTTTTTTTTTTTTTTTTTTTTTTTTTTTTTTTTTTTTTTTTTTTTTTTTTTTTTTTTTTTTTTTTTTTTTTTTTTTTTTTTTTTTTTTTTTTTTTTTTTTTTTTTTTTTTTTTTTTTTTTTTTTTTTTTTTTTTTTTTTTTTTTTTTTTTTTTTTTTTTTTTTTTTTTTTTTTT
EEEEEEEEEEEEEEEEEEEEEEEEEEEEEEEEEEEEEEEEEEEEEEEEEEEEEEEEEEEEEEEEEEEEEEEEEEEEEEEEEEEEEEEEEEEEEEEEEEEEEEEEEEEEEEEEEEEEEEEEEEEEEEEEEEEEEEEEEEEEEEEEEEEEEEEEEEEEEEEEEEEEEEEEEEEEEEEEEEE
RRRRRRRRRRRRRRRRRRRRRRRRRRRRRRRRRRRRRRRRRRRRRRRRRRRRRRRRRRRRRRRRRRRRRRRRRRRRRRRRRRRRRRRRRRRRRRRRRRRRRRRRRRRRRRRRRRRRRRRRRRRRRRRRRRRRRRRRRRRRRRRRRRRRRRRRRRRRRRRRRRRRRRRRRRRRRRRR
FFFFFFFFFFFFFFFFFFFFFFFFFFFFFFFFFFFFFFFFFFFFFFFFFFFFFFFFFFFFFFFFFFFFFFFFFFFFFFFFFFFFFFFFFFFFFFFFFFFFFFFFFFFFFFFFFFFFFFFFFFFFFFFFFFFFFFFFFFFFFFFFFFFFFFFFFFFFFFFFFFFFFFFFFFFFFFFFFFFF
AAAAAAAAAAAAAAAAAAAAAAAAAAAAAAAAAAAAAAAAAAAAAAAAAAAAAAAAAAAAAAAAAAAAAAAAAAAAAAAAAAAAAAAAAAAAAAAAAAAAAAAAAAAAAAAAAAAAAAAAAAAAAAAAAAAAAAAAAAAAAAAAAAAAAAAAAAAAAAAAAAAAAAAAAAAAAAAAAAAA
SSSSSSSSSSSSSSSSSSSSSSSSSSSSSSSSSSSSSSSSSSSSSSSSSSSSSSSSSSSSSSSSSSSSSSSSSSSSSSSSSSSSSSSSSSSSSSSSSSSSSSSSSSSSSSSSSSSSSSSSSSSSSSSSSSSSSSSSSSSSSSSSSSSSSSSSSSSSSSSSSSSSSSSSSSSSSSSSSS
TTTTTTTTTTTTTTTTTTTTTTTTTTTTTTTTTTTTTTTTTTTTTTTTTTTTTTTTTTTTTTTTTTTTTTTTTTTTTTTTTTTTTTTTTTTTTTTTTTTTTTTTTTTTTTTTTTTTTTTTTTTTTTTTTTTTTTTTTTTTTTTTTTTTTTTTTTTTTTTTTTTTTTTTTTTTTTTTTTTT
EEEEEEEEEEEEEEEEEEEEEEEEEEEEEEEEEEEEEEEEEEEEEEEEEEEEEEEEEEEEEEEEEEEEEEEEEEEEEEEEEEEEEEEEEEEEEEEEEEEEEEEEEEEEEEEEEEEEEEEEEEEEEEEEEEEEEEEEEEEEEEEEEEEEEEEEEEEEEEEEEEEEEEEEEEEEEEEEEEE
RRRRRRRRRRRRRRRRRRRRRRRRRRRRRRRRRRRRRRRRRRRRRRRRRRRRRRRRRRRRRRRRRRRRRRRRRRRRRRRRRRRRRRRRRRRRRRRRRRRRRRRRRRRRRRRRRRRRRRRRRRRRRRRRRRRRRRRRRRRRRRRRRRRRRRRRRRRRRRRRRRRRRRRRRRRRRRRR
FFFFFFFFFFFFFFFFFFFFFFFFFFFFFFFFFFFFFFFFFFFFFFFFFFFFFFFFFFFFFFFFFFFFFFFFFFFFFFFFFFFFFFFFFFFFFFFFFFFFFFFFFFFFFFFFFFFFFFFFFFFFFFFFFFFFFFFFFFFFFFFFFFFFFFFFFFFFFFFFFFFFFFFFFFFFFFFFFFFF
AAAAAAAAAAAAAAAAAAAAAAAAAAAAAAAAAAAAAAAAAAAAAAAAAAAAAAAAAAAAAAAAAAAAAAAAAAAAAAAAAAAAAAAAAAAAAAAAAAAAAAAAAAAAAAAAAAAAAAAAAAAAAAAAAAAAAAAAAAAAAAAAAAAAAAAAAAAAAAAAAAAAAAAAAAAAAAAAAAAA
SSSSSSSSSSSSSSSSSSSSSSSSSSSSSSSSSSSSSSSSSSSSSSSSSSSSSSSSSSSSSSSSSSSSSSSSSSSSSSSSSSSSSSSSSSSSSSSSSSSSSSSSSSSSSSSSSSSSSSSSSSSSSSSSSSSSSSSSSSSSSSSSSSSSSSSSSSSSSSSSSSSSSSSSSSSSSSSSSS
TTTTTTTTTTTTTTTTTTTTTTTTTTTTTTTTTTTTTTTTTTTTTTTTTTTTTTTTTTTTTTTTTTTTTTTTTTTTTTTTTTTTTTTTTTTTTTTTTTTTTTTTTTTTTTTTTTTTTTTTTTTTTTTTTTTTTTTTTTTTTTTTTTTTTTTTTTTTTTTTTTTTTTTTTTTTTTTTTTTT
EEEEEEEEEEEEEEEEEEEEEEEEEEEEEEEEEEEEEEEEEEEEEEEEEEEEEEEEEEEEEEEEEEEEEEEEEEEEEEEEEEEEEEEEEEEEEEEEEEEEEEEEEEEEEEEEEEEEEEEEEEEEEEEEEEEEEEEEEEEEEEEEEEEEEEEEEEEEEEEEEEEEEEEEEEEEEEEEEEE
RRRRRRRRRRRRRRRRRRRRRRRRRRRRRRRRRRRRRRRRRRRRRRRRRRRRRRRRRRRRRRRRRRRRRRRRRRRRRRRRRRRRRRRRRRRRRRRRRRRRRRRRRRRRRRRRRRRRRRRRRRRRRRRRRRRRRRRRRRRRRRRRRRRRRRRRRRRRRRRRRRRRRRRRRRRRRRRR
FFFFFFFFFFFFFFFFFFFFFFFFFFFFFFFFFFFFFFFFFFFFFFFFFFFFFFFFFFFFFFFFFFFFFFFFFFFFFFFFFFFFFFFFFFFFFFFFFFFFFFFFFFFFFFFFFFFFFFFFFFFFFFFFFFFFFFFFFFFFFFFFFFFFFFFFFFFFFFFFFFFFFFFFFFFFFFFFFFFF
AAAAAAAAAAAAAAAAAAAAAAAAAAAAAAAAAAAAAAAAAAAAAAAAAAAAAAAAAAAAAAAAAAAAAAAAAAAAAAAAAAAAAAAAAAAAAAAAAAAAAAAAAAAAAAAAAAAAAAAAAAAAAAAAAAAAAAAAAAAAAAAAAAAAAAAAAAAAAAAAAAAAAAAAAAAAAAAAAAAA
SSSSSSSSSSSSSSSSSSSSSSSSSSSSSSSSSSSSSSSSSSSSSSSSSSSSSSSSSSSSSSSSSSSSSSSSSSSSSSSSSSSSSSSSSSSSSSSSSSSSSSSSSSSSSSSSSSSSSSSSSSSSSSSSSSSSSSSSSSSSSSSSSSSSSSSSSSSSSSSSSSSSSSSSSSSSSSSSSS
TTTTTTTTTTTTTTTTTTTTTTTTTTTTTTTTTTTTTTTTTTTTTTTTTTTTTTTTTTTTTTTTTTTTTTTTTTTTTTTTTTTTTTTTTTTTTTTTTTTTTTTTTTTTTTTTTTTTTTTTTTTTTTTTTTTTTTTTTTTTTTTTTTTTTTTTTTTTTTTTTTTTTTTTTTTTTTTTTTTT
EEEEEEEEEEEEEEEEEEEEEEEEEEEEEEEEEEEEEEEEEEEEEEEEEEEEEEEEEEEEEEEEEEEEEEEEEEEEEEEEEEEEEEEEEEEEEEEEEEEEEEEEEEEEEEEEEEEEEEEEEEEEEEEEEEEEEEEEEEEEEEEEEEEEEEEEEEEEEEEEEEEEEEEEEEEEEEEEEEE
RRRRRRRRRRRRRRRRRRRRRRRRRRRRRRRRRRRRRRRRRRRRRRRRRRRRRRRRRRRRRRRRRRRRRRRRRRRRRRRRRRRRRRRRRRRRRRRRRRRRRRRRRRRRRRRRRRRRRRRRRRRRRRRRRRRRRRRRRRRRRRRRRRRRRRRRRRRRRRRRRRRRRRRRRRRRRRRR
FFFFFFFFFFFFFFFFFFFFFFFFFFFFFFFFFFFFFFFFFFFFFFFFFFFFFFFFFFFFFFFFFFFFFFFFFFFFFFFFFFFFFFFFFFFFFFFFFFFFFFFFFFFFFFFFFFFFFFFFFFFFFFFFFFFFFFFFFFFFFFFFFFFFFFFFFFFFFFFFFFFFFFFFFFFFFFFFFFFF
AAAAAAAAAAAAAAAAAAAAAAAAAAAAAAAAAAAAAAAAAAAAAAAAAAAAAAAAAAAAAAAAAAAAAAAAAAAAAAAAAAAAAAAAAAAAAAAAAAAAAAAAAAAAAAAAAAAAAAAAAAAAAAAAAAAAAAAAAAAAAAAAAAAAAAAAAAAAAAAAAAAAAAAAAAAAAAAAAAAA
SSSSSSSSSSSSSSSSSSSSSSSSSSSSSSSSSSSSSSSSSSSSSSSSSSSSSSSSSSSSSSSSSSSSSSSSSSSSSSSSSSSSSSSSSSSSSSSSSSSSSSSSSSSSSSSSSSSSSSSSSSSSSSSSSSSSSSSSSSSSSSSSSSSSSSSSSSSSSSSSSSSSSSSSSSSSSSSSSS
TTTTTTTTTTTTTTTTTTTTTTTTTTTTTTTTTTTTTTTTTTTTTTTTTTTTTTTTTTTTTTTTTTTTTTTTTTTTTTTTTTTTTTTTTTTTTTTTTTTTTTTTTTTTTTTTTTTTTTTTTTTTTTTTTTTTTTTTTTTTTTTTTTTTTTTTTTTTTTTTTTTTTTTTTTTTTTTTTTTT
EEEEEEEEEEEEEEEEEEEEEEEEEEEEEEEEEEEEEEEEEEEEEEEEEEEEEEEEEEEEEEEEEEEEEEEEEEEEEEEEEEEEEEEEEEEEEEEEEEEEEEEEEEEEEEEEEEEEEEEEEEEEEEEEEEEEEEEEEEEEEEEEEEEEEEEEEEEEEEEEEEEEEEEEEEEEEEEEEEE
RRRRRRRRRRRRRRRRRRRRRRRRRRRRRRRRRRRRRRRRRRRRRRRRRRRRRRRRRRRRRRRRRRRRRRRRRRRRRRRRRRRRRRRRRRRRRRRRRRRRRRRRRRRRRRRRRRRRRRRRRRRRRRRRRRRRRRRRRRRRRRRRRRRRRRRRRRRRRRRRRRRRRRRRRRRRRRRR
FFFFFFFFFFFFFFFFFFFFFFFFFFFFFFFFFFFFFFFFFFFFFFFFFFFFFFFFFFFFFFFFFFFFFFFFFFFFFFFFFFFFFFFFFFFFFFFFFFFFFFFFFFFFFFFFFFFFFFFFFFFFFFFFFFFFFFFFFFFFFFFFFFFFFFFFFFFFFFFFFFFFFFFFFFFFFFFFFFFF
AAAAAAAAAAAAAAAAAAAAAAAAAAAAAAAAAAAAAAAAAAAAAAAAAAAAAAAAAAAAAAAAAAAAAAAAAAAAAAAAAAAAAAAAAAAAAAAAAAAAAAAAAAAAAAAAAAAAAAAAAAAAAAAAAAAAAAAAAAAAAAAAAAAAAAAAAAAAAAAAAAAAAAAAAAAAAAAAAAAA
SSSSSSSSSSSSSSSSSSSSSSSSSSSSSSSSSSSSSSSSSSSSSSSSSSSSSSSSSSSSSSSSSSSSSSSSSSSSSSSSSSSSSSSSSSSSSSSSSSSSSSSSSSSSSSSSSSSSSSSSSSSSSSSSSSSSSSSSSSSSSSSSSSSSSSSSSSSSSSSSSSSSSSSSSSSSSSSSSS
TTTTTTTTTTTTTTTTTTTTTTTTTTTTTTTTTTTTTTTTTTTTTTTTTTTTTTTTTTTTTTTTTTTTTTTTTTTTTTTTTTTTTTTTTTTTTTTTTTTTTTTTTTTTTTTTTTTTTTTTTTTTTTTTTTTTTTTTTTTTTTTTTTTTTTTTTTTTTTTTTTTTTTTTTTTTTTTTTTT
EEEEEEEEEEEEEEEEEEEEEEEEEEEEEEEEEEEEEEEEEEEEEEEEEEEEEEEEEEEEEEEEEEEEEEEEEEEEEEEEEEEEEEEEEEEEEEEEEEEEEEEEEEEEEEEEEEEEEEEEEEEEEEEEEEEEEEEEEEEEEEEEEEEEEEEEEEEEEEEEEEEEEEEEEEEEEEEEEE
RRRRRRRRRRRRRRRRRRRRRRRRRRRRRRRRRRRRRRRRRRRRRRRRRRRRRRRRRRRRRRRRRRRRRRRRRRRRRRRRRRRRRRRRRRRRRRRRRRRRRRRRRRRRRRRRRRRRRRRRRRRRRRRRRRRRRRRRRRRRRRRRRRRRRRRRRRRRRRRRRRRRRRRRRRRRRRRR
FFFFFFFFFFFFFFFFFFFFFFFFFFFFFFFFFFFFFFFFFFFFFFFFFFFFFFFFFFFFFFFFFFFFFFFFFFFFFFFFFFFFFFFFFFFFFFFFFFFFFFFFFFFFFFFFFFFFFFFFFFFFFFFFFFFFFFFFFFFFFFFFFFFFFFFFFFFFFFFFFFFFFFFFFFFFFFFFFFFF
AAAAAAAAAAAAAAAAAAAAAAAAAAAAAAAAAAAAAAAAAAAAAAAAAAAAAAAAAAAAAAAAAAAAAAAAAAAAAAAAAAAAAAAAAAAAAAAAAAAAAAAAAAAAAAAAAAAAAAAAAAAAAAAAAAAAAAAAAAAAAAAAAAAAAAAAAAAAAAAAAAAAAAAAAAAAAAAAAAAA
SSSSSSSSSSSSSSSSSSSSSSSSSSSSSSSSSSSSSSSSSSSSSSSSSSSSSSSSSSSSSSSSSSSSSSSSSSSSSSSSSSSSSSSSSSSSSSSSSSSSSSSSSSSSSSSSSSSSSSSSSSSSSSSSSSSSSSSSSSSSSSSSSSSSSSSSSSSSSSSSSSSSSSSSSSSSSSSSSS
TTTTTTTTTTTTTTTTTTTTTTTTTTTTTTTTTTTTTTTTTTTTTTTTTTTTTTTTTTTTTTTTTTTTTTTTTTTTTTTTTTTTTTTTTTTTTTTTTTTTTTTTTTTTTTTTTTTTTTTTTTTTTTTTTTTTTTTTTTTTTTTTTTTTTTTTTTTTTTTTTTTTTTTTTTTTTTTTTTT
EEEEEEEEEEEEEEEEEEEEEEEEEEEEEEEEEEEEEEEEEEEEEEEEEEEEEEEEEEEEEEEEEEEEEEEEEEEEEEEEEEEEEEEEEEEEEEEEEEEEEEEEEEEEEEEEEEEEEEEEEEEEEEEEEEEEEEEEEEEEEEEEEEEEEEEEEEEEEEEEEEEEEEEEEEEEEEEEEE
RRRRRRRRRRRRRRRRRRRRRRRRRRRRRRRRRRRRRRRRRRRRRRRRRRRRRRRRRRRRRRRRRRRRRRRRRRRRRRRRRRRRRRRRRRRRRRRRRRRRRRRRRRRRRRRRRRRRRRRRRRRRRRRRRRRRRRRRRRRRRRRRRRRRRRRRRRRRRRRRRRRRRRRRRRRRRRRR
FFFFFFFFFFFFFFFFFFFFFFFFFFFFFFFFFFFFFFFFFFFFFFFFFFFFFFFFFFFFFFFFFFFFFFFFFFFFFFFFFFFFFFFFFFFFFFFFFFFFFFFFFFFFFFFFFFFFFFFFFFFFFFFFFFFFFFFFFFFFFFFFFFFFFFFFFFFFFFFFFFFFFFFFFFFFFFFFFFFF
AAAAAAAAAAAAAAAAAAAAAAAAAAAAAAAAAAAAAAAAAAAAAAAAAAAAAAAAAAAAAAAAAAAAAAAAAAAAAAAAAAAAAAAAAAAAAAAAAAAAAAAAAAAAAAAAAAAAAAAAAAAAAAAAAAAAAAAAAAAAAAAAAAAAAAAAAAAAAAAAAAAAAAAAAAAAAAAAAAAA
SSSSSSSSSSSSSSSSSSSSSSSSSSSSSSSSSSSSSSSSSSSSSSSSSSSSSSSSSSSSSSSSSSSSSSSSSSSSSSSSSSSSSSSSSSSSSSSSSSSSSSSSSSSSSSSSSSSSSSSSSSSSSSSSSSSSSSSSSSSSSSSSSSSSSSSSSSSSSSSSSSSSSSSSSSSSSSSSSS
TTTTTTTTTTTTTTTTTTTTTTTTTTTTTTTTTTTTTTTTTTTTTTTTTTTTTTTTTTTTTTTTTTTTTTTTTTTTTTTTTTTTTTTTTTTTTTTTTTTTTTTTTTTTTTTTTTTTTTTTTTTTTTTTTTTTTTTTTTTTTTTTTTTTTTTTTTTTTTTTTTTTTTTTTTTTTTTTTTT
EEEEEEEEEEEEEEEEEEEEEEEEEEEEEEEEEEEEEEEEEEEEEEEEEEEEEEEEEEEEEEEEEEEEEEEEEEEEEEEEEEEEEEEEEEEEEEEEEEEEEEEEEEEEEEEEEEEEEEEEEEEEEEEEEEEEEEEEEEEEEEEEEEEEEEEEEEEEEEEEEEEEEEEEEEEEEEEEEE
RRRRRRRRRRRRRRRRRRRRRRRRRRRRRRRRRRRRRRRRRRRRRRRRRRRRRRRRRRRRRRRRRRRRRRRRRRRRRRRRRRRRRRRRRRRRRRRRRRRRRRRRRRRRRRRRRRRRRRRRRRRRRRRRRRRRRRRRRRRRRRRRRRRRRRRRRRRRRRRRRRRRRRRRRRRRRRRR
FFFFFFFFFFFFFFFFFFFFFFFFFFFFFFFFFFFFFFFFFFFFFFFFFFFFFFFFFFFFFFFFFFFFFFFFFFFFFFFFFFFFFFFFFFFFFFFFFFFFFFFFFFFFFFFFFFFFFFFFFFFFFFFFFFFFFFFFFFFFFFFFFFFFFFFFFFFFFFFFFFFFFFFFFFFFFFFFFFF
AAAAAAAAAAAAAAAAAAAAAAAAAAAAAAAAAAAAAAAAAAAAAAAAAAAAAAAAAAAAAAAAAAAAAAAAAAAAAAAAAAAAAAAAAAAAAAAAAAAAAAAAAAAAAAAAAAAAAAAAAAAAAAAAAAAAAAAAAAAAAAAAAAAAAAAAAAAAAAAAAAAAAAAAAAAAAAAAAAAA
SSSSSSSSSSSSSSSSSSSSSSSSSSSSSSSSSSSSSSSSSSSSSSSSSSSSSSSSSSSSSSSSSSSSSSSSSSSSSSSSSSSSSSSSSSSSSSSSSSSSSSSSSSSSSSSSSSSSSSSSSSSSSSSSSSSSSSSSSSSSSSSSSSSSSSSSSSSSSSSSSSSSSSSSSSSSSSSSS
TTTTTTTTTTTTTTTTTTTTTTTTTTTTTTTTTTTTTTTTTTTTTTTTTTTTTTTTTTTTTTTTTTTTTTTTTTTTTTTTTTTTTTTTTTTTTTTTTTTTTTTTTTTTTTTTTTTTTTTTTTTTTTTTTTTTTTTTTTTTTTTTTTTTTTTTTTTTTTTTTTTTTTTTTTTTTTTTTTT
EEEEEEEEEEEEEEEEEEEEEEEEEEEEEEEEEEEEEEEEEEEEEEEEEEEEEEEEEEEEEEEEEEEEEEEEEEEEEEEEEEEEEEEEEEEEEEEEEEEEEEEEEEEEEEEEEEEEEEEEEEEEEEEEEEEEEEEEEEEEEEEEEEEEEEEEEEEEEEEEEEEEEEEEEEEEEEEEEE
RRRRRRRRRRRRRRRRRRRRRRRRRRRRRRRRRRRRRRRRRRRRRRRRRRRRRRRRRRRRRRRRRRRRRRRRRRRRRRRRRRRRRRRRRRRRRRRRRRRRRRRRRRRRRRRRRRRRRRRRRRRRRRRRRRRRRRRRRRRRRRRRRRRRRRRRRRRRRRRRRRRRRRRRRRRRRRRR
FFFFFFFFFFFFFFFFFFFFFFFFFFFFFFFFFFFFFFFFFFFFFFFFFFFFFFFFFFFFFFFFFFFFFFFFFFFFFFFFFFFFFFFFFFFFFFFFFFFFFFFFFFFFFFFFFFFFFFFFFFFFFFFFFFFFFFFFFFFFFFFFFFFFFFFFFFFFFFFFFFFFFFFFFFFFFFFFFFF
AAAAAAAAAAAAAAAAAAAAAAAAAAAAAAAAAAAAAAAAAAAAAAAAAAAAAAAAAAAAAAAAAAAAAAAAAAAAAAAAAAAAAAAAAAAAAAAAAAAAAAAAAAAAAAAAAAAAAAAAAAAAAAAAAAAAAAAAAAAAAAAAAAAAAAAAAAAAAAAAAAAAAAAAAAAAAAAAAAAA
SSSSSSSSSSSSSSSSSSSSSSSSSSSSSSSSSSSSSSSSSSSSSSSSSSSSSSSSSSSSSSSSSSSSSSSSSSSSSSSSSSSSSSSSSSSSSSSSSSSSSSSSSSSSSSSSSSSSSSSSSSSSSSSSSSSSSSSSSSSSSSSSSSSSSSSSSSSSSSSSSSSSSSSSSSSSSSSSS
TTTTTTTTTTTTTTTTTTTTTTTTTTTTTTTTTTTTTTTTTTTTTTTTTTTTTTTTTTTTTTTTTTTTTTTTTTTTTTTTTTTTTTTTTTTTTTTTTTTTTTTTTTTTTTTTTTTTTTTTTTTTTTTTTTTTTTTTTTTTTTTTTTTTTTTTTTTTTTTTTTTTTTTTTTTTTTTTTTT
EEEEEEEEEEEEEEEEEEEEEEEEEEEEEEEEEEEEEEEEEEEEEEEEEEEEEEEEEEEEEEEEEEEEEEEEEEEEEEEEEEEEEEEEEEEEEEEEEEEEEEEEEEEEEEEEEEEEEEEEEEEEEEEEEEEEEEEEEEEEEEEEEEEEEEEEEEEEEEEEEEEEEEEEEEEEEEEEEE
RRRRRRRRRRRRRRRRRRRRRRRRRRRRRRRRRRRRRRRRRRRRRRRRRRRRRRRRRRRRRRRRRRRRRRRRRRRRRRRRRRRRRRRRRRRRRRRRRRRRRRRRRRRRRRRRRRRRRRRRRRRRRRRRRRRRRRRRRRRRRRRRRRRRRRRRRRRRRRRRRRRRRRRRRRRRRRRR
FFFFFFFFFFFFFFFFFFFFFFFFFFFFFFFFFFFFFFFFFFFFFFFFFFFFFFFFFFFFFFFFFFFFFFFFFFFFFFFFFFFFFFFFFFFFFFFFFFFFFFFFFFFFFFFFFFFFFFFFFFFFFFFFFFFFFFFFFFFFFFFFFFFFFFFFFFFFFFFFFFFFFFFFFFFFFFFFFFFF
AAAAAAAAAAAAAAAAAAAAAAAAAAAAAAAAAAAAAAAAAAAAAAAAAAAAAAAAAAAAAAAAAAAAAAAAAAAAAAAAAAAAAAAAAAAAAAAAAAAAAAAAAAAAAAAAAAAAAAAAAAAAAAAAAAAAAAAAAAAAAAAAAAAAAAAAAAAAAAAAAAAAAAAAAAAAAAAAAAAA
SSSSSSSSSSSSSSSSSSSSSSSSSSSSSSSSSSSSSSSSSSSSSSSSSSSSSSSSSSSSSSSSSSSSSSSSSSSSSSSSSSSSSSSSSSSSSSSSSSSSSSSSSSSSSSSSSSSSSSSSSSSSSSSSSSSSSSSSSSSSSSSSSSSSSSSSSSSSSSSSSSSSSSSSSSSSSSSSSS
TTTTTTTTTTTTTTTTTTTTTTTTTTTTTTTTTTTTTTTTTTTTTTTTTTTTTTTTTTTTTTTTTTTTTTTTTTTTTTTTTTTTTTTTTTTTTTTTTTTTTTTTTTTTTTTTTTTTTTTTTTTTTTTTTTTTTTTTTTTTTTTTTTTTTTTTTTTTTTTTTTTTTTTTTTTTTTTTTTT
EEEEEEEEEEEEEEEEEEEEEEEEEEEEEEEEEEEEEEEEEEEEEEEEEEEEEEEEEEEEEEEEEEEEEEEEEEEEEEEEEEEEEEEEEEEEEEEEEEEEEEEEEEEEEEEEEEEEEEEEEEEEEEEEEEEEEEEEEEEEEEEEEEEEEEEEEEEEEEEEEEEEEEEEEEEEEEEEEE
RRRRRRRRRRRRRRRRRRRRRRRRRRRRRRRRRRRRRRRRRRRRRRRRRRRRRRRRRRRRRRRRRRRRRRRRRRRRRRRRRRRRRRRRRRRRRRRRRRRRRRRRRRRRRRRRRRRRRRRRRRRRRRRRRRRRRRRRRRRRRRRRRRRRRRRRRRRRRRRRRRRRRRRRRRRRRRRR
FFFFFFFFFFFFFFFFFFFFFFFFFFFFFFFFFFFFFFFFFFFFFFFFFFFFFFFFFFFFFFFFFFFFFFFFFFFFFFFFFFFFFFFFFFFFFFFFFFFFFFFFFFFFFFFFFFFFFFFFFFFFFFFFFFFFFFFFFFFFFFFFFFFFFFFFFFFFFFFFFFFFFFFFFFFFFFFFFFFF
AAAAAAAAAAAAAAAAAAAAAAAAAAAAAAAAAAAAAAAAAAAAAAAAAAAAAAAAAAAAAAAAAAAAAAAAAAAAAAAAAAAAAAAAAAAAAAAAAAAAAAAAAAAAAAAAAAAAAAAAAAAAAAAAAAAAAAAAAAAAAAAAAAAAAAAAAAAAAAAAAAAAAAAAAAAAAAAAAAAA
SSSSSSSSSSSSSSSSSSSSSSSSSSSSSSSSSSSSSSSSSSSSSSSSSSSSSSSSSSSSSSSSSSSSSSSSSSSSSSSSSSSSSSSSSSSSSSSSSSSSSSSSSSSSSSSSSSSSSSSSSSSSSSSSSSSSSSSSSSSSSSSSSSSSSSSSSSSSSSSSSSSSSSSSSSSSSSSSS
TTTTTTTTTTTTTTTTTTTTTTTTTTTTTTTTTTTTTTTTTTTTTTTTTTTTTTTTTTTTTTTTTTTTTTTTTTTTTTTTTTTTTTTTTTTTTTTTTTTTTTTTTTTTTTTTTTTTTTTTTTTTTTTTTTTTTTTTTTTTTTTTTTTTTTTTTTTTTTTTTTTTTTTTTTTTTTTTTTT
EEEEEEEEEEEEEEEEEEEEEEEEEEEEEEEEEEEEEEEEEEEEEEEEEEEEEEEEEEEEEEEEEEEEEEEEEEEEEEEEEEEEEEEEEEEEEEEEEEEEEEEEEEEEEEEEEEEEEEEEEEEEEEEEEEEEEEEEEEEEEEEEEEEEEEEEEEEEEEEEEEEEEEEEEEEEEEEEEE
RRRRRRRRRRRRRRRRRRRRRRRRRRRRRRRRRRRRRRRRRRRRRRRRRRRRRRRRRRRRRRRRRRRRRRRRRRRRRRRRRRRRRRRRRRRRRRRRRRRRRRRRRRRRRRRRRRRRRRRRRRRRRRRRRRRRRRRRRRRRRRRRRRRRRRRRRRRRRRRRRRRRRRRRRRRRRRRR
FFFFFFFFFFFFFFFFFFFFFFFFFFFFFFFFFFFFFFFFFFFFFFFFFFFFFFFFFFFFFFFFFFFFFFFFFFFFFFFFFFFFFFFFFFFFFFFFFFFFFFFFFFFFFFFFFFFFFFFFFFFFFFFFFFFFFFFFFFFFFFFFFFFFFFFFFFFFFFFFFFFFFFFFFFFFFFFFFFFF
AAAAAAAAAAAAAAAAAAAAAAAAAAAAAAAAAAAAAAAAAAAAAAAAAAAAAAAAAAAAAAAAAAAAAAAAAAAAAAAAAAAAAAAAAAAAAAAAAAAAAAAAAAAAAAAAAAAAAAAAAAAAAAAAAAAAAAAAAAAAAAAAAAAAAAAAAAAAAAAAAAAAAAAAAAAAAAAAAAAA
SSSSSSSSSSSSSSSSSSSSSSSSSSSSSSSSSSSSSSSSSSSSSSSSSSSSSSSSSSSSSSSSSSSSSSSSSSSSSSSSSSSSSSSSSSSSSSSSSSSSSSSSSSSSSSSSSSSSSSSSSSSSSSSSSSSSSSSSSSSSSSSSSSSSSSSSSSSSSSSSSSSSSSSSSSSSSSSSS
TTTTTTTTTTTTTTTTTTTTTTTTTTTTTTTTTTTTTTTTTTTTTTTTTTTTTTTTTTTTTTTTTTTTTTTTTTTTTTTTTTTTTTTTTTTTTTTTTTTTTTTTTTTTTTTTTTTTTTTTTTTTTTTTTTTTTTTTTTTTTTTTTTTTTTTTTTTTTTTTTTTTTTTTTTTTTTTTTTT
EEEEEEEEEEEEEEEEEEEEEEEEEEEEEEEEEEEEEEEEEEEEEEEEEEEEEEEEEEEEEEEEEEEEEEEEEEEEEEEEEEEEEEEEEEEEEEEEEEEEEEEEEEEEEEEEEEEEEEEEEEEEEEEEEEEEEEEEEEEEEEEEEEEEEEEEEEEEEEEEEEEEEEEEEEEEEEEEEE
RRRRRRRRRRRRRRRRRRRRRRRRRRRRRRRRRRRRRRRRRRRRRRRRRRRRRRRRRRRRRRRRRRRRRRRRRRRRRRRRRRRRRRRRRRRRRRRRRRRRRRRRRRRRRRRRRRRRRRRRRRRRRRRRRRRRRRRRRRRRRRRRRRRRRRRRRRRRRRRRRRRRRRRRRRRRRRRR
FFFFFFFFFFFFFFFFFFFFFFFFFFFFFFFFFFFFFFFFFFFFFFFFFFFFFFFFFFFFFFFFFFFFFFFFFFFFFFFFFFFFFFFFFFFFFFFFFFFFFFFFFFFFFFFFFFFFFFFFFFFFFFFFFFFFFFFFFFFFFFFFFFFFFFFFFFFFFFFFFFFFFFFFFFFFFFFFFFFF
AAAAAAAAAAAAAAAAAAAAAAAAAAAAAAAAAAAAAAAAAAAAAAAAAAAAAAAAAAAAAAAAAAAAAAAAAAAAAAAAAAAAAAAAAAAAAAAAAAAAAAAAAAAAAAAAAAAAAAAAAAAAAAAAAAAAAAAAAAAAAAAAAAAAAAAAAAAAAAAAAAAAAAAAAAAAAAAAAAAA
SSSSSSSSSSSSSSSSSSSSSSSSSSSSSSSSSSSSSSSSSSSSSSSSSSSSSSSSSSSSSSSSSSSSSSSSSSSSSSSSSSSSSSSSSSSSSSSSSSSSSSSSSSSSSSSSSSSSSSSSSSSSSSSSSSSSSSSSSSSSSSSSSSSSSSSSSSSSSSSSSSSSSSSSSSSSSSSSSS
TTTTTTTTTTTTTTTTTTTTTTTTTTTTTTTTTTTTTTTTTTTTTTTTTTTTTTTTTTTTTTTTTTTTTTTTTTTTTTTTTTTTTTTTTTTTTTTTTTTTTTTTTTTTTTTTTTTTTTTTTTTTTTTTTTTTTTTTTTTTTTTTTTTTTTTTTTTTTTTTTTTTTTTTTTTTTTTTTTT
EEEEEEEEEEEEEEEEEEEEEEEEEEEEEEEEEEEEEEEEEEEEEEEEEEEEEEEEEEEEEEEEEEEEEEEEEEEEEEEEEEEEEEEEEEEEEEEEEEEEEEEEEEEEEEEEEEEEEEEEEEEEEEEEEEEEEEEEEEEEEEEEEEEEEEEEEEEEEEEEEEEEEEEEEEEEEEEEEE
RRRRRRRRRRRRRRRRRRRRRRRRRRRRRRRRRRRRRRRRRRRRRRRRRRRRRRRRRRRRRRRRRRRRRRRRRRRRRRRRRRRRRRRRRRRRRRRRRRRRRRRRRRRRRRRRRRRRRRRRRRRRRRRRRRRRRRRRRRRRRRRRRRRRRRRRRRRRRRRRRRRRRRRRRRRRRRRR
FFFFFFFFFFFFFFFFFFFFFFFFFFFFFFFFFFFFFFFFFFFFFFFFFFFFFFFFFFFFFFFFFFFFFFFFFFFFFFFFFFFFFFFFFFFFFFFFFFFFFFFFFFFFFFFFFFFFFFFFFFFFFFFFFFFFFFFFFFFFFFFFFFFFFFFFFFFFFFFFFFFFFFFFFFFFFFFFFFFF

RRRRRRRRRRRRRRRRRRRRRRRRRRRRRRRRRRRRRRRRRRRRRRRRRRRRRRRRRRRRRRRRRRRRRRRRRRRRRRRRRRRRRRRRRRRRRRRRRRRRRRRRRRRRRRRRRRRRRRRRRRRRRRRRRRRRRRRRRRRRRRRRRRRRRRRRRRRRRR
FFFFFFFFFFFFFFFFFFFFFFFFFFFFFFFFFFFFFFFFFFFFFFFFFFFFFFFFFFFFFFFFFFFFFFFFFFFFFFFFFFFFFFFFFFFFFFFFFFFFFFFFFFFFFFFFFFFFFFFFFFFFFFFFFFFFFFFFFFFFFFFFFFFFFFFFFFFFFFFFFF
AAAAAAAAAAAAAAAAAAAAAAAAAAAAAAAAAAAAAAAAAAAAAAAAAAAAAAAAAAAAAAAAAAAAAAAAAAAAAAAAAAAAAAAAAAAAAAAAAAAAAAAAAAAAAAAAAAAAAAAAAAAAAAAAAAAAAAAAAAAAAAAAAAAAAAAAAAAAAAAA
SSSSSSSSSSSSSSSSSSSSSSSSSSSSSSSSSSSSSSSSSSSSSSSSSSSSSSSSSSSSSSSSSSSSSSSSSSSSSSSSSSSSSSSSSSSSSSSSSSSSSSSSSSSSSSSSSSSSSSSSSSSSSSSSSSSSSSSSSSSSSSSSSSSSSSSSSSSSSS
TTTTTTTTTTTTTTTTTTTTTTTTTTTTTTTTTTTTTTTTTTTTTTTTTTTTTTTTTTTTTTTTTTTTTTTTTTTTTTTTTTTTTTTTTTTTTTTTTTTTTTTTTTTTTTTTTTTTTTTTTTTTTTTTTTTTTTTTTTTTTTTTTTTTTTTTTTTTTTTT
EEEEEEEEEEEEEEEEEEEEEEEEEEEEEEEEEEEEEEEEEEEEEEEEEEEEEEEEEEEEEEEEEEEEEEEEEEEEEEEEEEEEEEEEEEEEEEEEEEEEEEEEEEEEEEEEEEEEEEEEEEEEEEEEEEEEEEEEEEEEEEEEEEEEEEEEEEEEEEEE
RRRRRRRRRRRRRRRRRRRRRRRRRRRRRRRRRRRRRRRRRRRRRRRRRRRRRRRRRRRRRRRRRRRRRRRRRRRRRRRRRRRRRRRRRRRRRRRRRRRRRRRRRRRRRRRRRRRRRRRRRRRRRRRRRRRRRRRRRRRRRRRRRRRRRRRRRRRRRR
FFFFFFFFFFFFFFFFFFFFFFFFFFFFFFFFFFFFFFFFFFFFFFFFFFFFFFFFFFFFFFFFFFFFFFFFFFFFFFFFFFFFFFFFFFFFFFFFFFFFFFFFFFFFFFFFFFFFFFFFFFFFFFFFFFFFFFFFFFFFFFFFFFFFFFFFFFFFFFFFFF
AAAAAAAAAAAAAAAAAAAAAAAAAAAAAAAAAAAAAAAAAAAAAAAAAAAAAAAAAAAAAAAAAAAAAAAAAAAAAAAAAAAAAAAAAAAAAAAAAAAAAAAAAAAAAAAAAAAAAAAAAAAAAAAAAAAAAAAAAAAAAAAAAAAAAAAAAAAAAAAA
SSSSSSSSSSSSSSSSSSSSSSSSSSSSSSSSSSSSSSSSSSSSSSSSSSSSSSSSSSSSSSSSSSSSSSSSSSSSSSSSSSSSSSSSSSSSSSSSSSSSSSSSSSSSSSSSSSSSSSSSSSSSSSSSSSSSSSSSSSSSSSSSSSSSSSSSSSSSSS
TTTTTTTTTTTTTTTTTTTTTTTTTTTTTTTTTTTTTTTTTTTTTTTTTTTTTTTTTTTTTTTTTTTTTTTTTTTTTTTTTTTTTTTTTTTTTTTTTTTTTTTTTTTTTTTTTTTTTTTTTTTTTTTTTTTTTTTTTTTTTTTTTTTTTTTTTTTTTTTT
EEEEEEEEEEEEEEEEEEEEEEEEEEEEEEEEEEEEEEEEEEEEEEEEEEEEEEEEEEEEEEEEEEEEEEEEEEEEEEEEEEEEEEEEEEEEEEEEEEEEEEEEEEEEEEEEEEEEEEEEEEEEEEEEEEEEEEEEEEEEEEEEEEEEEEEEEEEEEEEE
RRRRRRRRRRRRRRRRRRRRRRRRRRRRRRRRRRRRRRRRRRRRRRRRRRRRRRRRRRRRRRRRRRRRRRRRRRRRRRRRRRRRRRRRRRRRRRRRRRRRRRRRRRRRRRRRRRRRRRRRRRRRRRRRRRRRRRRRRRRRRRRRRRRRRRRRRRRRRR
FFFFFFFFFFFFFFFFFFFFFFFFFFFFFFFFFFFFFFFFFFFFFFFFFFFFFFFFFFFFFFFFFFFFFFFFFFFFFFFFFFFFFFFFFFFFFFFFFFFFFFFFFFFFFFFFFFFFFFFFFFFFFFFFFFFFFFFFFFFFFFFFFFFFFFFFFFFFFFFFF
AAAAAAAAAAAAAAAAAAAAAAAAAAAAAAAAAAAAAAAAAAAAAAAAAAAAAAAAAAAAAAAAAAAAAAAAAAAAAAAAAAAAAAAAAAAAAAAAAAAAAAAAAAAAAAAAAAAAAAAAAAAAAAAAAAAAAAAAAAAAAAAAAAAAAAAAAAAAAAAA
SSSSSSSSSSSSSSSSSSSSSSSSSSSSSSSSSSSSSSSSSSSSSSSSSSSSSSSSSSSSSSSSSSSSSSSSSSSSSSSSSSSSSSSSSSSSSSSSSSSSSSSSSSSSSSSSSSSSSSSSSSSSSSSSSSSSSSSSSSSSSSSSSSSSSSSSSSSSSS
TTTTTTTTTTTTTTTTTTTTTTTTTTTTTTTTTTTTTTTTTTTTTTTTTTTTTTTTTTTTTTTTTTTTTTTTTTTTTTTTTTTTTTTTTTTTTTTTTTTTTTTTTTTTTTTTTTTTTTTTTTTTTTTTTTTTTTTTTTTTTTTTTTTTTTTTTTTTTTTT
EEEEEEEEEEEEEEEEEEEEEEEEEEEEEEEEEEEEEEEEEEEEEEEEEEEEEEEEEEEEEEEEEEEEEEEEEEEEEEEEEEEEEEEEEEEEEEEEEEEEEEEEEEEEEEEEEEEEEEEEEEEEEEEEEEEEEEEEEEEEEEEEEEEEEEEEEEEEEEEE
RRRRRRRRRRRRRRRRRRRRRRRRRRRRRRRRRRRRRRRRRRRRRRRRRRRRRRRRRRRRRRRRRRRRRRRRRRRRRRRRRRRRRRRRRRRRRRRRRRRRRRRRRRRRRRRRRRRRRRRRRRRRRRRRRRRRRRRRRRRRRRRRRRRRRRRRRRRRRR
FFFFFFFFFFFFFFFFFFFFFFFFFFFFFFFFFFFFFFFFFFFFFFFFFFFFFFFFFFFFFFFFFFFFFFFFFFFFFFFFFFFFFFFFFFFFFFFFFFFFFFFFFFFFFFFFFFFFFFFFFFFFFFFFFFFFFFFFFFFFFFFFFFFFFFFFFFFFFFFFFF
AAAAAAAAAAAAAAAAAAAAAAAAAAAAAAAAAAAAAAAAAAAAAAAAAAAAAAAAAAAAAAAAAAAAAAAAAAAAAAAAAAAAAAAAAAAAAAAAAAAAAAAAAAAAAAAAAAAAAAAAAAAAAAAAAAAAAAAAAAAAAAAAAAAAAAAAAAAAAAAA
SSSSSSSSSSSSSSSSSSSSSSSSSSSSSSSSSSSSSSSSSSSSSSSSSSSSSSSSSSSSSSSSSSSSSSSSSSSSSSSSSSSSSSSSSSSSSSSSSSSSSSSSSSSSSSSSSSSSSSSSSSSSSSSSSSSSSSSSSSSSSSSSSSSSSSSSSSSSSS
TTTTTTTTTTTTTTTTTTTTTTTTTTTTTTTTTTTTTTTTTTTTTTTTTTTTTTTTTTTTTTTTTTTTTTTTTTTTTTTTTTTTTTTTTTTTTTTTTTTTTTTTTTTTTTTTTTTTTTTTTTTTTTTTTTTTTTTTTTTTTTTTTTTTTTTTTTTTTTTT
EEEEEEEEEEEEEEEEEEEEEEEEEEEEEEEEEEEEEEEEEEEEEEEEEEEEEEEEEEEEEEEEEEEEEEEEEEEEEEEEEEEEEEEEEEEEEEEEEEEEEEEEEEEEEEEEEEEEEEEEEEEEEEEEEEEEEEEEEEEEEEEEEEEEEEEEEEEEEEEE
RRRRRRRRRRRRRRRRRRRRRRRRRRRRRRRRRRRRRRRRRRRRRRRRRRRRRRRRRRRRRRRRRRRRRRRRRRRRRRRRRRRRRRRRRRRRRRRRRRRRRRRRRRRRRRRRRRRRRRRRRRRRRRRRRRRRRRRRRRRRRRRRRRRRRRRRRRRRRR
FFFFFFFFFFFFFFFFFFFFFFFFFFFFFFFFFFFFFFFFFFFFFFFFFFFFFFFFFFFFFFFFFFFFFFFFFFFFFFFFFFFFFFFFFFFFFFFFFFFFFFFFFFFFFFFFFFFFFFFFFFFFFFFFFFFFFFFFFFFFFFFFFFFFFFFFFFFFFFFFFF
AAAAAAAAAAAAAAAAAAAAAAAAAAAAAAAAAAAAAAAAAAAAAAAAAAAAAAAAAAAAAAAAAAAAAAAAAAAAAAAAAAAAAAAAAAAAAAAAAAAAAAAAAAAAAAAAAAAAAAAAAAAAAAAAAAAAAAAAAAAAAAAAAAAAAAAAAAAAAAAA
SSSSSSSSSSSSSSSSSSSSSSSSSSSSSSSSSSSSSSSSSSSSSSSSSSSSSSSSSSSSSSSSSSSSSSSSSSSSSSSSSSSSSSSSSSSSSSSSSSSSSSSSSSSSSSSSSSSSSSSSSSSSSSSSSSSSSSSSSSSSSSSSSSSSSSSSSSSSSS
TTTTTTTTTTTTTTTTTTTTTTTTTTTTTTTTTTTTTTTTTTTTTTTTTTTTTTTTTTTTTTTTTTTTTTTTTTTTTTTTTTTTTTTTTTTTTTTTTTTTTTTTTTTTTTTTTTTTTTTTTTTTTTTTTTTTTTTTTTTTTTTTTTTTTTTTTTTTTTT
EEEEEEEEEEEEEEEEEEEEEEEEEEEEEEEEEEEEEEEEEEEEEEEEEEEEEEEEEEEEEEEEEEEEEEEEEEEEEEEEEEEEEEEEEEEEEEEEEEEEEEEEEEEEEEEEEEEEEEEEEEEEEEEEEEEEEEEEEEEEEEEEEEEEEEEEEEEEEEEE
RRRRRRRRRRRRRRRRRRRRRRRRRRRRRRRRRRRRRRRRRRRRRRRRRRRRRRRRRRRRRRRRRRRRRRRRRRRRRRRRRRRRRRRRRRRRRRRRRRRRRRRRRRRRRRRRRRRRRRRRRRRRRRRRRRRRRRRRRRRRRRRRRRRRRRRRRRRRRR
FFFFFFFFFFFFFFFFFFFFFFFFFFFFFFFFFFFFFFFFFFFFFFFFFFFFFFFFFFFFFFFFFFFFFFFFFFFFFFFFFFFFFFFFFFFFFFFFFFFFFFFFFFFFFFFFFFFFFFFFFFFFFFFFFFFFFFFFFFFFFFFFFFFFFFFFFFFFFFFFFF
AAAAAAAAAAAAAAAAAAAAAAAAAAAAAAAAAAAAAAAAAAAAAAAAAAAAAAAAAAAAAAAAAAAAAAAAAAAAAAAAAAAAAAAAAAAAAAAAAAAAAAAAAAAAAAAAAAAAAAAAAAAAAAAAAAAAAAAAAAAAAAAAAAAAAAAAAAAAAAAA
SSSSSSSSSSSSSSSSSSSSSSSSSSSSSSSSSSSSSSSSSSSSSSSSSSSSSSSSSSSSSSSSSSSSSSSSSSSSSSSSSSSSSSSSSSSSSSSSSSSSSSSSSSSSSSSSSSSSSSSSSSSSSSSSSSSSSSSSSSSSSSSSSSSSSSSSSSSSSS
TTTTTTTTTTTTTTTTTTTTTTTTTTTTTTTTTTTTTTTTTTTTTTTTTTTTTTTTTTTTTTTTTTTTTTTTTTTTTTTTTTTTTTTTTTTTTTTTTTTTTTTTTTTTTTTTTTTTTTTTTTTTTTTTTTTTTTTTTTTTTTTTTTTTTTTTTTTTTTT
EEEEEEEEEEEEEEEEEEEEEEEEEEEEEEEEEEEEEEEEEEEEEEEEEEEEEEEEEEEEEEEEEEEEEEEEEEEEEEEEEEEEEEEEEEEEEEEEEEEEEEEEEEEEEEEEEEEEEEEEEEEEEEEEEEEEEEEEEEEEEEEEEEEEEEEEEEEEEEEE
RRRRRRRRRRRRRRRRRRRRRRRRRRRRRRRRRRRRRRRRRRRRRRRRRRRRRRRRRRRRRRRRRRRRRRRRRRRRRRRRRRRRRRRRRRRRRRRRRRRRRRRRRRRRRRRRRRRRRRRRRRRRRRRRRRRRRRRRRRRRRRRRRRRRRRRRRRRRRR
FFFFFFFFFFFFFFFFFFFFFFFFFFFFFFFFFFFFFFFFFFFFFFFFFFFFFFFFFFFFFFFFFFFFFFFFFFFFFFFFFFFFFFFFFFFFFFFFFFFFFFFFFFFFFFFFFFFFFFFFFFFFFFFFFFFFFFFFFFFFFFFFFFFFFFFFFFFFFFFFFF
AAAAAAAAAAAAAAAAAAAAAAAAAAAAAAAAAAAAAAAAAAAAAAAAAAAAAAAAAAAAAAAAAAAAAAAAAAAAAAAAAAAAAAAAAAAAAAAAAAAAAAAAAAAAAAAAAAAAAAAAAAAAAAAAAAAAAAAAAAAAAAAAAAAAAAAAAAAAAAAA
SSSSSSSSSSSSSSSSSSSSSSSSSSSSSSSSSSSSSSSSSSSSSSSSSSSSSSSSSSSSSSSSSSSSSSSSSSSSSSSSSSSSSSSSSSSSSSSSSSSSSSSSSSSSSSSSSSSSSSSSSSSSSSSSSSSSSSSSSSSSSSSSSSSSSSSSSSSSSS
TTTTTTTTTTTTTTTTTTTTTTTTTTTTTTTTTTTTTTTTTTTTTTTTTTTTTTTTTTTTTTTTTTTTTTTTTTTTTTTTTTTTTTTTTTTTTTTTTTTTTTTTTTTTTTTTTTTTTTTTTTTTTTTTTTTTTTTTTTTTTTTTTTTTTTTTTTTTTTT
EEEEEEEEEEEEEEEEEEEEEEEEEEEEEEEEEEEEEEEEEEEEEEEEEEEEEEEEEEEEEEEEEEEEEEEEEEEEEEEEEEEEEEEEEEEEEEEEEEEEEEEEEEEEEEEEEEEEEEEEEEEEEEEEEEEEEEEEEEEEEEEEEEEEEEEEEEEEEEEE
RRRRRRRRRRRRRRRRRRRRRRRRRRRRRRRRRRRRRRRRRRRRRRRRRRRRRRRRRRRRRRRRRRRRRRRRRRRRRRRRRRRRRRRRRRRRRRRRRRRRRRRRRRRRRRRRRRRRRRRRRRRRRRRRRRRRRRRRRRRRRRRRRRRRRRRRRRRRRR
FFFFFFFFFFFFFFFFFFFFFFFFFFFFFFFFFFFFFFFFFFFFFFFFFFFFFFFFFFFFFFFFFFFFFFFFFFFFFFFFFFFFFFFFFFFFFFFFFFFFFFFFFFFFFFFFFFFFFFFFFFFFFFFFFFFFFFFFFFFFFFFFFFFFFFFFFFFFFFFFFF
AAAAAAAAAAAAAAAAAAAAAAAAAAAAAAAAAAAAAAAAAAAAAAAAAAAAAAAAAAAAAAAAAAAAAAAAAAAAAAAAAAAAAAAAAAAAAAAAAAAAAAAAAAAAAAAAAAAAAAAAAAAAAAAAAAAAAAAAAAAAAAAAAAAAAAAAAAAAAAAA
SSSSSSSSSSSSSSSSSSSSSSSSSSSSSSSSSSSSSSSSSSSSSSSSSSSSSSSSSSSSSSSSSSSSSSSSSSSSSSSSSSSSSSSSSSSSSSSSSSSSSSSSSSSSSSSSSSSSSSSSSSSSSSSSSSSSSSSSSSSSSSSSSSSSSSSSSSSSSS
TTTTTTTTTTTTTTTTTTTTTTTTTTTTTTTTTTTTTTTTTTTTTTTTTTTTTTTTTTTTTTTTTTTTTTTTTTTTTTTTTTTTTTTTTTTTTTTTTTTTTTTTTTTTTTTTTTTTTTTTTTTTTTTTTTTTTTTTTTTTTTTTTTTTTTTTTTTTTTT
EEEEEEEEEEEEEEEEEEEEEEEEEEEEEEEEEEEEEEEEEEEEEEEEEEEEEEEEEEEEEEEEEEEEEEEEEEEEEEEEEEEEEEEEEEEEEEEEEEEEEEEEEEEEEEEEEEEEEEEEEEEEEEEEEEEEEEEEEEEEEEEEEEEEEEEEEEEEEEEE
RRRRRRRRRRRRRRRRRRRRRRRRRRRRRRRRRRRRRRRRRRRRRRRRRRRRRRRRRRRRRRRRRRRRRRRRRRRRRRRRRRRRRRRRRRRRRRRRRRRRRRRRRRRRRRRRRRRRRRRRRRRRRRRRRRRRRRRRRRRRRRRRRRRRRRRRRRRRRR
FFFFFFFFFFFFFFFFFFFFFFFFFFFFFFFFFFFFFFFFFFFFFFFFFFFFFFFFFFFFFFFFFFFFFFFFFFFFFFFFFFFFFFFFFFFFFFFFFFFFFFFFFFFFFFFFFFFFFFFFFFFFFFFFFFFFFFFFFFFFFFFFFFFFFFFFFFFFFFFFFF
AAAAAAAAAAAAAAAAAAAAAAAAAAAAAAAAAAAAAAAAAAAAAAAAAAAAAAAAAAAAAAAAAAAAAAAAAAAAAAAAAAAAAAAAAAAAAAAAAAAAAAAAAAAAAAAAAAAAAAAAAAAAAAAAAAAAAAAAAAAAAAAAAAAAAAAAAAAAAAAA
SSSSSSSSSSSSSSSSSSSSSSSSSSSSSSSSSSSSSSSSSSSSSSSSSSSSSSSSSSSSSSSSSSSSSSSSSSSSSSSSSSSSSSSSSSSSSSSSSSSSSSSSSSSSSSSSSSSSSSSSSSSSSSSSSSSSSSSSSSSSSSSSSSSSSSSSSSSSSS
TTTTTTTTTTTTTTTTTTTTTTTTTTTTTTTTTTTTTTTTTTTTTTTTTTTTTTTTTTTTTTTTTTTTTTTTTTTTTTTTTTTTTTTTTTTTTTTTTTTTTTTTTTTTTTTTTTTTTTTTTTTTTTTTTTTTTTTTTTTTTTTTTTTTTTTTTTTTTTT
EEEEEEEEEEEEEEEEEEEEEEEEEEEEEEEEEEEEEEEEEEEEEEEEEEEEEEEEEEEEEEEEEEEEEEEEEEEEEEEEEEEEEEEEEEEEEEEEEEEEEEEEEEEEEEEEEEEEEEEEEEEEEEEEEEEEEEEEEEEEEEEEEEEEEEEEEEEEEEEE
RRRRRRRRRRRRRRRRRRRRRRRRRRRRRRRRRRRRRRRRRRRRRRRRRRRRRRRRRRRRRRRRRRRRRRRRRRRRRRRRRRRRRRRRRRRRRRRRRRRRRRRRRRRRRRRRRRRRRRRRRRRRRRRRRRRRRRRRRRRRRRRRRRRRRRRRRRRRRR
FFFFFFFFFFFFFFFFFFFFFFFFFFFFFFFFFFFFFFFFFFFFFFFFFFFFFFFFFFFFFFFFFFFFFFFFFFFFFFFFFFFFFFFFFFFFFFFFFFFFFFFFFFFFFFFFFFFFFFFFFFFFFFFFFFFFFFFFFFFFFFFFFFFFFFFFFFFFFFFFFF
AAAAAAAAAAAAAAAAAAAAAAAAAAAAAAAAAAAAAAAAAAAAAAAAAAAAAAAAAAAAAAAAAAAAAAAAAAAAAAAAAAAAAAAAAAAAAAAAAAAAAAAAAAAAAAAAAAAAAAAAAAAAAAAAAAAAAAAAAAAAAAAAAAAAAAAAAAAAAAAA
SSSSSSSSSSSSSSSSSSSSSSSSSSSSSSSSSSSSSSSSSSSSSSSSSSSSSSSSSSSSSSSSSSSSSSSSSSSSSSSSSSSSSSSSSSSSSSSSSSSSSSSSSSSSSSSSSSSSSSSSSSSSSSSSSSSSSSSSSSSSSSSSSSSSSSSSSSSSSS
TTTTTTTTTTTTTTTTTTTTTTTTTTTTTTTTTTTTTTTTTTTTTTTTTTTTTTTTTTTTTTTTTTTTTTTTTTTTTTTTTTTTTTTTTTTTTTTTTTTTTTTTTTTTTTTTTTTTTTTTTTTTTTTTTTTTTTTTTTTTTTTTTTTTTTTTTTTTTTT
EEEEEEEEEEEEEEEEEEEEEEEEEEEEEEEEEEEEEEEEEEEEEEEEEEEEEEEEEEEEEEEEEEEEEEEEEEEEEEEEEEEEEEEEEEEEEEEEEEEEEEEEEEEEEEEEEEEEEEEEEEEEEEEEEEEEEEEEEEEEEEEEEEEEEEEEEEEEEEEE
RRRRRRRRRRRRRRRRRRRRRRRRRRRRRRRRRRRRRRRRRRRRRRRRRRRRRRRRRRRRRRRRRRRRRRRRRRRRRRRRRRRRRRRRRRRRRRRRRRRRRRRRRRRRRRRRRRRRRRRRRRRRRRRRRRRRRRRRRRRRRRRRRRRRRRRRRRRRRR
FFFFFFFFFFFFFFFFFFFFFFFFFFFFFFFFFFFFFFFFFFFFFFFFFFFFFFFFFFFFFFFFFFFFFFFFFFFFFFFFFFFFFFFFFFFFFFFFFFFFFFFFFFFFFFFFFFFFFFFFFFFFFFFFFFFFFFFFFFFFFFFFFFFFFFFFFFFFFFFFFF
AAAAAAAAAAAAAAAAAAAAAAAAAAAAAAAAAAAAAAAAAAAAAAAAAAAAAAAAAAAAAAAAAAAAAAAAAAAAAAAAAAAAAAAAAAAAAAAAAAAAAAAAAAAAAAAAAAAAAAAAAAAAAAAAAAAAAAAAAAAAAAAAAAAAAAAAAAAAAAAA
SSSSSSSSSSSSSSSSSSSSSSSSSSSSSSSSSSSSSSSSSSSSSSSSSSSSSSSSSSSSSSSSSSSSSSSSSSSSSSSSSSSSSSSSSSSSSSSSSSSSSSSSSSSSSSSSSSSSSSSSSSSSSSSSSSSSSSSSSSSSSSSSSSSSSSSSSSSSSS
TTTTTTTTTTTTTTTTTTTTTTTTTTTTTTTTTTTTTTTTTTTTTTTTTTTTTTTTTTTTTTTTTTTTTTTTTTTTTTTTTTTTTTTTTTTTTTTTTTTTTTTTTTTTTTTTTTTTTTTTTTTTTTTTTTTTTTTTTTTTTTTTTTTTTTTTTTTTTTT
EEEEEEEEEEEEEEEEEEEEEEEEEEEEEEEEEEEEEEEEEEEEEEEEEEEEEEEEEEEEEEEEEEEEEEEEEEEEEEEEEEEEEEEEEEEEEEEEEEEEEEEEEEEEEEEEEEEEEEEEEEEEEEEEEEEEEEEEEEEEEEEEEEEEEEEEEEEEEEEE
RRRRRRRRRRRRRRRRRRRRRRRRRRRRRRRRRRRRRRRRRRRRRRRRRRRRRRRRRRRRRRRRRRRRRRRRRRRRRRRRRRRRRRRRRRRRRRRRRRRRRRRRRRRRRRRRRRRRRRRRRRRRRRRRRRRRRRRRRRRRRRRRRRRRRRRRRRRRRR
FFFFFFFFFFFFFFFFFFFFFFFFFFFFFFFFFFFFFFFFFFFFFFFFFFFFFFFFFFFFFFFFFFFFFFFFFFFFFFFFFFFFFFFFFFFFFFFFFFFFFFFFFFFFFFFFFFFFFFFFFFFFFFFFFFFFFFFFFFFFFFFFFFFFFFFFFFFFFFFFFF
AAAAAAAAAAAAAAAAAAAAAAAAAAAAAAAAAAAAAAAAAAAAAAAAAAAAAAAAAAAAAAAAAAAAAAAAAAAAAAAAAAAAAAAAAAAAAAAAAAAAAAAAAAAAAAAAAAAAAAAAAAAAAAAAAAAAAAAAAAAAAAAAAAAAAAAAAAAAAAAA
SSSSSSSSSSSSSSSSSSSSSSSSSSSSSSSSSSSSSSSSSSSSSSSSSSSSSSSSSSSSSSSSSSSSSSSSSSSSSSSSSSSSSSSSSSSSSSSSSSSSSSSSSSSSSSSSSSSSSSSSSSSSSSSSSSSSSSSSSSSSSSSSSSSSSSSSSSSSSS
TTTTTTTTTTTTTTTTTTTTTTTTTTTTTTTTTTTTTTTTTTTTTTTTTTTTTTTTTTTTTTTTTTTTTTTTTTTTTTTTTTTTTTTTTTTTTTTTTTTTTTTTTTTTTTTTTTTTTTTTTTTTTTTTTTTTTTTTTTTTTTTTTTTTTTTTTTTTTTT
EEEEEEEEEEEEEEEEEEEEEEEEEEEEEEEEEEEEEEEEEEEEEEEEEEEEEEEEEEEEEEEEEEEEEEEEEEEEEEEEEEEEEEEEEEEEEEEEEEEEEEEEEEEEEEEEEEEEEEEEEEEEEEEEEEEEEEEEEEEEEEEEEEEEEEEEEEEEEEEE
RRRRRRRRRRRRRRRRRRRRRRRRRRRRRRRRRRRRRRRRRRRRRRRRRRRRRRRRRRRRRRRRRRRRRRRRRRRRRRRRRRRRRRRRRRRRRRRRRRRRRRRRRRRRRRRRRRRRRRRRRRRRRRRRRRRRRRRRRRRRRRRRRRRRRRRRRRRRRR
FFFFFFFFFFFFFFFFFFFFFFFFFFFFFFFFFFFFFFFFFFFFFFFFFFFFFFFFFFFFFFFFFFFFFFFFFFFFFFFFFFFFFFFFFFFFFFFFFFFFFFFFFFFFFFFFFFFFFFFFFFFFFFFFFFFFFFFFFFFFFFFFFFFFFFFFFFFFFFFFFF
AAAAAAAAAAAAAAAAAAAAAAAAAAAAAAAAAAAAAAAAAAAAAAAAAAAAAAAAAAAAAAAAAAAAAAAAAAAAAAAAAAAAAAAAAAAAAAAAAAAAAAAAAAAAAAAAAAAAAAAAAAAAAAAAAAAAAAAAAAAAAAAAAAAAAAAAAAAAAAAA
SSSSSSSSSSSSSSSSSSSSSSSSSSSSSSSSSSSSSSSSSSSSSSSSSSSSSSSSSSSSSSSSSSSSSSSSSSSSSSSSSSSSSSSSSSSSSSSSSSSSSSSSSSSSSSSSSSSSSSSSSSSSSSSSSSSSSSSSSSSSSSSSSSSSSSSSSSSSSS
TTTTTTTTTTTTTTTTTTTTTTTTTTTTTTTTTTTTTTTTTTTTTTTTTTTTTTTTTTTTTTTTTTTTTTTTTTTTTTTTTTTTTTTTTTTTTTTTTTTTTTTTTTTTTTTTTTTTTTTTTTTTTTTTTTTTTTTTTTTTTTTTTTTTTTTTTTTTTTT
EEEEEEEEEEEEEEEEEEEEEEEEEEEEEEEEEEEEEEEEEEEEEEEEEEEEEEEEEEEEEEEEEEEEEEEEEEEEEEEEEEEEEEEEEEEEEEEEEEEEEEEEEEEEEEEEEEEEEEEEEEEEEEEEEEEEEEEEEEEEEEEEEEEEEEEEEEEEEEEE
RRRRRRRRRRRRRRRRRRRRRRRRRRRRRRRRRRRRRRRRRRRRRRRRRRRRRRRRRRRRRRRRRRRRRRRRRRRRRRRRRRRRRRRRRRRRRRRRRRRRRRRRRRRRRRRRRRRRRRRRRRRRRRRRRRRRRRRRRRRRRRRRRRRRRRRRRRRRRR
FFFFFFFFFFFFFFFFFFFFFFFFFFFFFFFFFFFFFFFFFFFFFFFFFFFFFFFFFFFFFFFFFFFFFFFFFFFFFFFFFFFFFFFFFFFFFFFFFFFFFFFFFFFFFFFFFFFFFFFFFFFFFFFFFFFFFFFFFFFFFFFFFFFFFFFFFFFFFFFFFF
AAAAAAAAAAAAAAAAAAAAAAAAAAAAAAAAAAAAAAAAAAAAAAAAAAAAAAAAAAAAAAAAAAAAAAAAAAAAAAAAAAAAAAAAAAAAAAAAAAAAAAAAAAAAAAAAAAAAAAAAAAAAAAAAAAAAAAAAAAAAAAAAAAAAAAAAAAAAAAAA
SSSSSSSSSSSSSSSSSSSSSSSSSSSSSSSSSSSSSSSSSSSSSSSSSSSSSSSSSSSSSSSSSSSSSSSSSSSSSSSSSSSSSSSSSSSSSSSSSSSSSSSSSSSSSSSSSSSSSSSSSSSSSSSSSSSSSSSSSSSSSSSSSSSSSSSSSSSSSS
TTTTTTTTTTTTTTTTTTTTTTTTTTTTTTTTTTTTTTTTTTTTTTTTTTTTTTTTTTTTTTTTTTTTTTTTTTTTTTTTTTTTTTTTTTTTTTTTTTTTTTTTTTTTTTTTTTTTTTTTTTTTTTTTTTTTTTTTTTTTTTTTTTTTTTTTTTTTTTT
EEEEEEEEEEEEEEEEEEEEEEEEEEEEEEEEEEEEEEEEEEEEEEEEEEEEEEEEEEEEEEEEEEEEEEEEEEEEEEEEEEEEEEEEEEEEEEEEEEEEEEEEEEEEEEEEEEEEEEEEEEEEEEEEEEEEEEEEEEEEEEEEEEEEEEEEEEEEEEEE
RRRRRRRRRRRRRRRRRRRRRRRRRRRRRRRRRRRRRRRRRRRRRRRRRRRRRRRRRRRRRRRRRRRRRRRRRRRRRRRRRRRRRRRRRRRRRRRRRRRRRRRRRRRRRRRRRRRRRRRRRRRRRRRRRRRRRRRRRRRRRRRRRRRRRRRRRRRRRR
FFFFFFFFFFFFFFFFFFFFFFFFFFFFFFFFFFFFFFFFFFFFFFFFFFFFFFFFFFFFFFFFFFFFFFFFFFFFFFFFFFFFFFFFFFFFFFFFFFFFFFFFFFFFFFFFFFFFFFFFFFFFFFFFFFFFFFFFFFFFFFFFFFFFFFFFFFFFFFFFFF
AAAAAAAAAAAAAAAAAAAAAAAAAAAAAAAAAAAAAAAAAAAAAAAAAAAAAAAAAAAAAAAAAAAAAAAAAAAAAAAAAAAAAAAAAAAAAAAAAAAAAAAAAAAAAAAAAAAAAAAAAAAAAAAAAAAAAAAAAAAAAAAAAAAAAAAAAAAAAAAA
SSSSSSSSSSSSSSSSSSSSSSSSSSSSSSSSSSSSSSSSSSSSSSSSSSSSSSSSSSSSSSSSSSSSSSSSSSSSSSSSSSSSSSSSSSSSSSSSSSSSSSSSSSSSSSSSSSSSSSSSSSSSSSSSSSSSSSSSSSSSSSSSSSSSSSSSSSSSSS
TTTTTTTTTTTTTTTTTTTTTTTTTTTTTTTTTTTTTTTTTTTTTTTTTTTTTTTTTTTTTTTTTTTTTTTTTTTTTTTTTTTTTTTTTTTTTTTTTTTTTTTTTTTTTTTTTTTTTTTTTTTTTTTTTTTTTTTTTTTTTTTTTTTTTTTTTTTTTT
EEEEEEEEEEEEEEEEEEEEEEEEEEEEEEEEEEEEEEEEEEEEEEEEEEEEEEEEEEEEEEEEEEEEEEEEEEEEEEEEEEEEEEEEEEEEEEEEEEEEEEEEEEEEEEEEEEEEEEEEEEEEEEEEEEEEEEEEEEEEEEEEEEEEEEEEEEEEEEEE
RRRRRRRRRRRRRRRRRRRRRRRRRRRRRRRRRRRRRRRRRRRRRRRRRRRRRRRRRRRRRRRRRRRRRRRRRRRRRRRRRRRRRRRRRRRRRRRRRRRRRRRRRRRRRRRRRRRRRRRRRRRRRRRRRRRRRRRRRRRRRRRRRRRRRRRRRRRRRR
FFFFFFFFFFFFFFFFFFFFFFFFFFFFFFFFFFFFFFFFFFFFFFFFFFFFFFFFFFFFFFFFFFFFFFFFFFFFFFFFFFFFFFFFFFFFFFFFFFFFFFFFFFFFFFFFFFFFFFFFFFFFFFFFFFFFFFFFFFFFFFFFFFFFFFFFFFFFFFFFFF
AAAAAAAAAAAAAAAAAAAAAAAAAAAAAAAAAAAAAAAAAAAAAAAAAAAAAAAAAAAAAAAAAAAAAAAAAAAAAAAAAAAAAAAAAAAAAAAAAAAAAAAAAAAAAAAAAAAAAAAAAAAAAAAAAAAAAAAAAAAAAAAAAAAAAAAAAAAAAAAA
SSSSSSSSSSSSSSSSSSSSSSSSSSSSSSSSSSSSSSSSSSSSSSSSSSSSSSSSSSSSSSSSSSSSSSSSSSSSSSSSSSSSSSSSSSSSSSSSSSSSSSSSSSSSSSSSSSSSSSSSSSSSSSSSSSSSSSSSSSSSSSSSSSSSSSSSSSSSSS
TTTTTTTTTTTTTTTTTTTTTTTTTTTTTTTTTTTTTTTTTTTTTTTTTTTTTTTTTTTTTTTTTTTTTTTTTTTTTTTTTTTTTTTTTTTTTTTTTTTTTTTTTTTTTTTTTTTTTTTTTTTTTTTTTTTTTTTTTTTTTTTTTTTTTTTTTTTTTT
EEEEEEEEEEEEEEEEEEEEEEEEEEEEEEEEEEEEEEEEEEEEEEEEEEEEEEEEEEEEEEEEEEEEEEEEEEEEEEEEEEEEEEEEEEEEEEEEEEEEEEEEEEEEEEEEEEEEEEEEEEEEEEEEEEEEEEEEEEEEEEEEEEEEEEEEEEEEEEEE
RRRRRRRRRRRRRRRRRRRRRRRRRRRRRRRRRRRRRRRRRRRRRRRRRRRRRRRRRRRRRRRRRRRRRRRRRRRRRRRRRRRRRRRRRRRRRRRRRRRRRRRRRRRRRRRRRRRRRRRRRRRRRRRRRRRRRRRRRRRRRRRRRRRRRRRRRRRRRR
FFFFFFFFFFFFFFFFFFFFFFFFFFFFFFFFFFFFFFFFFFFFFFFFFFFFFFFFFFFFFFFFFFFFFFFFFFFFFFFFFFFFFFFFFFFFFFFFFFFFFFFFFFFFFFFFFFFFFFFFFFFFFFFFFFFFFFFFFFFFFFFFFFFFFFFFFFFFFFFFFF
AAAAAAAAAAAAAAAAAAAAAAAAAAAAAAAAAAAAAAAAAAAAAAAAAAAAAAAAAAAAAAAAAAAAAAAAAAAAAAAAAAAAAAAAAAAAAAAAAAAAAAAAAAAAAAAAAAAAAAAAAAAAAAAAAAAAAAAAAAAAAAAAAAAAAAAAAAAAAAAA
SSSSSSSSSSSSSSSSSSSSSSSSSSSSSSSSSSSSSSSSSSSSSSSSSSSSSSSSSSSSSSSSSSSSSSSSSSSSSSSSSSSSSSSSSSSSSSSSSSSSSSSSSSSSSSSSSSSSSSSSSSSSSSSSSSSSSSSSSSSSSSSSSSSSSSSSSSSSSS
TTTTTTTTTTTTTTTTTTTTTTTTTTTTTTTTTTTTTTTTTTTTTTTTTTTTTTTTTTTTTTTTTTTTTTTTTTTTTTTTTTTTTTTTTTTTTTTTTTTTTTTTTTTTTTTTTTTTTTTTTTTTTTTTTTTTTTTTTTTTTTTTTTTTTTTTTTTTTTT
EEEEEEEEEEEEEEEEEEEEEEEEEEEEEEEEEEEEEEEEEEEEEEEEEEEEEEEEEEEEEEEEEEEEEEEEEEEEEEEEEEEEEEEEEEEEEEEEEEEEEEEEEEEEEEEEEEEEEEEEEEEEEEEEEEEEEEEEEEEEEEEEEEEEEEEEEEEEEEEE
RRRRRRRRRRRRRRRRRRRRRRRRRRRRRRRRRRRRRRRRRRRRRRRRRRRRRRRRRRRRRRRRRRRRRRRRRRRRRRRRRRRRRRRRRRRRRRRRRRRRRRRRRRRRRRRRRRRRRRRRRRRRRRRRRRRRRRRRRRRRRRRRRRRRRRRRRRRRRR
FFFFFFFFFFFFFFFFFFFFFFFFFFFFFFFFFFFFFFFFFFFFFFFFFFFFFFFFFFFFFFFFFFFFFFFFFFFFFFFFFFFFFFFFFFFFFFFFFFFFFFFFFFFFFFFFFFFFFFFFFFFFFFFFFFFFFFFFFFFFFFFFFFFFFFFFFFFFFFFFFF
AAAAAAAAAAAAAAAAAAAAAAAAAAAAAAAAAAAAAAAAAAAAAAAAAAAAAAAAAAAAAAAAAAAAAAAAAAAAAAAAAAAAAAAAAAAAAAAAAAAAAAAAAAAAAAAAAAAAAAAAAAAAAAAAAAAAAAAAAAAAAAAAAAAAAAAAAAAAAAAA
SSSSSSSSSSSSSSSSSSSSSSSSSSSSSSSSSSSSSSSSSSSSSSSSSSSSSSSSSSSSSSSSSSSSSSSSSSSSSSSSSSSSSSSSSSSSSSSSSSSSSSSSSSSSSSSSSSSSSSSSSSSSSSSSSSSSSSSSSSSSSSSSSSSSSSSSSSSSSS
TTTTTTTTTTTTTTTTTTTTTTTTTTTTTTTTTTTTTTTTTTTTTTTTTTTTTTTTTTTTTTTTTTTTTTTTTTTTTTTTTTTTTTTTTTTTTTTTTTTTTTTTTTTTTTTTTTTTTTTTTTTTTTTTTTTTTTTTTTTTTTTTTTTTTTTTTTTTTTT
EEEEEEEEEEEEEEEEEEEEEEEEEEEEEEEEEEEEEEEEEEEEEEEEEEEEEEEEEEEEEEEEEEEEEEEEEEEEEEEEEEEEEEEEEEEEEEEEEEEEEEEEEEEEEEEEEEEEEEEEEEEEEEEEEEEEEEEEEEEEEEEEEEEEEEEEEEEEEEEE
RRRRRRRRRRRRRRRRRRRRRRRRRRRRRRRRRRRRRRRRRRRRRRRRRRRRRRRRRRRRRRRRRRRRRRRRRRRRRRRRRRRRRRRRRRRRRRRRRRRRRRRRRRRRRRRRRRRRRRRRRRRRRRRRRRRRRRRRRRRRRRRRRRRRRRRRRRRRRR
FFFFFFFFFFFFFFFFFFFFFFFFFFFFFFFFFFFFFFFFFFFFFFFFFFFFFFFFFFFFFFFFFFFFFFFFFFFFFFFFFFFFFFFFFFFFFFFFFFFFFFFFFFFFFFFFFFFFFFFFFFFFFFFFFFFFFFFFFFFFFFFFFFFFFFFFFFFFFFFFFF

USA
50014
ESMERALDA
ESMERALDA

fig. 1
*Esmeralda* Farr 50
Designed by Farr Yacht Design, built by Goetz Custom Boats
U.S.A., 2000

Sailing the finished boat

Matilda McQuaid

# AN INTERVIEW WITH ERIC GOETZ

Advanced composite technology has completely transformed many areas of the design world—from aeronautics to sports to architecture. As athletes and others strive for speed and lightness to achieve ultimate performance, carbon-fiber composites have become ever more integral to their success.

Eric Goetz, a master boat builder, has been experimenting with carbon-fiber composites since the late 1970s. His company, Goetz Custom Boats, has made over one hundred boats in its thirty-year history, ranging from high-performance America's Cup sailboats to recreational powerboats. All exemplify a profound knowledge of advanced materials and techniques, and demonstrate that technology can be effectively and innovatively integrated to produce lighter, faster, and more efficiently made boats. In the following interview, Goetz discusses the materials and processes he uses in his workshop, and how textiles are fundamental to creating the fastest boats on the water today.

fig. 2
Forming the hull skin by laying the unidirectional carbon-fiber tape onto the mold

MM: I'm interested in how you ultimately got into boat building. In many ways, it is typical for designers or craftsmen to begin as the user, and then become the maker, like you did. What was it about your experiences as a sailor that led you into this profession?
EG: Well, it wasn't so much the sailing part, which was wonderful for me anyway, but the fact that my father likes to make things. He's a pretty good amateur woodworker, so it was that, coupled with the fact that we were always working on, or sailing, the family boat. My father made furniture as a hobby, and when I was a kid, he guided me through all kinds of little projects. So, combining sailing and making things was where I landed.
MM: What was your connection to the Rhode Island School of Design (RISD) when you were studying at Brown University?
EG: I took industrial design classes at RISD, which helped me to understand process. And I found I was good at process.
MM: What was the focus of the program? Was it on materials? Were they working with composites at RISD, or were they experimenting with other materials?
EG: The industrial design course I took centered around furniture. The interesting thing is that a lot of that activity took place in the woodshop at RISD, but many of the students were putting together all kinds of different materials, such as sheet metal and laminated veneer. So even though metal and wood are conventional building materials, they were combined in unconventional ways. The first boats I built were all veneer wooden boats. Cold-molded is what we called them. You build a three-dimensional form and drape the veneer over it—like creating plywood over a curved form—and then everything is held together with epoxy, making it strong.
MM: In an article about your company in *Professional Boatbuilder* magazine, the writer, Paul Lazarus, notes that when you opened your shop in 1975 the "crew members of your operation shared a certain cultural spirit of the era which manifested itself in an experimental approach to technology." This new technology was wood-epoxy composite construction. Can you talk about this type of construction?
EG: We started with straight cold-molding. A forty-foot boat would have five layers of three-millimeter or one-eighth-inch veneer. After awhile, with competitive pressures to build a lighter, stronger, and stiffer boat, this forced us to combine things. We began putting foam core in our boats, and a honeycomb core. We kept advancing this composite with a wood composite, but it used epoxy-resin, high-tech cores, and vacuum-bag technology. So in 1983, when we couldn't sell another wood boat, or a wood-composite boat, our next step was to sell fabric boats. We used different fibers, like carbon, instead of wood, but the techniques were the same. Because we developed our process in wood, when it came time to change, we just used different materials.
MM: But there must have been some

fig. 3
Laying out the honeycomb core of the hull

adjustments in terms of working style, going from a wood-based composite to a textile-based composite?
EG: Not as much as you might think. We still built a wooden form, and we still used a vacuum bag. The biggest problem was introducing the resin to the wood, which you do with a roller, or the resin to the composite, which we first did with a hand roller, and then eventually with a machine that would feed dry fabric through a resin bath, like a grandmother's wringer washer. And out would come a piece of fabric with about the right amount of resin on it. That's what we call wet-preg.
MM: So with the wet-preg, was it cured over time?
EG: It's room-temperature cure, so you have to get it organized quickly, and under the bag quickly, because you only have three to five hours of working time, and then, all of a sudden, things start to gel.
MM: What is prepreg?
EG: Prepreg is like suspended animation, it's frozen. The fabric is impregnated with resin and put in the freezer so it can be shipped around the country. You unfreeze it when you actually want to use it. Once we're ready, we bring it out of the freezer, and get it to room

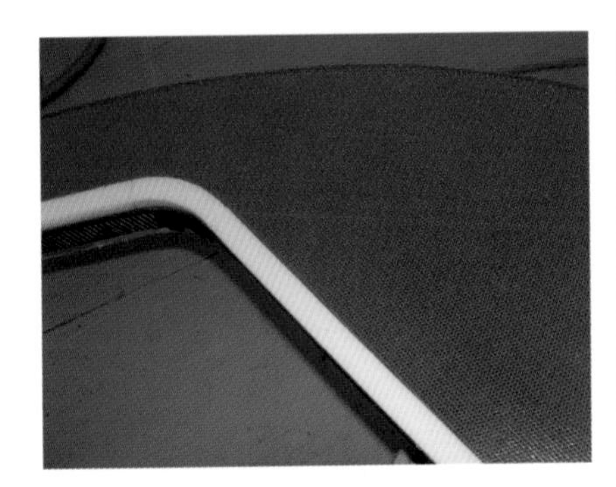

fig. 4
Cutting the structural frames (bulkheads) from sheets of honeycomb core

temperature. It has what we call a long "out-time," which means you can leave it out for six or eight weeks before you have to cook it. So you have a lot of time to get things where you want them.

MM: What were other differences?

EG: As we advanced into the wet-preg and prepreg—with prepreg you have to introduce heat as a way of curing your resin systems—there was a whole other layer of processing that needed to be applied.

MM: Were you looking toward any of the other industries that were using composite construction for inspiration and/or research, or were they looking at you?

EG: Basically, we were following airplane manufacture. There is one branch of the airplane designers who work primarily with metal, but there are others who are more experimental and can expand the craft's range—whether it's the amount of payload, speed, or distance. The main driver of such research is the military, but there is also the Experimental Airplane Society, which includes such design leaders as Burt Rutan, who is a pioneer in the composites industry and has designed record-breaking composite aircrafts.

MM: Are there counterparts to this in the boating world?

EG: One of our owners, Bill Koch, for whom we built many boats, including the 1992 America's Cup winner and also an eighty-five-footer that was world champion in offshore racing, was instrumental because of his interest in technology. He was from MIT and very influenced by technology, not just because "I want it to go faster," but more out of an interest to experiment and push the envelope. One time he flew us out to the experimental air show in Oshkosh, Wisconsin. When we were building the boats for the 1992 America's Cup, he had us working with the Department of Aeronautics and Astrophysics from Stanford University.

MM: In what other ways has your company been an experimental leader in terms of material and process developments?

EG: Once we started on this experimental trail with Koch, we were introduced to many people. We always had an experimental flavor to our company, so we did some interesting projects, including a wooden boat in 1982 whose keel moves from side-to-side in order to actually stabilize the boat while it's sailing. The boat is still in existence today and is the grandfather of all these wonderful, modern boats that the French are building. Now that we are recognized as experimental leaders, we are working with our suppliers to try new combinations of materials or try to help them tweak the chemistry of their resins—to provide more time out of the freezer or a slightly different cure, or to hold two dissimilar materials together. Right now we have numerous test samples off at the lab that will help us examine a new group of resins that we'd like to include in our arsenal. We continue to experiment, and we have to, in order to stay ahead of the pack. If we don't experiment, and

fig. 5
Curing the laminate with localized vacuum bagging

fig. 6
Structural frames, now with their carbon skin, complete and ready to go in the boat

fig. 7
Completed hull taped out to receive the structural frames

think, and push the envelope, whether it's chemistry or processing or various fabrics, we're going to be left behind.

MM: When did you start using advanced composites?

EG: Early 1980s, late 1970s—late 1970s actually.

MM: That seems early, because the only industry using carbon-fiber composites was aerospace. The sports industry was not even using it much at that point.

EG: No, it wasn't. It was really the aerospace domain. It's interesting, I went to Lockheed Aerospace, in Marietta, Georgia, around 1993. I was looking at a rudder assembly for an airplane, and said, "Wow, this is pretty old-fashioned." It was made with a titanium rudder stock [axle] with welded titanium frames [ribs] and carbon skins glued and riveted to the frames. The rudders we build are made with carbon stocks, foam or honeycomb core, and carbon skins bonded to the post and core. The composite monocoque rudders that we use do not have dissimilar materials held together by brittle joints. And they couldn't understand my comment, as it was a fighter plane. So we found that we had pushed the envelope a little further than some of the aerospace engineers. In the 1992 America's Cup, Bill Koch's first boat was built by Hercules Aerospace in Magna, Utah, and it took them twice as long in calendar time and in man-hours as we normally take.

MM: And why is that?

EG: It's just a different culture; ours is very experimental, which is, "Let's build it and get it on the water and get it sailing and in service as quickly as possible." Even though they are building carbon-fiber rocket motor casings, they'll do prototypes, and then pre-production prototypes. It takes them four or five years to get them up and running. And for us, for an America's Cup boat, we have to have it sailing in no time at all—twenty-two weeks.

MM: What's the difference between fiberglass boat hulls and carbon-composite boat hulls, apart from the fiber?

EG: There again, that's a matter of philosophy. We do build fiberglass boat hulls, but we use our high-tech resins and our high-tech techniques. We take advantage of the properties of the materials. When one talks about fiberglass boat hulls in a sort of generic sense, or production sense, those builders don't use the same resins or techniques we use, and, while they build perfectly viable vehicles, they are not race or high-performance vehicles. Right now, for instance, we are building a high-performance boat for someone who has a record of two hundred miles per hour on the water. We are building the structure for his next boat so that he can keep the record.

MM: And what is that made out of?

EG: It's a combination of Kevlar and carbon and aluminum honeycomb, very high-tech stuff. Something that the normal boat manufacture cannot conceive of using.

MM: Besides being heavier, how is fiberglass different from carbon fiber?

fig. 8
Structural frames bonded in place with high-strength rubber and toughened epoxy adhesive

EG: It's a very good material. It's very strong and economical, and it's more elastic than carbon fiber. In the old days, a boat would have 65% resin and 35% glass, and the resin is not the strong part of the matrix, fiber is. When we build fiberglass boats, we have about 70% to 72% fiber and about 28% to 30% resin.

MM: Why do you have a lot more fiber? Is it because the resin weighs more?

EG: We've taken advantage of the fibers. The resin weighs more, but its main function is to hold the bundles of fiber together and in columns, so that when you push or pull on them, they don't buckle. If they do buckle, then you're going to have a failure. In high-tech applications, one wants just enough resin to hold the bundles together; any more is wasted weight.

MM: Are you experimenting with other types of fibers?

EG: Actually, we are experimenting mostly with other types of resins. We do use different fibers, but the fibers themselves are so much stronger than the resins. It's the resins that tend to fail first.

MM: What changes have you seen in composite construction since you entered the field, and do you think textiles have played a central role in the changes?

EG: I started working with wood fiber, and then epoxy resins, and vacuum bags, and foam cores. And then, when demand was for faster and lighter, we had to change the materials, so we went from wood fiber to glass, carbon, and Kevlar fibers. My next education was in weave styles. I've learned a lot about what different weave styles can do for us, and what they can do to the fibers. We have woven carbon and Kevlar made on everyday looms that can also weave cotton, rayon, or other fibers. However, when you start examining fiber, whether it's wool or cotton or carbon, you find that a normal weaving process can degrade the fibers by as much as 40% to 50%. That's not very good, so we went from woven to knitted or stitched fabrics, and then to unidirectional. "Uni" is straightforward, as all the fiber runs in the same direction. Then you put layer after layer on top of each other like plywood, and suddenly, it behaves like wood.

MM: Are you mainly implementing unidirectional?

EG: We are using combinations of fabrics and materials, because there are different uses for different weave styles. For instance, even though you may degrade the strength of a fabric by weaving it, a woven material is a safety net, a reinforcement against impact. The unidirectional is strong, but it has interlaminar shear, causing the layers to separate and peel off from each other. A fabric, on the other hand, will hold together like a safety net.

MM: What are some of the applications for different weave structures?

EG: When we build some of our more complicated shapes, like sinks, we want to use something like a twill, or a two- or five-harness satin weave, that is able to make compound curves more easily than

fig. 9
Rolling the boat into its upright position

unidirectional or even wovens. When we construct a staircase, such as the one for architect Toshiko Mori, we are using a combination of three different weave structures (fig. 12). We use a very fine woven material as a surface, then unidirectional, because we want to carry the load across the staircase to the edges, and we're using a quadraxial, a material that is knitted on four axes—0°, +45°, –45°, and 90°—so it's like a big net. We are building a simple staircase, but we are using various materials and weave structures to do different jobs within the laminate stack.

MM: How thick is the staircase?

EG: The skin of the staircase is about two millimeters, and then there's about twenty-five millimeters of core, and another two millimeters of fabric. Within the two millimeters, we have seven different directions of fiber.

MM: In boats you use a twill because it can be manipulated three-dimensionally more than other woven structures. Would you ever use knitted structure in boat hulls for the same reason?

EG: Absolutely. We're building a fifty-six-footer right now, and this boat is a combination of high-tech techniques and high-tech resins. But the owner is price-conscious, so we're using a prepreg that is not optimal for load carrying, but will be very strong nevertheless, and slightly less expensive, because less labor will go into it. The laminate stack is such that we are

fig. 10
Structural frames fully installed

putting down one layer of material, which has four different axes of knit material. The owner gets a carbon-fiber boat, but we put in less labor. The compromise is that it's slightly heavier, but still very strong.

MM: What was your first America's Cup boat?

EG: That was a boat called *Stars and Stripes* for Dennis Conner, for the 1992 Cup.

MM: How many have you made since then?

EG: Nine since then, nine and a half actually, because we built the tool or the mold that Hercules Aerospace used to build the 1992 Cup.

MM: Can you describe the process of making the America's Cup boat? How much time do you have when you are first approached, and what happens, more or less sequentially, in its making?

EG: The ideal is that we would be involved from the onset, as a member of the design-and-build team. In highly designed projects such as these, the designers go off and fiddle with the shape, meanwhile the engineers get together with the builders and say, "Okay, how are we going to build this thing?" And then we spend some time testing. For example, in the 1992 Cup, Bill Koch had custom-made carbon fiber built to the limits of the rules, we then tested it to see if in fact it was worth the dollars. We also tested

fig. 11
Lowering the deck onto the hull

with the bonding to see how to stick the whole boat together—the bulkhead to the hull. We went through a series of tests before even starting to build the boat. The designers never stop working, so management has to say, "You have to deliver your shape on such-and-such date so the boat builders have enough time to build it." Once they deliver that shape, we have normally twenty-two weeks to build. This is the longest single piece of the process, so that time needs to be minimized.

When we get the final shape of the boat, we send it to our back room, where our CNC (Computer Numeric Control) machine cuts that shape at full size. We then erect our tool or mold. Once the tool is surfaced and very smooth, we begin laminating. Meanwhile, as we are constructing the hull, we are also making the deck and the interior structure—the keel box, maststep, bulkheads, and any other components. First you make the molds, followed by the pieces, and then you get to the assembly phase, and you take all these elements and glue them together—that's a bit like model airplanes. And when that's done, you have to go to the outfitting stage, where you put on sail-handling equipment, winches and tracks, and steering gear, and paint it—finally, it's ready to go sailing.

MM: Does it get outfitted at your workshop?

EG: Yes, when the boat leaves the shop, the winches are on and the steering is in; you just have to put the mast in and the keel on.

MM: Can you talk about laying out the fiber on the mold?

EG: We get a laminate schedule, which is essentially a guide to the loads on the boat characterized by high load, low load, or no load. Next, the engineers locate areas that need extra carbon, around the keel, for instance, or under the mast. They also specify the number of layers and direction of material—three plies running fore and aft, for example, and then four plies running athwartship. Each ply is anywhere from two-tenths to three-tenths of a millimeter thick, with around seven or eight plies in the whole skin. In the high-load areas, there can be as many as fifteen or even twenty plies of material.

MM: How much back-and-forth is there? Since you have built so many boats, do you give the engineers guidance about how many plies?

EG: No, we don't tell them how many plies. The engineers have to figure out the exact number and orientation of the plies. We give them guidance on process. They have to tell us how much to put there, but this is how we anticipate putting it together, and how we cure it, and these are the properties that can be expected from something that we build. They get a pretty complete picture of what they are going to get from us.

MM: Once they give you all those directions, you lay all the fibers out, and then what happens after that?

EG: After we lay it out, we cook it. All of these boats are cured at 100°C (212°F). We have a big oven.

The largest boat that we have cooked was 105 feet long, 22 feet wide, and 16 feet high. First we have to get it to temperature, which takes about four hours, then it stays at temperature for another four hours. We turn off the oven, let it cool down, open up the package, and, assuming everything went as planned, we are ready for the next step. The hull is taken off the tool and the assembly phase begins.

MM: How much of an America's Cup boat is composite-based?

EG: To give you an idea, when we built the Maxi boat *The Card* in 1988 or 1989, there was about $40,000 worth of metal in the boat—and that was in 1988 dollars! Now we are down to probably $5,000 worth of metal in a boat that size. Everything is made out of composites. For example, the chain plates, which are the anchors for the wires that hold the mast up, would have been made of metal and bolted on twenty years ago; today, they are all carbon fiber, and you just glue them on and walk away.

MM: What are some interesting collaborations or projects that have come out of your experience with America's Cup boat building, as well as boat building in general, with composites? You've mentioned Toshiko Mori's project with the fiberglass stairs, but are there others that are spinoffs of what you've been doing for so many years?

EG: There's the *Ark* in Detroit, which is 165 feet long, 45 feet wide, and 16 or 18 feet high. It's in a library and hangs from the fourth floor into an atrium. We built it out of fiberglass and covered it in mahogany. The company who commissioned the project houses its technical library in the *Ark*. There's another project we did for Lockheed Aerospace, in which we built a radar range reflector, which was about 60 feet long, 22 feet wide, and 5 feet high, and looks like a manta ray. They use it in Palmdale, California, at their radar range. The reflector sits on top of a pole and is studied in the radar, becoming a known shape. They calibrate the radar and replace the reflector with an airplane. They can then compare what the radar sees in the airplane against their shape, which is known.

MM: I understand large aerospace companies are in awe of how you manage to construct such large parts out of composites efficiently as well as economically. Are you collaborating with them on any projects?

EG: When we collaborated with Lockheed Aerospace, they thought it was amazing that the only part of this calibration project that was on time and on budget was our piece. And they asked us to do another one, and when we couldn't do it in the timeframe they asked for, they went to someone else; and, of course, it turned out misshapen and late, and the rework was huge.

MM: Do you think there is a backlash or a revisiting of the use of carbon fiber now in design? For example, do you really want an extremely light chair if it's not meant to be transportable?

EG: Well, that really depends. I have a set of chairs in my house that are pretty awesome chairs.

fig. 12
Fiberglass stair for a residence in Sarasota, Florida
Toshiko Mori, architect, built by Goetz Custom Boats
U.S.A., designed 2002, completed 2004
Continuous fiberglass sandwich construction stair with molded guardrail, consisting of end-grain balsa core and e-glass fiberglass skin

figs. 13, 14 (facing page)
*Esmeralda* Farr 50, interior
Built by Goetz Custom Boats
U.S.A., 2000
Nomex® core, sandwiched between layers of woven aramid fiber

They were made in the late 1950s or early 1960s in Copenhagen, out of ash. If you pick one up, you would say, "I've never felt a chair as light as this." But there are other times I would agree with you, that you just want to flop down in a chair and get absorbed by the cushions, and don't necessarily need the lightest thing. There is a carbon-fiber chic—people want their briefcases made out of carbon, and their cars, and those fake carbon sunglass frames. There are all kinds of places where you see fake and real carbon.

MM: Do you see that happening at all in the boat industry?

EG: Absolutely. We just finished a seventy-footer, and the interior is a real mix of textures, visually as well as tactilely. For instance, the table, desk, and sink in the bathroom are all carbon fiber. There are other areas that are natural wood or have linen wallpaper on the bulkhead, so you can feel this woven texture with your hand. There are still other areas where we put natural wood and very lightly painted it, so it has color but you can see the grain. It's a really wonderful mix of textures.

MM: Are there future projects that are particularly exciting that you can mention?

EG: We would like to get involved in one of the Volvo Around-the-World boats. We have built boats for the Whitbread/Volvo Around-the-World Race in the past, but for this race, they have a new, exciting specification for the boats which makes them technically very complex and interesting. We're talking to several groups of people about it.

MM: And in terms of expanding outside the boating industry?

EG: Well, we're sort of pushing the envelope. Toshiko Mori has been a real stalwart in helping us do this. We did a project with Jamie Carpenter last year, and we've built several cupolas for buildings in Newport, Rhode Island, and some sculptures that hung in the Pan Am–Met Life building for many years by the architect Warren Platner. So, really, what we do is build sculptural shapes—it doesn't matter if they're decorative or for sailing or cruising or racing. Actually, one of the most intriguing projects that are pending is for an artist in New York who was commissioned to do a sculpture for a college campus in Florida. She wanted to make a balloon that was about four feet in diameter and suspended twenty feet in the air. I had worked out the way we would make her balloon out of fiberglass, and attach it to a stainless-steel rod that looked like a coil of twine so that it would seem lighter than air.

MM: You're a problem-solver, and you seem to enjoy the challenges of your profession. Your work crosses all disciplines and enables you to collaborate with artists, architects, designers, all types of engineers, etc. It's great that you have that kind of openness.

EG: Well, that's the fun of it.

Interview conducted July 2004

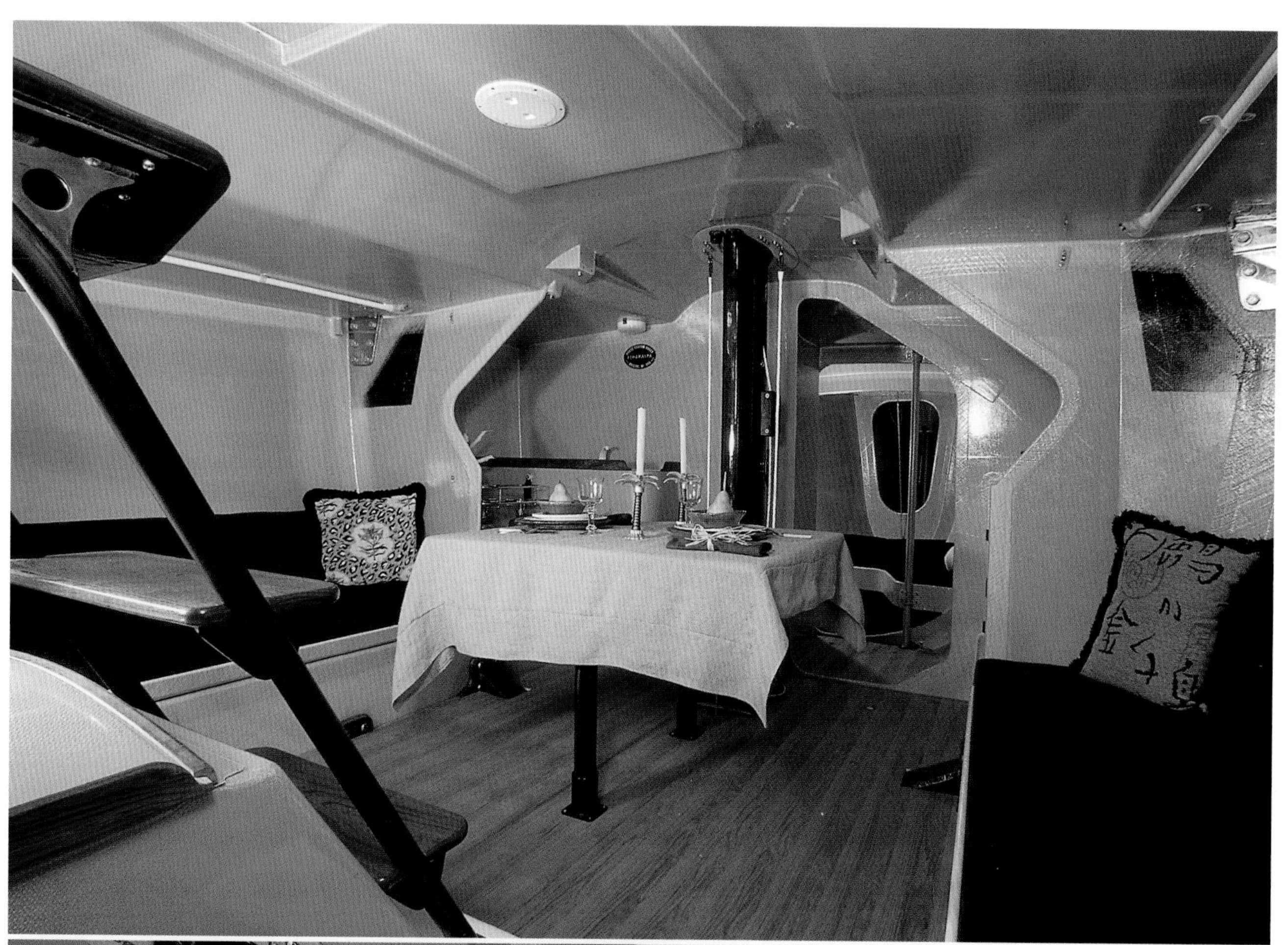

M-
-M

figs. 15, 16
Vanguard *Vector* racing dinghy
Designed by Bob Ames, manufactured by Vanguard Sailboats, engineering partner Steve Clark
U.S.A., designed 2000, manufactured 2004
Hull and deck of woven, knitted, unidirectional, and random glass fiber, polyester resin, PVC foam core; mast of aluminum; topmast of glass and carbon fiber, epoxy
464.8 x 182.9 cm (15 ft. 3 in. x 6 ft.)

Vanguard's *Vector* is among the fastest of the small racing boats or dinghies commercially available today. The rounded and narrow hull, made out of polyester resin and fiberglass, has been designed to reduce waterline length and water displacement, resulting in a skiff that can plane or skim over the surface of the water with less resistance and greater speed. The two-person crew trapezes or hikes to keep the sail vertical, while maintaining as long as possible the planing on the surface of the water. The ergonomics of the deck create comfortable hiking and foot positions in all wind conditions.

fig. 17 (facing page)
Cheetah™ Flex-Foot® prosthetic
Manufactured by Ossur North America
Designed by Van Phillips, engineered by Hilary Pouchak, textile manufactured by Newport Adhesives & Composites
U.S.A., designed prior to 2000, manufactured 2005
Plain-woven carbon-fiber composite

Flex-Foot® prosthetics were invented by amputee runner Van Phillips in the early 1980s. Dissatisfied with the prosthetics available to him, he began to explore strong, lightweight, and flexible carbon-fiber composites. His creations are designed for specialized use in a variety of activities. Today, some form of the Flex-Foot® is worn by over 90% of elite amputee athletes. The result of a close collaboration between engineer and athlete, Flex-Foot® has taken Tom Whittaker to the top of Mount Everest, and Paralympic gold medalist Marlon Shirley to world-record-breaking times. His time in the 100-meter dash is just tenths of a second over Florence Griffith-Joyner's. The design of the Cheetah™ Flex-Foot®, a sprinting foot, is based on the rear leg of a cheetah. Woven carbon fiber is molded into a shape that acts as a spring-board. Unlike traditional prosthetics, the inclined mid-section directs the ground contact toward the front two inches of the "toe." A C-shaped segment stores energy, compressing and springing back, propelling the runner forward.

figs. 18 (left)
Marlon Shirley, Paralympic gold medalist (Sydney, 2000), wearing the Cheetah™ Flex-Foot®

fig. 19
Jan Ullrich riding ADA Carbon Wheels in the 1999 World Championships, Verona, Italy

figs. 20–22
ADA Carbon Wheels
Designed by Cees Beers, manufactured by ADA
The Netherlands, designed 1999, manufactured 2005
Tubular rim of woven and press-molded carbon and Kevlar® K49 fiber, braided spokes of Kevlar® K49, carbon, and Twaron®
Rim width: 21.5 mm (0.8 in.); weight: 410–690 grams (0.9–1.5 lbs)

ADA Carbon Wheels, engineered for road and track racing, are handmade exclusively by Dutch wheel builder Cees Beers and his family. The custom-designed rim is press-molded of a single piece of twill or satin woven carbon and Kevlar® K49 aerospace certified fibers. Number and weave of carbon fibers is customized for the rider and cycling discipline — time trial, sprinting, climbing, etc. The spokes are handmade from a core of carbon, Kevlar® K49, and Twaron® fibers, with a braided cover of Kevlar®. In addition to incredible strength and stiffness, these wheels have outstanding aerodynamic properties due to their narrow rim width of eight-tenths of an inch, and low-profile spoke pattern. ADA wheels have been ridden to victory in World and National Championships, the Olympics, and the Tour de France.

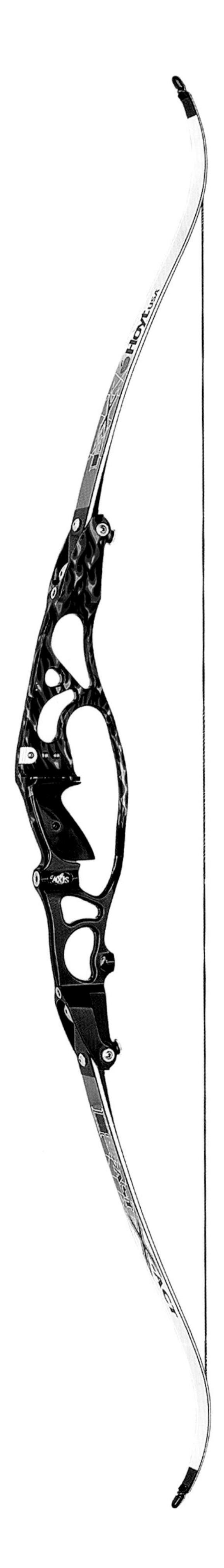


fig. 23 (left)
Axis recurve bow with FX limbs
Designed by Randy Walk, engineered by George Tekmitchov and Gideon Jolly, manufactured by Hoyt U.S.A., textile designed by Gordon Composites
U.S.A., designed 1999, manufactured 2004
Unidirectional fiberglass, Zero-90 fiberglass, unidirectional and bias carbon-fiber composite limbs; Dyneema® string

fig. 24 (facing page)
Flow bicycle saddle
Designed by Tylor Garland and Aldis Rauda, engineered by Dahti Technology, manufactured by SaddleCo., textile designed by Bob Benko
U.S.A., designed 2001, manufactured 2003
Leno woven mesh of elastomeric polyester monofilaments molded onto a thermoplastic perimeter; hollow titanium rail

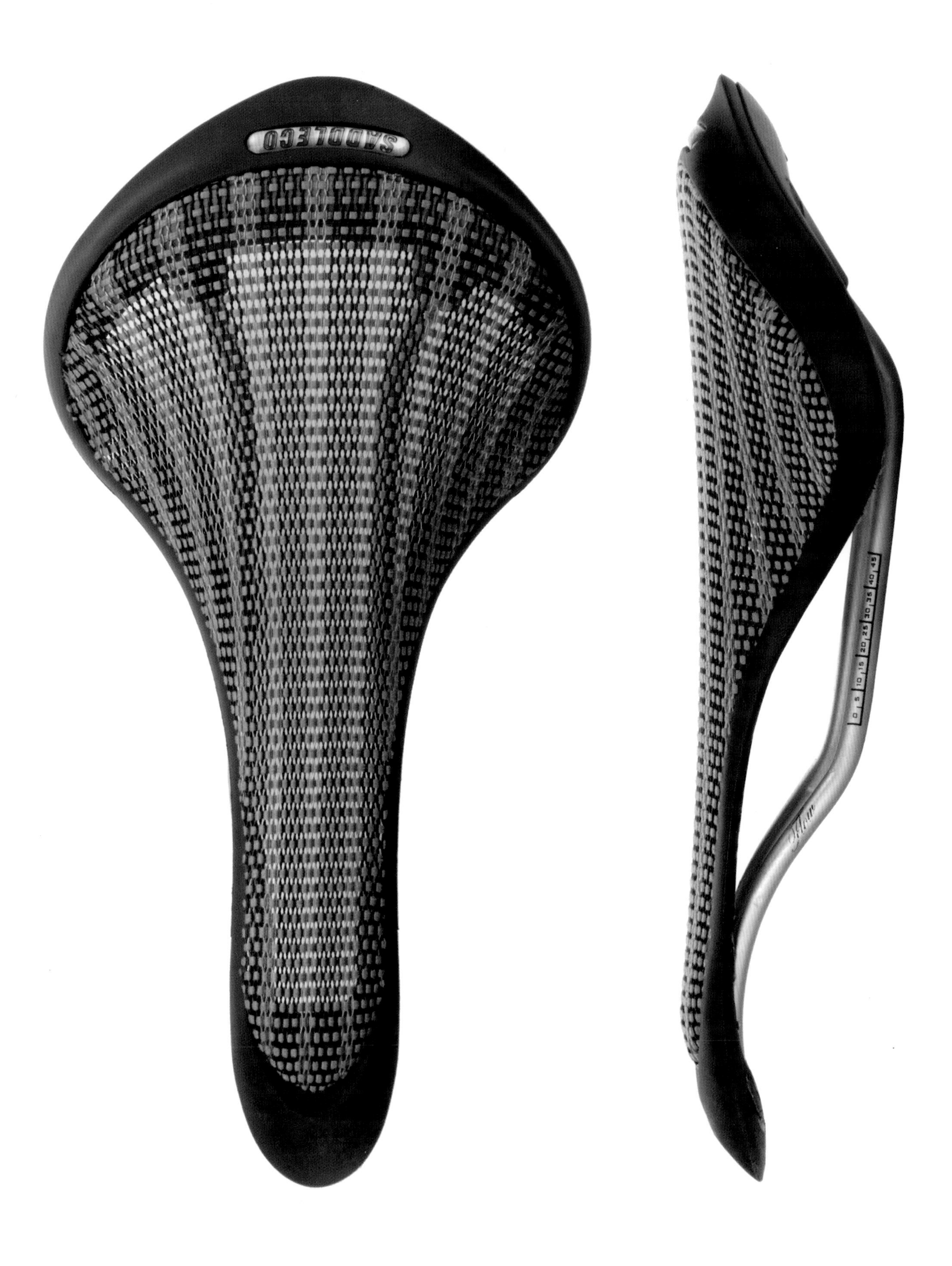
SADDLECO
0 5 10 15 20 25 30 35 40 45

fig. 1
Cladding detail of Swiss Re Headquarters
Foster and Partners, architects;
Ove Arup and Partners, engineers
England, 1997–2004

The mullion system is organized as continuous structural strands placed on bias, providing bracing for wind loads.

Philip Beesley and Sean Hanna

# A TRANSFORMED ARCHITECTURE

A new generation of giant-scale textiles is at the core of a revolution in architecture. Soft textile foundations are fundamentally changing the way we think about our built environment. Textile-based building concepts range from flexible skeletons and meshwork skins to structures that move and respond to their occupants. These structures replace traditional views of solid, gravity-bound building with an interwoven, floating new world.

In a 1913 drawing published in *Scientific American* magazine, Harvey Wiley Corbett imagined a city that dissolves into a complex of layers: buildings, bridges, and roads rise up to the sky and stretch deep underground supported by airy lattices made of steel frames and meshworks (fig. 2). Just a few years after Corbett, the eminent American architect and engineer R. Buckminster Fuller began drawing his own early visions of an interconnected infrastructure built from an immersive web of lattices (fig. 3). Today, giant textiles are being used to realize these radiant structures.

Since the beginning of the Industrial Revolution, leading structural engineers have been fascinated with the potential of intermeshed, lightweight, flexible structures. These evolving structures have steadily increased the role of tension forces, replacing the dense masses of compression-based

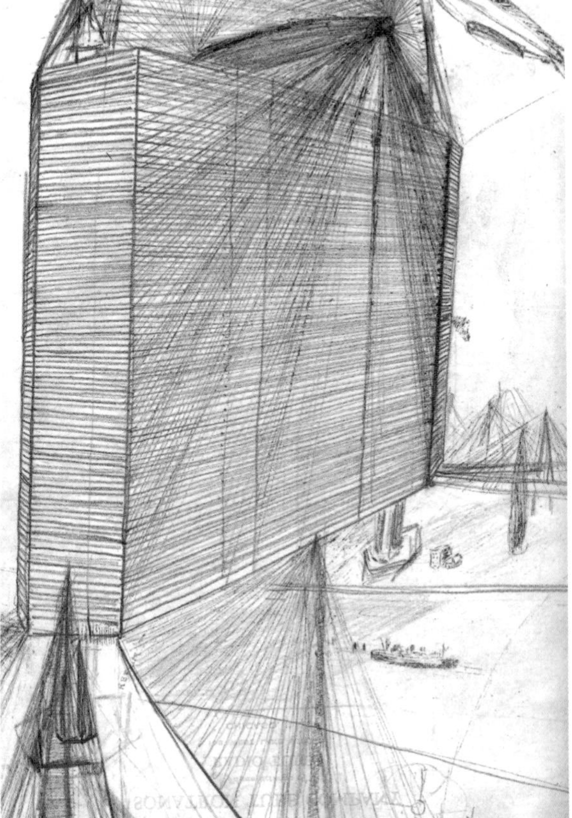

fig. 2 (left)
Harvey Wiley Corbett, 1913, in *Scientific American* magazine

Intermeshed skeletons support the multiple levels of Corbett's imaginary view of a city of the future.

fig. 3 (right)
Suspension bridge and office building, 1928
Sketch by R. Buckminster Fuller

Fuller attempted to integrate the structural systems of the Brooklyn Bridge and a Ferris Wheel in this visualization.

traditional buildings with open, more efficient systems. Some of these early experiments hid their slender interior structures. Alexandre-Gustave Eiffel's Statue of Liberty of 1886, for example, presides over New York Harbor as a presumably massive figure. Underneath the dress, however, is a light framework composed of continuous, supple bands of iron woven together into a basketwork that supports the sculptural drapery, made from thin overlapping tiles of bronze (figs. 4, 5).

While much of twentieth-century architecture continued to wear a mask of heavy cladding derived from ancient Roman and Greek construction, structural cores were being transformed into resilient skeletons that flex and respond to the dynamic loads of wind and earthquakes.[1] Instead of being built from the ground up, new building skins hung from above and became thinner and thinner. The development of the curtain wall, early in the century, was a turning point for architecture. In a break from previous construction methods, the enveloping jacket of a curtain-wall building is composed of a metal and glass cladding system formed from a continuous network of structural strands. The skin is able to support itself as a unified fabric, requiring only intermittent fastening to carry its weight (fig. 6).

fig. 4 (right)
*Liberty Enlightening the World*
Frederic-Auguste Bartholdi and
Alexandre-Gustave Eiffel
U.S.A., 1886

fig. 5 (below)
Interior view of *Liberty*, showing forged
meshwork supporting the folded profiles
of the bronze drapery skin

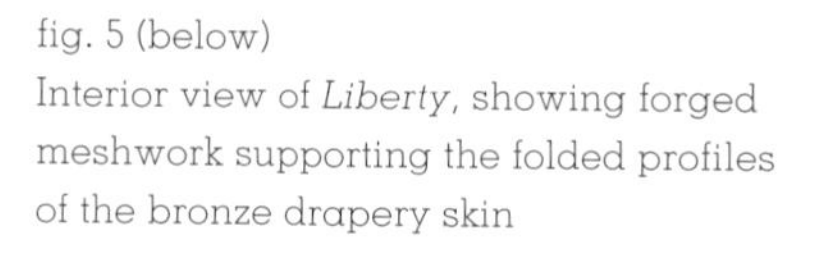

fig. 6
View of contemporary London, showing modern curtain-wall towers, with eighteenth- and nineteenth-century masonry buildings in foreground; Swiss Re Headquarters by Foster and Partners in distance

Recent projects continue the evolution of the curtain wall. Foster and Partners' tower for the Swiss Re Headquarters uses a unique structural skin made by placing members on a bias that dissolves the distinction between vertical and horizontal. The curved structural shell bulges gently in the middle and converges at the top, formed by an intersecting helical structure—essentially a skewed curtain wall. Similarly, the Seattle Public Library, designed by Rem Koolhaas's Office for Metropolitan Architecture, consists of a series of stacked, sloped floor plates, held in place by an angled, tight fishnet structure. The angled intersections of the exterior members and floor planes brace each other, keeping the structure stable (fig. 7). Pursuing even greater lightness, the face of Richard Rogers's Channel Four headquarters in London is composed of plates of glass assembled entirely without mullions, instead supported by a network of cables. This cable-net produces a hovering, agile presence that seems opposite to the classical paradigm of permanence (fig. 8).

## ANCIENT ROOTS

While many of these examples depend on highly advanced materials and technology, the principles have ancient roots using humble fabrics. Recent archaeological finds date the beginnings of permanent construction almost immediately after the last ice age, approximately seven thousand years ago, contemporary with the earliest evidence of woven textiles. As the nomadic existence of the Paleolithic age gave way to the first settlements, and transportable tentlike huts clad in animal skins were replaced by architecture designed to last for generations, the first building materials to emerge were

fig. 7 (top)
Seattle Public Library
Office for Metropolitan Architecture, architects; Front Inc., envelope consultants
U.S.A., 2004

External meshwork provides primary structural system, resisting wind and lateral forces

fig. 8 (bottom)
Channel Four Television Headquarters
Richard Rogers, architect
England, 1991–94

Interior view of front façade showing cable-stayed structural glass façade

not masonry, but woven. A meshwork of small, flexible branches formed the underlayer of cladding and served to brace the larger structural members, stiffening the then-circular house.

Thatch, which is the binding of straw or grass fibers together as a roofing material, and wattle, a lattice of flexible twigs and small branches woven horizontally through a series of vertical wooden stakes, were the standards for a building's exterior surface. The wattle provided excellent tensile strength, held fast by clay daub—the combination formed an efficient structure that made an integrated fabric.

The technologies emerging at the beginning of this new millennium return to these traditions and share many of the underlying principles.[2] Carbon-fiber and resin matrices are being fabricated to outperform the steel and concrete buildings of today. The basic concepts of ancient wattle-and-daub and thatch techniques still apply to these lightweight building systems.

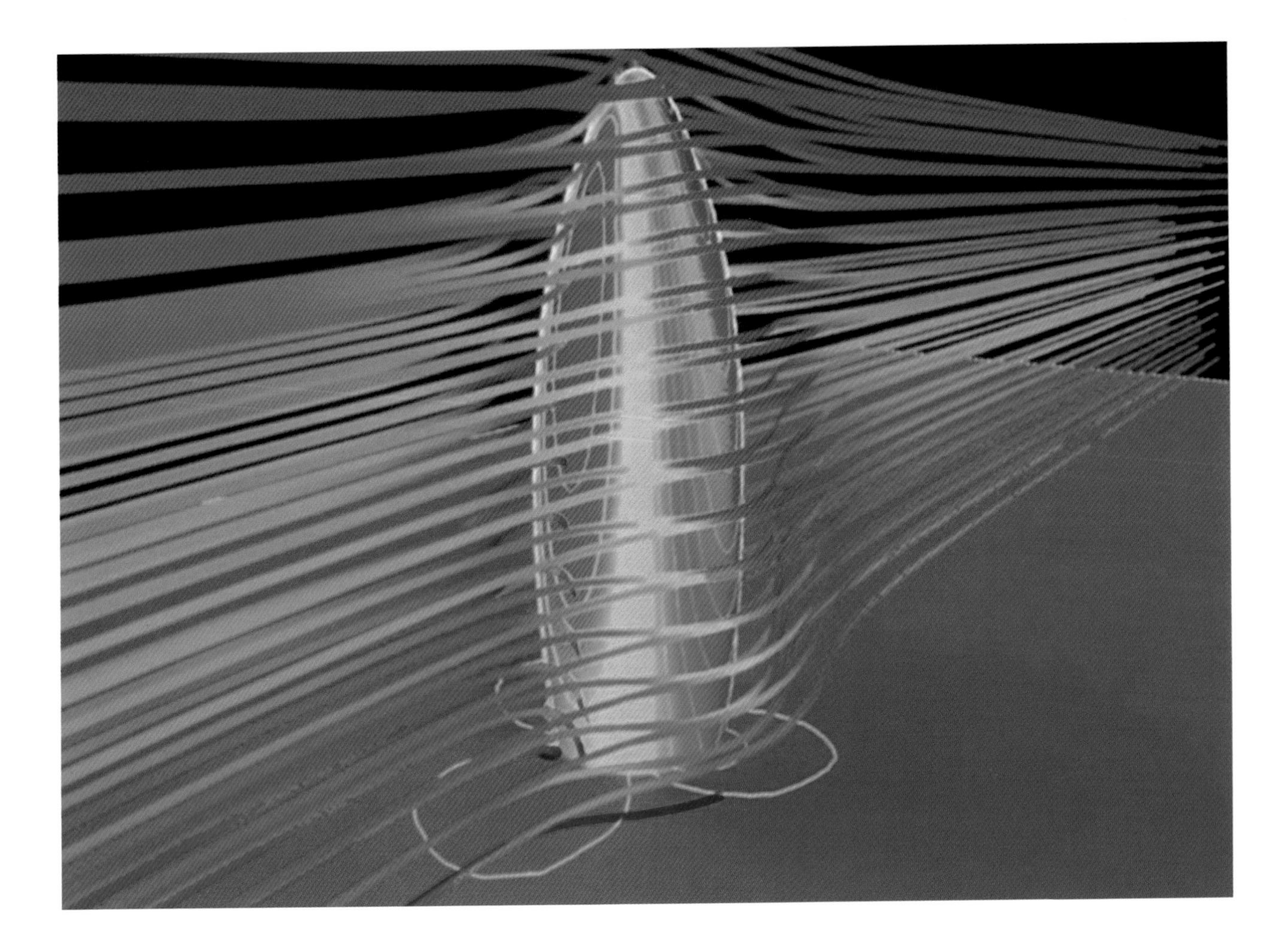

fig. 9
Swiss Re Headquarters
Foster and Partners, architects
England, 1997–2004

Illustration of wind loads on tower; structures that act in tension can handle these forces more effectively than compression-based masonry systems.

New fibers used in architecture include composites of glass and carbon that are stronger and more efficient than traditional steel and glass assemblies. These textiles require new methods of construction. This kind of building uses continuous chains of components and distributed structures that take advantage of a meshwork of woven elements. Because these materials are constantly evolving, specialists responsible for building systems are obliged to rethink fundamental systems of design and means of construction.

## THE PROBLEM OF EFFICIENCY

A prime criterion for engineering is the concept of efficiency. Heavy materials naturally tend to take more energy to transport and form than lightweight materials, but for much of human history they seemed efficient because the only forces that they had to serve were the downward-directed forces of compression and gravity. It is a relatively simple matter to work with gravity using unskilled labor by piling one heavy block on top of another. However, traditional masonry buildings are quite vulnerable to shifting and buckling forces. Ancient history is marked by a succession of catastrophes in which entire cities were destroyed by earthquakes and floods. The challenge of building an effective structure dramatically increases when considering the lateral and upward-pulling pressures from wind and earthquakes, because these forces are much more complex than compression in the way they behave. These factors require tension, and masonry is extremely inefficient in handling tension (fig. 9).

On the other hand, tensile forces can easily be handled by thin, continuous strands of resilient material. To deal with complex back-and-forth waves of forces, a lattice of long strands applied in opposite directions can be used. Confronting the pull of upward-lifting wind and the twisting, buckling forces of gravity has required a change in engineering practice. The force of tension, which previously played only a minor role in architecture, has become just as significant as compression.

## INTERCONNECTED SYSTEMS

Most traditional buildings engineered today use components of construction fabricated in a strict order. Primary elements support secondary, secondary supports tertiary, and so on. For example, a foundation and structural core in concrete might form the basis for a grid of steel columns between regular floor plates, and on these a skeletal grid of windows may be hung, with finishes added. The order of construction is roughly parallel to a steady decrease in size of each element, and in the hierarchy of support. Each must carry the combined load of every subsequent element that is later added to it. The first stages of construction therefore form the immovable, stable base that supports everything that is to come.[3]

In a textile, the process is quite different. Every fiber has an integral role in maintaining structure, each as important as its neighbor. The fibers are long, usually spanning the entire length or width of the textile. The structural properties are evenly distributed throughout the fabric, as each thread connects to the others. Instead of fixed, rigid connections based on compression, textile structures use tension. The binding of one fiber to the next is achieved through the tension exerted by the immediately adjacent fibers. Rather than relying on support from the previous, stronger member, the system is circular, holding itself in balance. The necessity of constructing components only after their supporting members are complete is removed, and a wide range of diverse elements can be built at the same time (fig. 10).

From 1947 to 1948, while working with artists at Black Mountain College in North Carolina, visionary Buckminster Fuller developed the concept of synergy, meaning the "behavior of whole systems unpredicated by the behavior of their parts taken separately."[4] During a career of pioneering work in engineering space-frame and tensegrity systems, Fuller explored complex interactions of structural elements that reinforce the whole. Using synergy, he described textiles as exemplary systems for architecture. A distribution of forces occurs as each thread joins a large number of similar threads. The whole collection can tolerate extensive damage, spreading this risk throughout many elements. If one thread snaps, the proximity of identical components, and their flexibility, allows the system to adapt dynamically to the new condition. However, the complexity and dynamic qualities of this behavior put it beyond the reach of standard nineteenth-century analysis methods and Fuller's time.

Another esteemed American scientist and engineer, John Argyris, invented Finite Element Analysis in the 1950s to study the attributes of complex systems and geometries. The method breaks down a continuous, dynamic structure into many simple, linked elements. Finite Element Analysis has become the standard method used today to design fabric structures. For example, a stretched membrane held in tension as a doubly curved surface can be simulated by a grid of individual, connected elements similar to the pattern of fibers in the fabric. Each of these components can be analyzed in relation to all elements to which it is connected to calculate the stresses, determine optimal curvature, plan the location of seams, and orient the warp and weft of the fabric (fig. 11).[5]

The textile examples that follow attempt to achieve tangible goals: lighter, stronger, more responsive, and more efficient. In pursuing these objectives, these projects present innovative ways of designing and looking at the world.

## CARBON TOWER, PETER TESTA

Perhaps no other project asserts Fuller's idea of synergy more than Peter Testa's Carbon Tower, an extraordinarily ambitious design based on advanced textile technology. The proposed tower is built of carbon fiber and composite materials. A prototype forty-story office building, the main structure is woven together, rather than assembled from a series of distinct parts. The building's shell consists of twenty-four helical bands thousands of feet long, winding in both directions around the cylindrical volume. Instead of relying on a rigid internal core and a series of columns for stability, these thin bands of carbon fiber, each a foot wide and an inch thick, run continuously from the bottom to the top of the building and take the entire vertical compressive load. The forty floor plates are tied in to the external structure, acting in tension. The floors keep the helix from collapsing while the helix, in turn, supports the floors. Both systems are interdependent; the helix, for example, would collapse in a scissor motion if the joints were not bonded together and the tension of the floors was removed (figs. 12–15).

The goal of Testa's project is to achieve an unparalleled synergy of elements, where multiple systems in the building—structural members, airflow, and circulation of people—act together. The Carbon Tower offers a strategy to eliminate joints and abrupt changes in material. This approach requires a break from the tradition of reductionism, and an almost complete abandonment of the principles of hierarchies in building systems. Massive calculation is required to model the complex interactions between elements and the environment. In the most recent generation of computing, this simulation and analysis have become possible. Connections between structural elements are crucial to the design, and the transformation between the floor and helix

figs. 10, 11 (facing page)
Swiss Re Headquarters, tower envelope
Foster and Partners, architects
England, 1997–2004

Layout system of structural elements showing fabric system of interconnected spirals and circular matrices of elements (top)

Analysis of arching structural fibers within tower; example of Finite Element Analysis software by Bentley Systems (bottom)

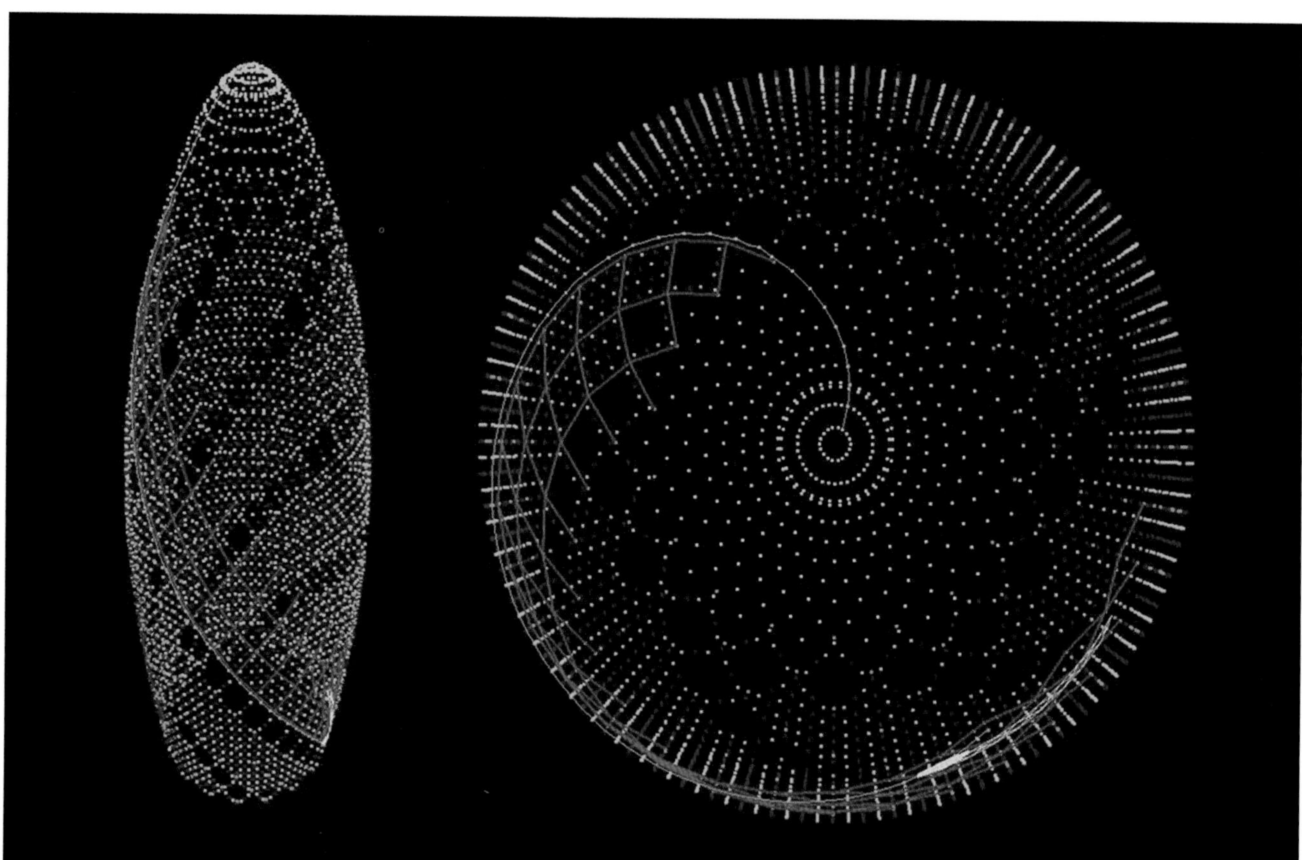

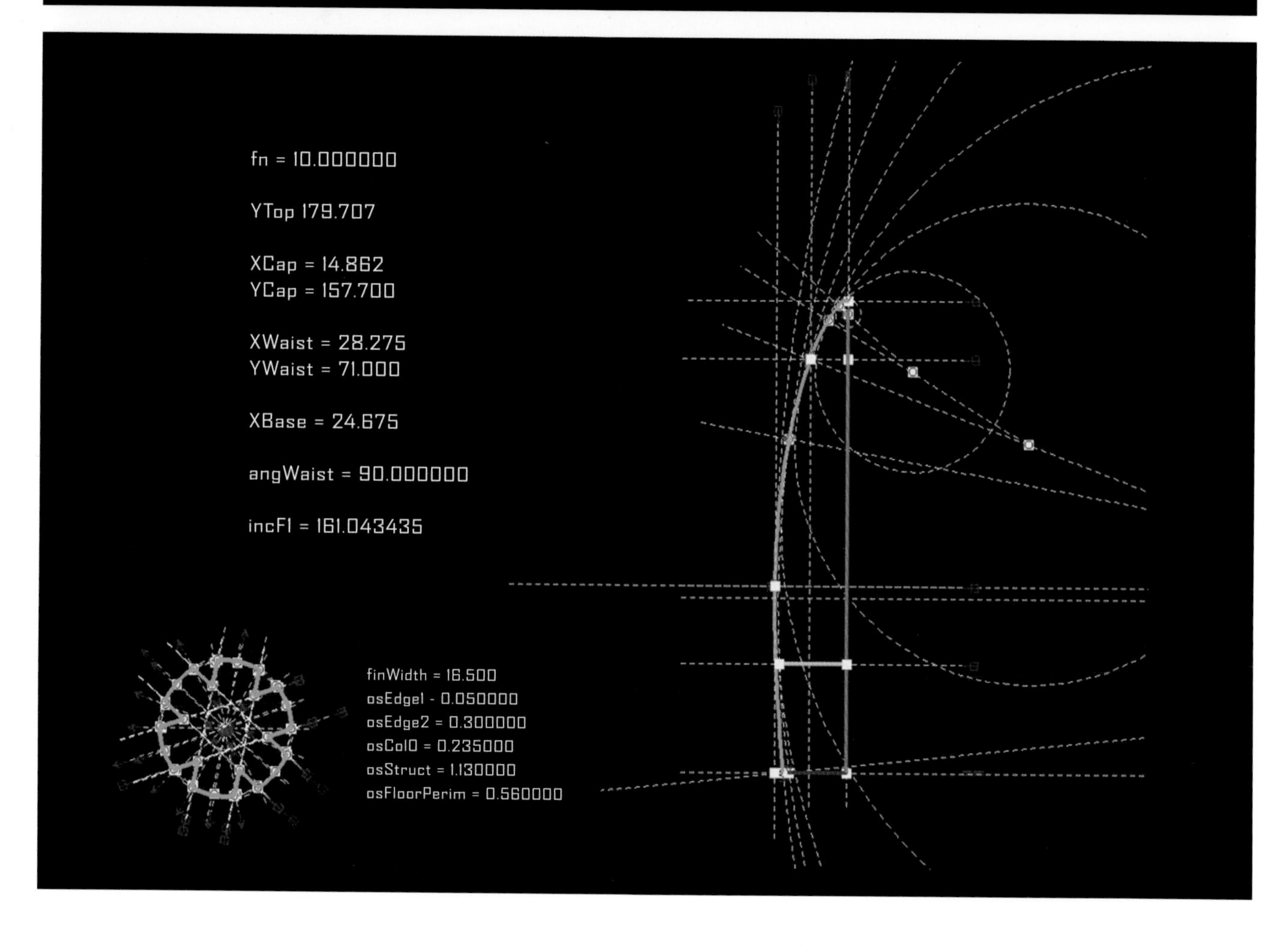
fn = 10.000000
YTop 179.707
XCap = 14.862
YCap = 157.700
XWaist = 28.275
YWaist = 71.000
XBase = 24.675
angWaist = 90.000000
incFl = 161.043435
finWidth = 16.500
osEdge1 - 0.050000
osEdge2 = 0.300000
osCol0 = 0.235000
osStruct = 1.130000
osFloorPerim = 0.560000

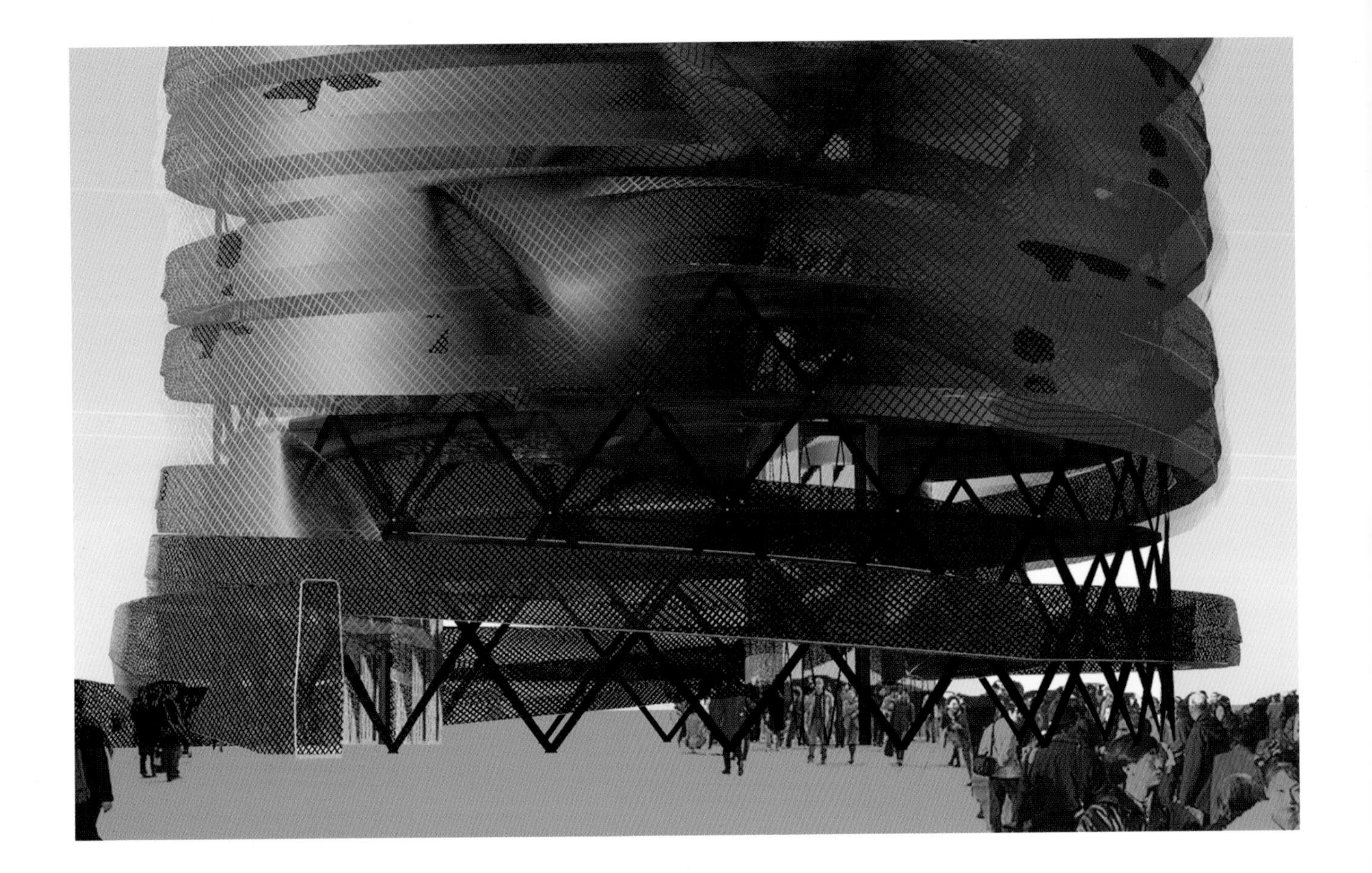

fig. 12
Carbon Tower, exterior
Testa Architecture and Design
Peter Testa and Devyn Weiser, principal architects; Ian Ferguson and Hans-Michael Foeldeak, project designers; Markus Schulte, principal structural engineer; Mahadev Raman, principal mechanical engineer; Ove Arup, New York, engineers; Simon Greenwold, Weaver software designer
U.S.A., 2004
Double-helix woven structure of twenty-four twisted strands of pultruded and braided carbon fiber, stabilized by continuous braided tendons within floor plates, two external filament-wound ramps providing lateral brace, exterior of a ventilating tensile membrane

features gradual movement of one component into the next. The same cables that form the helix also provide the basic structural framework for the floors.

A new construction method is used for making these cables from impregnated carbon fibers. Pultrusion is a method for producing continuous extrusions of composite materials. Carbon-fiber composite is formed by passing raw strands of material through a resin-impregnation bath and then through a die to shape it into the appropriate cross section. The resin cures while the material is passing through the die, and the final product emerges immediately. This technique is used to make a range of stiff rods or flexible fibers that can be twisted, braided, or bundled into cables. The equipment for the fabrication process is portable, and allows many of the materials to be made directly on site.

When the main structural members of the perimeter helix are pultruded, the fibers cross at the point of a floor plate. Some of the carbon strands are diverted from the main vertical member and are grouped to form cables that run to the opposite side of the helix, tying the external structure together. A floor structure is then woven into and layered onto this surface network of cables. The building is literally woven and braided together.

All fibers in the structure are continuous as they travel up and around the helix, spanning the full height of the building. The bands comprising the helix are constructed by two robotic devices working in tandem: a pultruder on each of the twenty-four vertically spiraling members, closely followed by

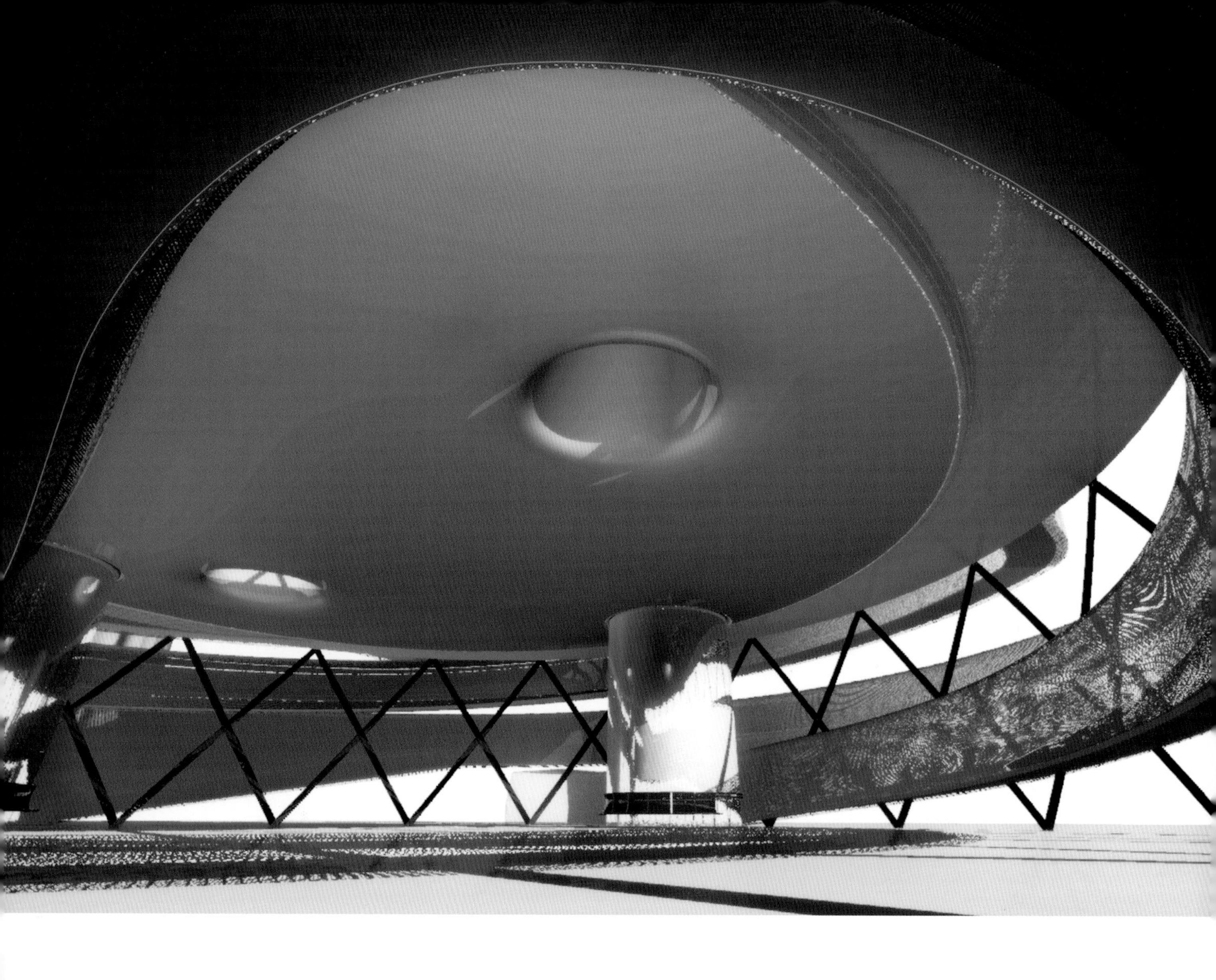

fig. 13
Carbon Tower
Testa Architecture and Design
U.S.A., 2004

Interior of lobby, with the "virtual duct" that runs the full height of the tower

a series of braiders that shape the same fibers into floors. The robots weave simultaneously, moving up the steadily rising building floor by floor.

Two central strands can be seen winding vertically through the middle of the building. These are air-distribution ducts, or "virtual ducts," which consist primarily of voids in the floor plates. A fabric-based duct system, which varies in weave density and strand thickness, hangs from the top of the building through these voids. The alterations in the weave are based on air pressure, such that the weave opens up as the building gets higher. The movement of air is handled by a dynamic and responsive system that allows the enclosure to contract and expand seasonally as ventilation conditions change. The floors through which the vertical voids pass are also hollow, allowing air to be drawn from the outside of the building, effectively integrating air flow with the structure. Standard ductwork as a separate, applied system is eliminated, increasing the floor-to-ceiling height and bringing more daylight into the building, offering reduction in energy consumption.

An external ramp system that follows the perimeter of the building performs several functions. By connecting to the structure at the floor plates, the ramp stiffens the entire building. It is shallow relative to the primary helix, and the

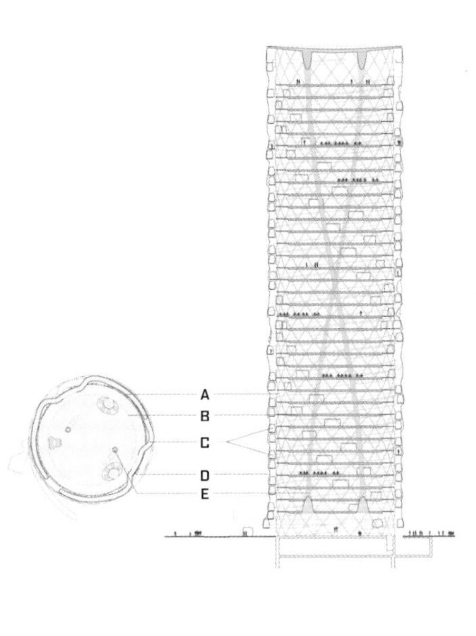


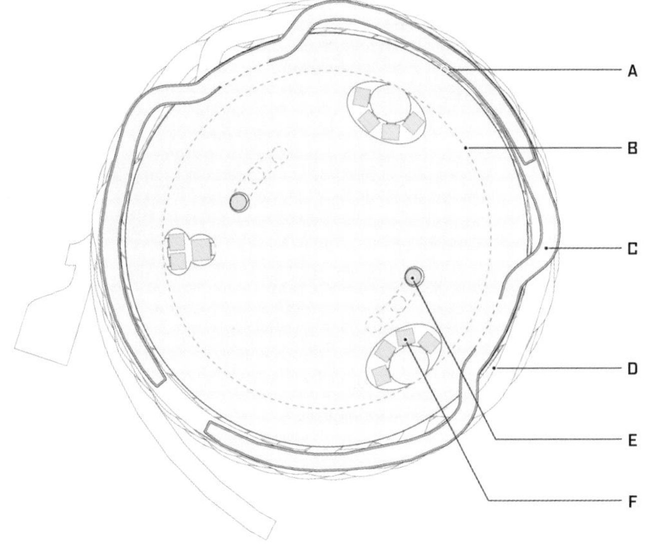


**A** Carbon Fiber Double-Helix Primary Structure
**B** Tensile Composite Floor Plates
**C** Filament-Wound Carbon Fiber Ramps
**D** Transparent Thin Film Enclosure
**E** Woven Ventilation Ducts
**F** Elevators

fig. 14 (right)
Carbon Tower
Testa Architecture and Design
U.S.A., 2004

Interior, showing filament wound ramp

fig. 15 (above)
Carbon Tower
Testa Architecture and Design
U.S.A., 2004

Section and typical floor plan

resulting increase in length allows it to create a substantial torque around the structure. The ramp is designed to contract with cables along its length to form active lateral bracing, which can be adjusted in relation to high winds, allowing the building to react to extreme weather.

Many of the structural problems of the Carbon Tower stretch the boundaries of standard design software. Testa developed a program titled Weaver to assist visualization of the structure. The program allows interactive play with geometry, mocking up woven patterns quickly by adjusting parameters of existing options. Because of the demanding nature of the new materials and the synergy between them, testing and analysis for this project lie outside established engineering practice. The practical limit on what can be designed is therefore based on what can be physically tested. No substantial precedents have been found for the use of composites in an architectural project of this magnitude and complexity.

The Carbon Tower suggests a number of future benefits for the construction industry. Carbon fiber offers many advantages over traditional materials. It is strong and light, and manufacturing carbon fiber and resin requires half the energy of steel. In the construction of a large building project, much of the cost of materials lies in the expense of transportation to the site, and here, too, there is a substantial saving; many components in the Carbon Tower, including the core elements, are manufactured on-site, and others are very light components that are easily transported. The advanced materials have the potential to last longer and require less maintenance than many standard materials currently in use.

## LEONA DRIVE RESIDENCE, MICHAEL MALTZAN

Exploiting the extraordinary visual qualities of carbon fiber, California architect Michael Maltzan has developed a design for a lightweight house made of a shimmering meshwork of translucent walls. The house, organized in three overlapping rectangular volumes, floats on a plateau overlooking the basin of Los Angeles. The public and private spaces pass over each other using long cantilevered structures. The stiff, light material, consisting of a woven carbon-fiber textile in a resin matrix, is ideal for supporting the loads of the long cantilevers (fig. 16).

Visually stunning, Maltzan has used a technique termed "ghost fiber" that binds aluminum powder to the carbon fibers, producing a reflection that dramatically decreases the apparent opacity of the gridded construction matrix. The overlapping layers of each space produce interference patterns that result from phase-relationships of the intersecting geometries of the fiber reinforcement. Wide spacing of the matrix fibers creates a distinct moiré effect, transforming the minimal, rigid geometry into a dynamic, open work.

A host of technical challenges accompany this project, raising provocative questions about the building industry. Carbon fiber is not an unusual material for transport vessels, but it is virtually unknown in the North American

fig. 16
Leona Drive Residence, Beverly Hills, California, model (southeast aerial view)
Michael Maltzan Architecture, architects; Michael Maltzan, Tom Goffigon, Tim Williams, Yong Kim, Gabriel Lopez, Bill Mowat, John Murphy, Nadine Quirmback, Ivy Yung, project team; Ove Arup, Los Angeles, structural and MEP (mechanical, electrical, and plumbing) engineers
U.S.A., designed 2002, completion date 2006
Carbon fiber, acrylic and epoxy resin, painted medium density fiberboard base
335.3 x 81.3 x 45.7 cm (11 ft. x 32 in. x 18 in.)

construction industry, where housing construction is normally managed with a catalog of predetermined parts. Intensive testing of each component has been required for the project. The thickness of the panels has generally been determined by structural performance tests, and a sandwich system is currently being developed to integrate the different needs of interior partitions and exterior areas requiring insulation and weather seals. The resin binding matrix has received special scrutiny because the longevity of the entire structure is restricted by the way the resin ages. Cleaning, resistance to scratching, and protection from the California sun have been particular issues requiring refinement of the resin formulation.

While a high level of experimentation characterizes the Leona Drive residence, the project carries the immediate reality of serving a family's daily life. This imperative has heightened the importance of practical solutions for maintenance and weather seals, while at the same time it has encouraged a highly personal approach to the sensual effects of the building material. Perhaps this suggests a working method for the next century: the combination of increasingly sophisticated materials created by a highly technological and remote industry and advanced structural applications will be balanced by a careful, hands-on approach. In his house, Maltzan is physically and intimately involved in the craft of building.

fig. 17
AirBeams by Vertigo™ inflatable support beam
Developed for U.S. Army Natick Soldier Center, U.S. Army Project Engineer Jean Hampel, designed by David Cronk (Vertigo™ Inc.), manufactured by Vertigo™ Inc.
U.S.A., designed 2001, manufactured 2004
Seamlessly braided Vectran™ fiber
Length: 36.6 x span: 25 m (120 x 82 ft.)

## RIGIDIZABLE STRUCTURES

Inflatable and rigidizable beam structures offer an architecture that can readily change its shape. The technologies in development at ILC Dover and Vertigo™ are intended for building in extreme conditions in space and on Earth. Designs range from low-gravity habitats on the Moon or Mars to deployable structures such as emergency shelters and army tents. Military and space astronautics are the first users of this technology; however, the potential performance and commercial applications of this adaptable system mean it will soon enter every neighborhood shopping center.

The AirBeam™ is a close relative of forms that are common today (fig. 17). Floating pool toys and air mattresses are small-scale cousins of inflatable architecture used for tennis domes, fairground buildings, and other temporary structures. Standard inflatables are usually made from coated textiles. A flexible polymer matrix covers the structure creating an airtight seal. Air is pumped into the form, acting as a compressive element within the tensile surface enclosure. This pressure must be sustained over time to maintain rigidity. While these inflatables offer an easily transportable and assembled architecture, they are prone to failure by puncture or power loss. Rigidizable structures are erected in the same way, but the high-performance matrix material in the surface becomes rigid and supports the structure without internal air pressure, removing the need for a constant power source. It suggests a new kind of semi-permanent structure that can be readily reused.

A typical rigidizable assembly from ILC Dover uses fiberglass fabric coated with epoxy resin and sealed with a polymer film to prevent the inactivated surface from being sticky to the touch. This initially flexible epoxy matrix becomes rigid and holds the fibers in place. It is the efficient combination of the flexible fabric and the rigid matrix that gives the structure its overall strength. Both are required in unison to hold the form—the matrix and textile act in compression, and the textile in tension. The stiffness and load-carrying capacities for a rigidizable beam are magnitudes higher than a similarly-sized inflatable beam.

A number of different methods can be used to make the structure rigid. Some of these result in a permanent shape, while others can be repeatedly deployed and retracted. The epoxy-coated fiberglass, used in cases of natural or artificial light exposure, forms a permanent, stable structure and can be made into specialized shapes, such as inflatable wings for aircraft. For forms intended to be repacked and transported, an alternative is based on a thermoplastic matrix made of "shape-memory" polymers that are imprinted with a memory of their fully-deployed shape. These polymers can change their states from soft to hard with heat (fig. 18).

Other methods differ in activation techniques. Some are set by a chemical reaction that generates foam inside the inflatable structure, while others are stiffened by a reaction with the chemical-laden gas used to inflate them. A variation of this structure uses thermosetting composites containing

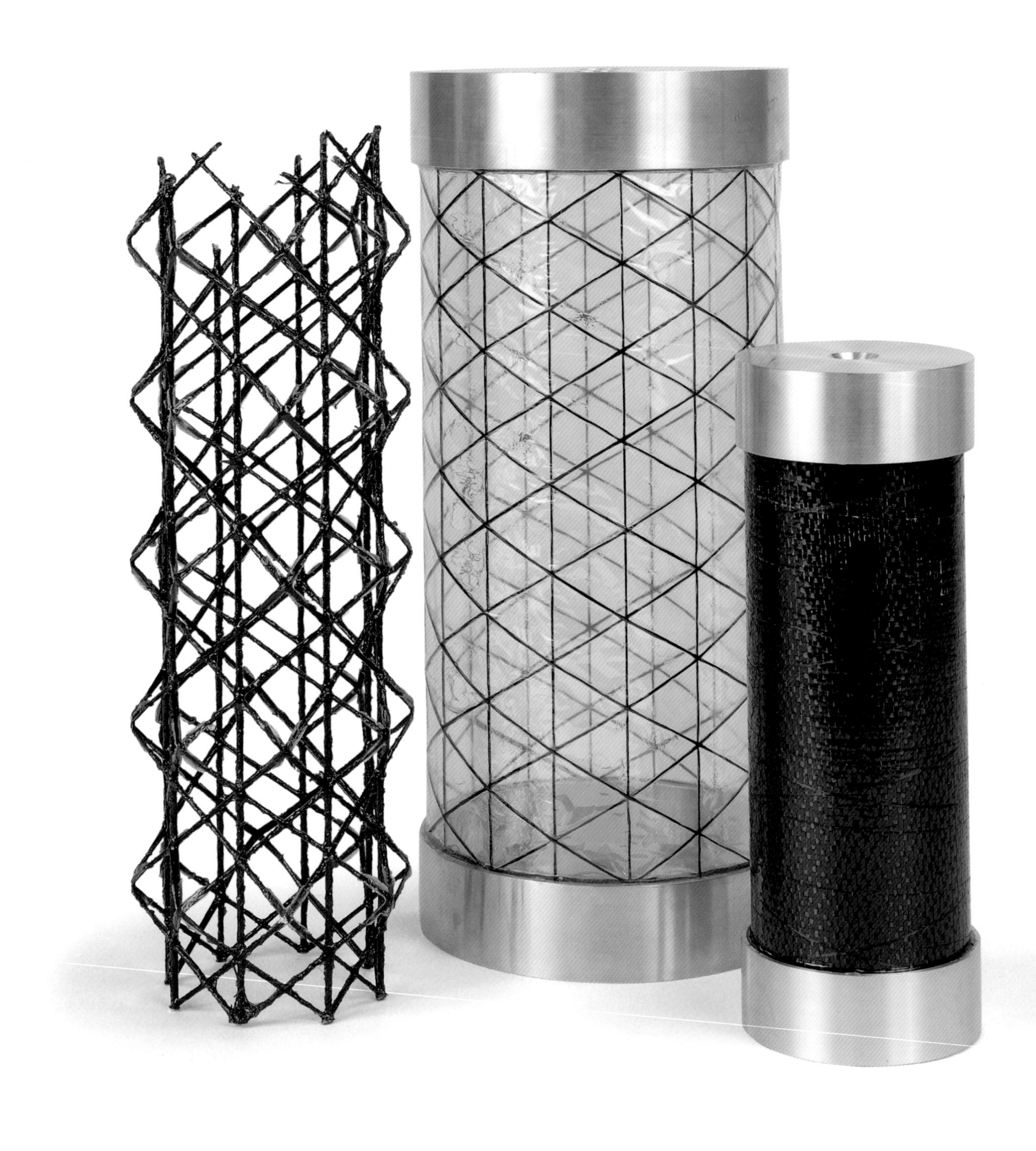

fig. 18
Examples of lightweight collapsible carbon/epoxy boom structures for space applications
Developed by ILC Dover Inc. and NASA
U.S.A., 1998–2001
Glass and carbon-fiber filaments; UV or shape-memory epoxy; Mylar® layers on both sides for gas retention; self-deploy when heated to a prescribed temperature

fig. 19
Carbon Isogrid rigidizable boom, deployed and packed
Developed by ILC Dover Inc. and NASA
U.S.A., 2001
Carbon filament wound structure with shape-memory epoxy; Kapton™ layers on both sides for gas retention; self-deploys when heated to a prescribed temperature

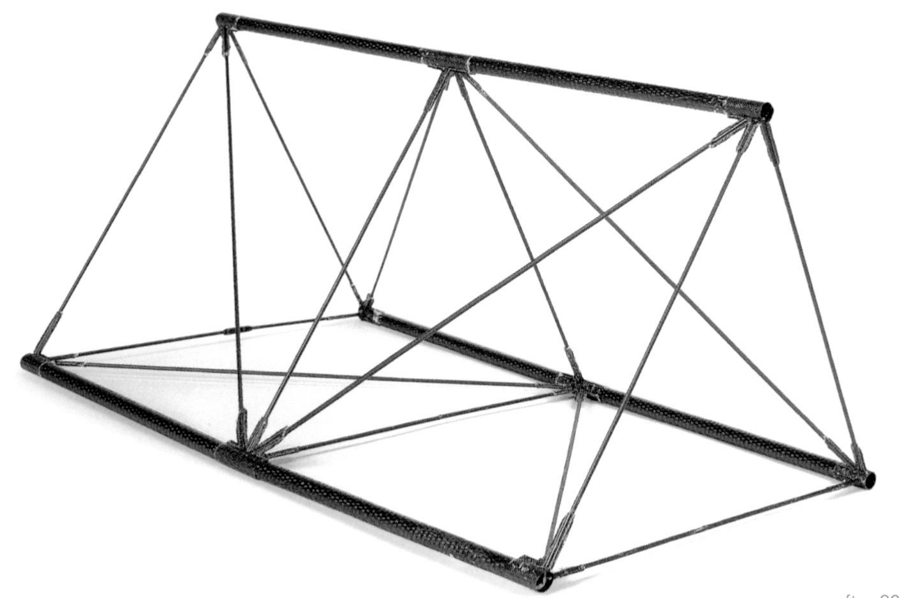

fig. 20
Space solar power rigidizable truss
Manufactured by ILC Dover Inc.
U.S.A., 2002
Carbon, epoxy

fig. 21
Wire gabions reinforce a slope in Tuscany, Italy

graphite that stiffens when subjected to a combined heat and pressure. In this case, a series of heaters is placed over the form while it inflates and sets. As an example, Isogrid booms, which are a class of structures, implement textile weave variations based on a low-density filament-wound construction (fig. 19). In contrast to the tight, regular pattern of woven reinforcement, these booms can be manufactured from loosely interlaced fiber bundles that resemble giant pieces of yarn.

Unlike inflatable structures, the strength of rigidizable beams can be readily analyzed, making them reliable as critical members of large assemblies. Standard inflatables derive their stability from a complex interaction of the exterior tensile membrane and the compressed, fluid air. While the interactions of the materials in the rigidizable structure are complex, their dynamic behavior can be analyzed quite simply by treating the components as traditional column, beam, and strut elements carrying tension and compression forces.

There are strong possibilities for these elements in buildings of the future. The range of scales for this technology is wide, from miniature rigidizable members measuring less than one-eighth of an inch in diameter to a nine-hundred-foot-long truss with triangular faces of ten to fifteen feet, intended for use in space (fig. 20). In a full-gravity environment there is a practical limit on vertical size, because thick walls are required to support the compounded loads that accumulate toward the base of large structures. Horizontally oriented components, however, can be arrayed along the ground indefinitely.

fig. 22
Pyramat® turf reinforcement mat
Manufactured by Synthetic Industries Corporation, Performance Fabrics Division/Geosolutions
U.S.A., 2004
Waffle weave (interlocked structure of uniform voids and projections) of UV radiation stabilized polypropylene monofilament yarns

## FLOATING CITIES

The ground that buildings stand on has been transformed. It is now common practice for civil engineers to include geotextiles—heavy-duty fabrics used for earthworks—in large-scale landscape construction. Filtering and drainage layers, reinforcing cable nets, beds of earth anchors, and arrays of wire gabions run beneath new cities (fig. 21).

Geotextiles are landscape-engineering technologies that are literally woven into the earth. Applied as a fabric to the surface, they can be integrated into the soil or root systems of vegetation, preserving existing fragile landforms and creating entirely new landscapes. Erosion protection is one very common use for the technology, providing a soft armor, as opposed to the hard erosion control of rock or concrete. New plantings and their embedded textiles grow together to form a single, integrated structure. The scale of these functional support systems is often enormous. Pyramat®, produced by Synthetic Industries, is used to prevent soil erosion and consists of a fabric woven of monofilament yarns formulated to resist breakdown under harsh sunlight (fig. 22). Pyramat's structure interlocks and combines with the existing ground. This three-dimensional geotextile has pyramid-like projections that capture and contain soil, holding it fast while water flows. The textile allows vegetation to take root in a now-stabilized soil. Enkamat®, made by Colbond, is another product that provides protection of embankments or slopes, both wet and dry (figs. 23–25). It is primarily used adjacent to roads and railways, and is seeded and filled with soil. The resulting textile-and-

figs. 23–25
Enkamat® turf reinforcement mat
Manufactured by Colbond Inc.
U.S.A., designed 1965, manufactured 2004
Three-dimensional matrix of entangled,
fused Nylon 6 filaments

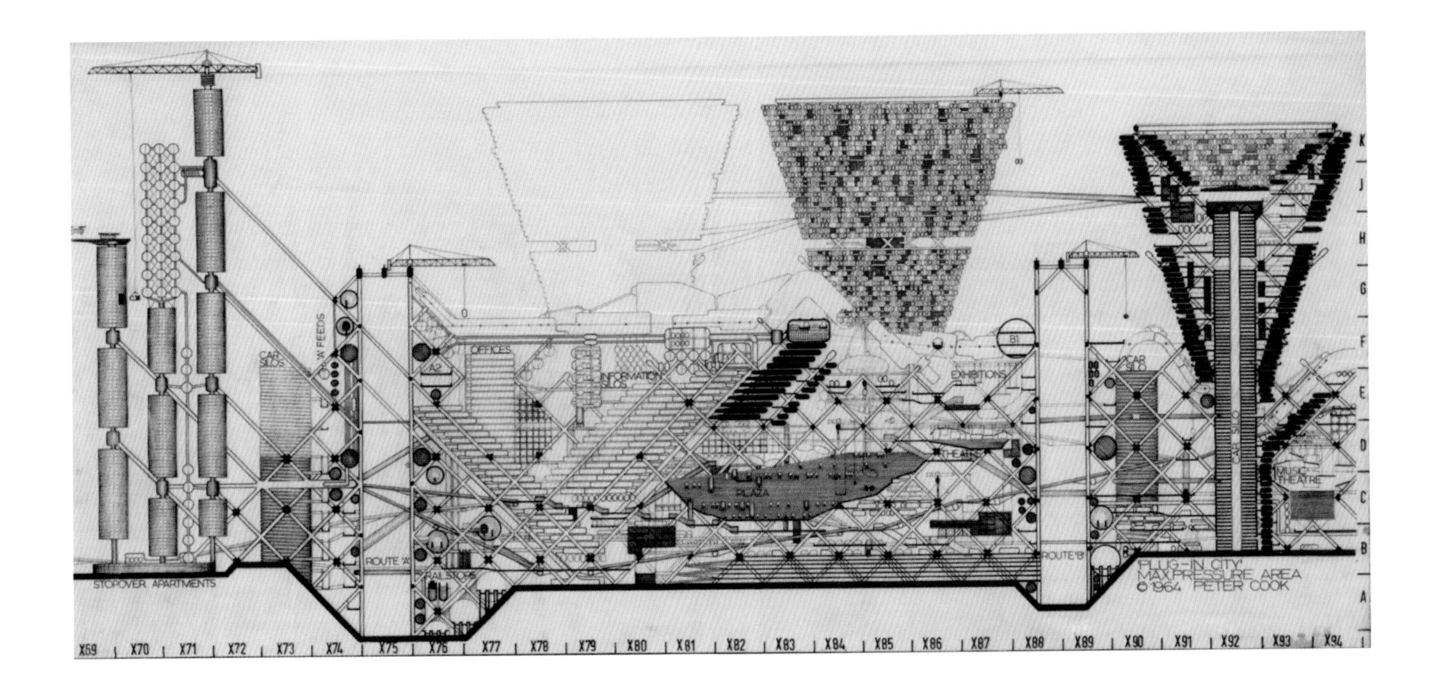


fig. 26
Plug-in City, visualization of the city of the future
Peter Cook, Archigram, 1964

earth combination is permeable by water and roots, and allows the growth of vegetation.

Repotex, manufactured by Huck, is a mat used for plant cultivation and repositioning in new environments, usually water and soft mud. The woven fibers form a base in which plant roots grow in the textile matrix, and the whole can float on top of the water's surface. Because the roots are held only to the textile, the entire living surface can be moved, forming islands of floating plants. The material itself is a coarse fabric structure made of either decomposable or non-decomposable materials, depending on the desired use. It can also be used to hold water and allow growth on flat and angled roofs. Purification of surface water can be accomplished by positioning plants on the surface. The plant textile can remove organic compounds from runoff from sewage farms and filter inorganic nutrients in intense agricultural areas.

## NOMADIC HUTS, OLD AND NEW

The nomadic huts of the Mesolithic age consisted of only a surface skin, stitched together of animal hides, transported from site to site. In the 1960s, speculative projects like Archigram's Plug-in City explored the possibility of rapid change in new architecture of the future for a society dedicated to speed and novelty (fig. 26). The recent development of disposable shops—filling retail vacancies in major cities for a month or two of intense promotional activity, and then disappearing—indicates that the fantastical proposals of two or three decades past are now a reality. A flexible structure using the technology demonstrated here by ILC Dover offers a means of reaching this vision while avoiding the waste of single-use temporary structures. Similarly,

fig. 27
In a 2002 view of Potsdamer Platz in the heart of Berlin, new towers stand on a hollow ground plane that covers a vast underground web of submerged expressways, subways, and utility corridors. The horizon is at the mid-point of this image.

with the implement of structural textiles, Buckminster Fuller's vision is coming true. Landscapes constructed from geotextiles and cities built from lightweight flexible skeletons are at the core of this new world.

A surge of possibilities accompanies this technology. These possibilities affect scales from molecular engineering to artificial landscapes that support entire cities (fig. 27). Peter Testa's explorations into self-assembling structures and lightweight fiber-producing robots offer tangible ways for the structures of the future to grow themselves. The Carbon Tower is an advanced illustration of the generation of buildings that will soon make up our surroundings. The dramatically expanded three-dimensional web that Corbett imagined is very nearly here.

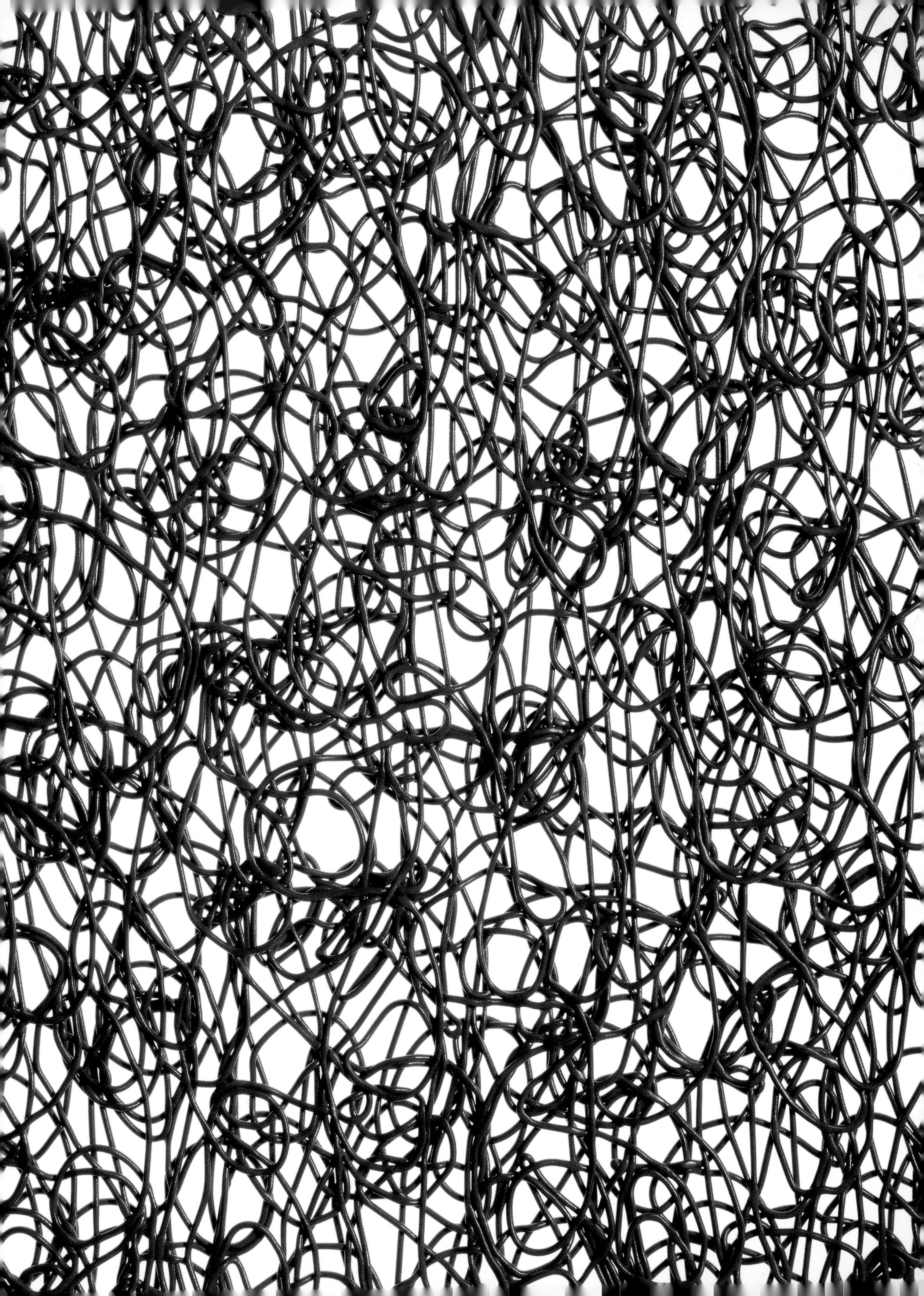

fig. 28 (facing page)
PEC-Mat® turf reinforcement mat
Designed by Thomas P. Duffy, manufactured by Greenstreak Inc.
U.S.A., designed 1986, manufactured 2004
Thermally welded PVC monofilament

fig. 29 (right)
Encircling fishing net
Manufactured by Toray Industries Inc.
Japan, designed 1996, manufactured 2003
Machine-made knotless netting of interconnected twisted polyester threads, heat treatment finish
4 cm (1 9/16 in.) compacted; expands to 226 cm (7 ft. 5 in.)
Cooper-Hewitt, National Design Museum, Gift of Toray Industries Inc., 2003-22-1

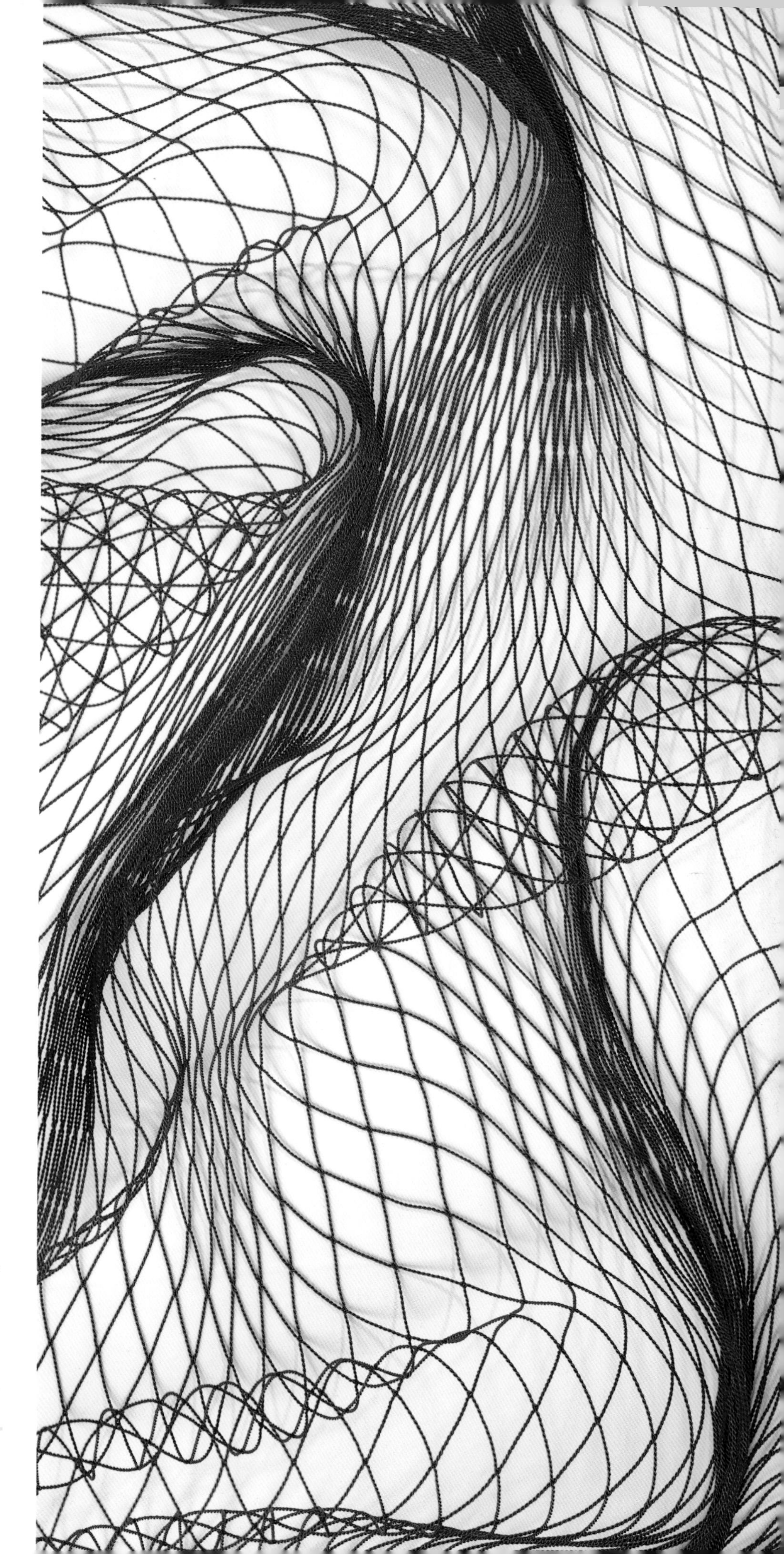

fig. 30
Rocket nozzle
Designed by Thiokol®, textile designed and manufactured by Foster-Miller Inc.
U.S.A., 1998
Braided carbon fiber, epoxy matrix
Length: 25.4 cm (10 in.); diameter at wide end: 22.9 cm (9 in.)

This prototype rocket nozzle exit cone is constructed from an advanced carbon-fiber braid. It is designed to replace significantly heavier metal exit cones and increase rocket payload.

fig. 31
Pi-braid
Designed and manufactured by
Foster-Miller Inc.
U.S.A., 2001
Three-dimensional braid of carbon fiber

This prototype three-dimensional pi-braid is designed to form strong yet lightweight joints between composite panels on aircraft fuselages.

fig. 32 (facing page)
Spacer fabric
Designed by Stefan Jung,
manufactured by Karl Mayer
Textilmaschinenfabrik Obertshausen
Germany, designed 2002,
manufactured 2005
Warp-knit polyester and
polyester monofilament

Spacer fabric can be used in a number of applications, including filtration and insoles for shoes, and as a substitute for foam in automobile-seat padding and upholstery.

fig. 33 (right above)
Pedestrian bridge for hiking trail
in Maui, Hawaii, 80' span
Designed by E. T. Techtonics, Inc.,
fabricated by Structural Fiberglass Inc.,
composite components manufactured
by Creative Pultrusions, Inc.
U.S.A., 1995
Pultruded fiberglass rovings and
non-woven mat, polyester resin
Courtesy of E. T. Techtonics, Inc.

Lightweight fiberglass components can be easily hand-carried to remote locations and assembled on-site. Fiberglass also resists rust, rot, salt corrosion, and termites, making it ideal for wilderness use.

fig. 34 (right below)
I-beam
Designed by Thomas Campbell, textile
designed and manufactured by Foster-
Miller Inc., manufactured by ACME
Fiberglass, sponsored by the National
Cooperative Highway Research Program
under the Innovations Deserving
Exploratory Analysis (IDEA) program
U.S.A., 1997
Carbon fiber composite
Courtesy of Foster-Miller Inc.

The composite I-beam was designed to extend an existing bridge by attaching a cantilevered pedestrian walkway as a retrofit, where traditional steel and concrete construction would be too heavy.

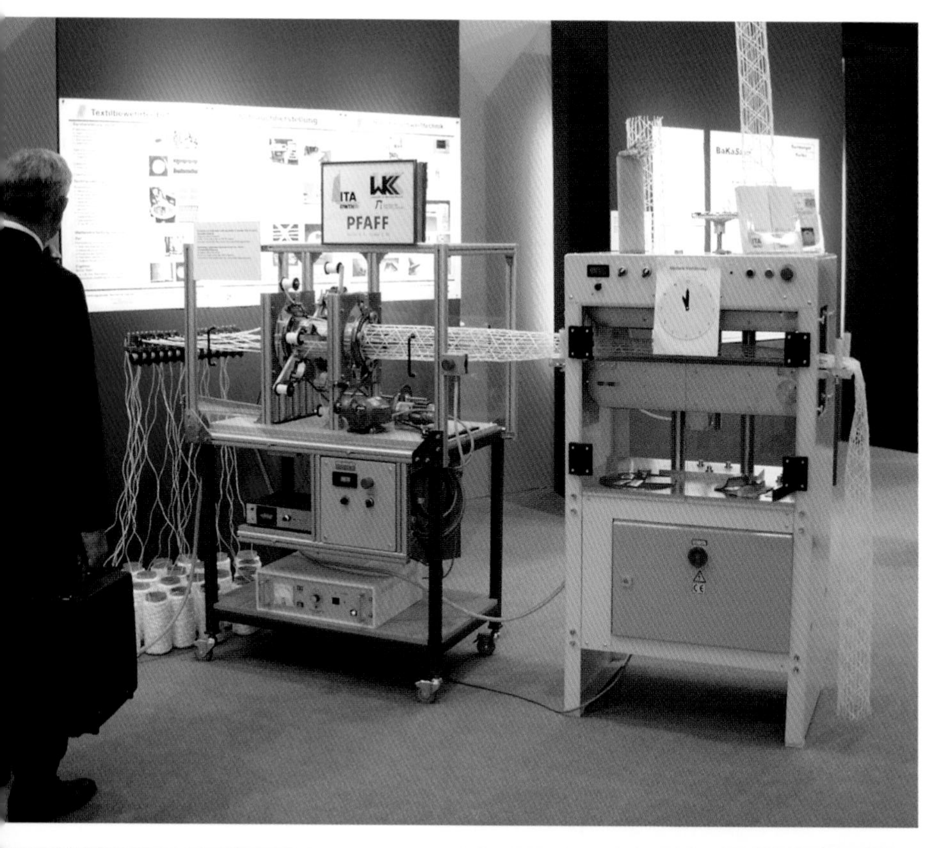


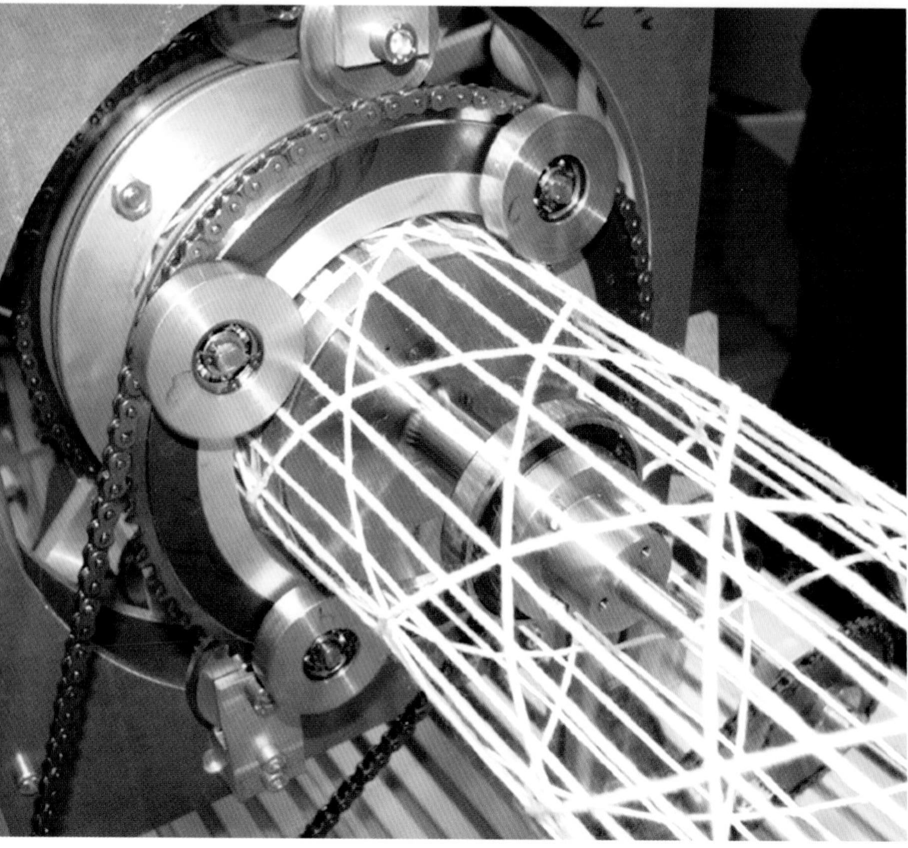

figs. 35–37
Ultrasonically welded tubular fabric
Designed by Prof. Dr. Thomas Gries, Dr. P. Stockmann, A. Roye, Prof. Dr. D. Eifler, Dr. G. Wagner, and S. Kruger; engineering partner Pfaff Industriemaschinen AG
Germany, designed 2001, manufactured 2005
Seamless circular non-crimp ultrasonic welded tube of stretched yarns of 90% AR (alkali-resistant) glass and 10% PP (polypropylene) fibers
Diameter: 12 cm (4¾ in.)

Winner of the Techtextil 2003 Innovation Prize, this ultrasonically welded tubular fabric is one example of a textile being used to enhance the safety and structural performance of building materials. Designed as a reinforcement textile for concrete, the multi-axial, multi-ply tubular fabric would give normally brittle concrete increased strength, energy absorption, and two-way bending ability. As a replacement for steel reinforcements, it drastically reduces weight, and can be fabricated on site. The plastic portion of the yarn is joined by softening the thermoplastic components at the intersections with an ultrasonic welding wheel.

SSSSSSSSSSSSSSSSSSSSSSSSSSSSSSSSSSSSSSSSSSSSSSSSSSSSSSSSSSSSSSSSSSSSSSSSSSSSSSSSSSSSSSSSSSSSSSSSSSSSSSSSSSSSSSSSSSSSSSSSSSSSSSSSSSSSSSSSSSSSSSSSSSSSSSSSSSSSSSSSSSSSSSSSSSSSSSSSSSSSSSSSSSSSSSSSSSSSSS
AAAAAAAAAAAAAAAAAAAAAAAAAAAAAAAAAAAAAAAAAAAAAAAAAAAAAAAAAAAAAAAAAAAAAAAAAAAAAAAAAAAAAAAAAAAAAAAAAAAAAAAAAAAAAAAAAAAAAAAAAAAAAAAAAAAAAAAAAAAAAAAAAAAAAAAAAAAAAAAAAAAAAAAAAAAAAAAAAAAAAAAAAAAA
FFFFFFFFFFFFFFFFFFFFFFFFFFFFFFFFFFFFFFFFFFFFFFFFFFFFFFFFFFFFFFFFFFFFFFFFFFFFFFFFFFFFFFFFFFFFFFFFFFFFFFFFFFFFFFFFFFFFFFFFFFFFFFFFFFFFFFFFFFFFFFFFFFFFFFFFFFFFFFFFFFFFFFFFFFFFFFFFFFFFFFFFFFFFFFFFFFFFFFFFFFFFFFFFFFFFFFFFF
EEEEEEEEEEEEEEEEEEEEEEEEEEEEEEEEEEEEEEEEEEEEEEEEEEEEEEEEEEEEEEEEEEEEEEEEEEEEEEEEEEEEEEEEEEEEEEEEEEEEEEEEEEEEEEEEEEEEEEEEEEEEEEEEEEEEEEEEEEEEEEEEEEEEEEEEEEEEEEEEEEEEEEEEEEEEEEEEEEEEEEEEEEEEEEEEEEEEEEEEEEEEEEEEE
RRRRRRRRRRRRRRRRRRRRRRRRRRRRRRRRRRRRRRRRRRRRRRRRRRRRRRRRRRRRRRRRRRRRRRRRRRRRRRRRRRRRRRRRRRRRRRRRRRRRRRRRRRRRRRRRRRRRRRRRRRRRRRRRRRRRRRRRRRRRRRRRRRRRRRRRRRRRRRRRRRRRRRRRRRRRRRRRRRRRRRRRRRRRRRRRRRRRR
SSSSSSSSSSSSSSSSSSSSSSSSSSSSSSSSSSSSSSSSSSSSSSSSSSSSSSSSSSSSSSSSSSSSSSSSSSSSSSSSSSSSSSSSSSSSSSSSSSSSSSSSSSSSSSSSSSSSSSSSSSSSSSSSSSSSSSSSSSSSSSSSSSSSSSSSSSSSSSSSSSSSSSSSSSSSSSSSSSSSSSSSSSSSSSSSSSSSSSS
AAAAAAAAAAAAAAAAAAAAAAAAAAAAAAAAAAAAAAAAAAAAAAAAAAAAAAAAAAAAAAAAAAAAAAAAAAAAAAAAAAAAAAAAAAAAAAAAAAAAAAAAAAAAAAAAAAAAAAAAAAAAAAAAAAAAAAAAAAAAAAAAAAAAAAAAAAAAAAAAAAAAAAAAAAAAAAAAAAAAAAAAAAAAA
FFFFFFFFFFFFFFFFFFFFFFFFFFFFFFFFFFFFFFFFFFFFFFFFFFFFFFFFFFFFFFFFFFFFFFFFFFFFFFFFFFFFFFFFFFFFFFFFFFFFFFFFFFFFFFFFFFFFFFFFFFFFFFFFFFFFFFFFFFFFFFFFFFFFFFFFFFFFFFFFFFFFFFFFFFFFFFFFFFFFFFFFFFFFFFFFFFFFFFFFFFFFFFFFFFFFFFFFFF
EEEEEEEEEEEEEEEEEEEEEEEEEEEEEEEEEEEEEEEEEEEEEEEEEEEEEEEEEEEEEEEEEEEEEEEEEEEEEEEEEEEEEEEEEEEEEEEEEEEEEEEEEEEEEEEEEEEEEEEEEEEEEEEEEEEEEEEEEEEEEEEEEEEEEEEEEEEEEEEEEEEEEEEEEEEEEEEEEEEEEEEEEEEEEEEEEEEEEEEEEEEEEEEE
RRRRRRRRRRRRRRRRRRRRRRRRRRRRRRRRRRRRRRRRRRRRRRRRRRRRRRRRRRRRRRRRRRRRRRRRRRRRRRRRRRRRRRRRRRRRRRRRRRRRRRRRRRRRRRRRRRRRRRRRRRRRRRRRRRRRRRRRRRRRRRRRRRRRRRRRRRRRRRRRRRRRRRRRRRRRRRRRRRRRRRRRRRRRRRRRRRRRR
SSSSSSSSSSSSSSSSSSSSSSSSSSSSSSSSSSSSSSSSSSSSSSSSSSSSSSSSSSSSSSSSSSSSSSSSSSSSSSSSSSSSSSSSSSSSSSSSSSSSSSSSSSSSSSSSSSSSSSSSSSSSSSSSSSSSSSSSSSSSSSSSSSSSSSSSSSSSSSSSSSSSSSSSSSSSSSSSSSSSSSSSSSSSSSSSSSSSS
AAAAAAAAAAAAAAAAAAAAAAAAAAAAAAAAAAAAAAAAAAAAAAAAAAAAAAAAAAAAAAAAAAAAAAAAAAAAAAAAAAAAAAAAAAAAAAAAAAAAAAAAAAAAAAAAAAAAAAAAAAAAAAAAAAAAAAAAAAAAAAAAAAAAAAAAAAAAAAAAAAAAAAAAAAAAAAAAAAAAAAAAAAAAA
FFFFFFFFFFFFFFFFFFFFFFFFFFFFFFFFFFFFFFFFFFFFFFFFFFFFFFFFFFFFFFFFFFFFFFFFFFFFFFFFFFFFFFFFFFFFFFFFFFFFFFFFFFFFFFFFFFFFFFFFFFFFFFFFFFFFFFFFFFFFFFFFFFFFFFFFFFFFFFFFFFFFFFFFFFFFFFFFFFFFFFFFFFFFFFFFFFFFFFFFFFFFFFFFFFFFFFFFFF
EEEEEEEEEEEEEEEEEEEEEEEEEEEEEEEEEEEEEEEEEEEEEEEEEEEEEEEEEEEEEEEEEEEEEEEEEEEEEEEEEEEEEEEEEEEEEEEEEEEEEEEEEEEEEEEEEEEEEEEEEEEEEEEEEEEEEEEEEEEEEEEEEEEEEEEEEEEEEEEEEEEEEEEEEEEEEEEEEEEEEEEEEEEEEEEEEEEEEEEEEEEEEEEEE
RRRRRRRRRRRRRRRRRRRRRRRRRRRRRRRRRRRRRRRRRRRRRRRRRRRRRRRRRRRRRRRRRRRRRRRRRRRRRRRRRRRRRRRRRRRRRRRRRRRRRRRRRRRRRRRRRRRRRRRRRRRRRRRRRRRRRRRRRRRRRRRRRRRRRRRRRRRRRRRRRRRRRRRRRRRRRRRRRRRRRRRRRRRRRRRRRRRRR
SSSSSSSSSSSSSSSSSSSSSSSSSSSSSSSSSSSSSSSSSSSSSSSSSSSSSSSSSSSSSSSSSSSSSSSSSSSSSSSSSSSSSSSSSSSSSSSSSSSSSSSSSSSSSSSSSSSSSSSSSSSSSSSSSSSSSSSSSSSSSSSSSSSSSSSSSSSSSSSSSSSSSSSSSSSSSSSSSSSSSSSSSSSSSSSSSSSSS
AAAAAAAAAAAAAAAAAAAAAAAAAAAAAAAAAAAAAAAAAAAAAAAAAAAAAAAAAAAAAAAAAAAAAAAAAAAAAAAAAAAAAAAAAAAAAAAAAAAAAAAAAAAAAAAAAAAAAAAAAAAAAAAAAAAAAAAAAAAAAAAAAAAAAAAAAAAAAAAAAAAAAAAAAAAAAAAAAAAAAAAAAAAAA
FFFFFFFFFFFFFFFFFFFFFFFFFFFFFFFFFFFFFFFFFFFFFFFFFFFFFFFFFFFFFFFFFFFFFFFFFFFFFFFFFFFFFFFFFFFFFFFFFFFFFFFFFFFFFFFFFFFFFFFFFFFFFFFFFFFFFFFFFFFFFFFFFFFFFFFFFFFFFFFFFFFFFFFFFFFFFFFFFFFFFFFFFFFFFFFFFFFFFFFFFFFFFFFFFFFFFFFFFF
EEEEEEEEEEEEEEEEEEEEEEEEEEEEEEEEEEEEEEEEEEEEEEEEEEEEEEEEEEEEEEEEEEEEEEEEEEEEEEEEEEEEEEEEEEEEEEEEEEEEEEEEEEEEEEEEEEEEEEEEEEEEEEEEEEEEEEEEEEEEEEEEEEEEEEEEEEEEEEEEEEEEEEEEEEEEEEEEEEEEEEEEEEEEEEEEEEEEEEEEEEEEEEEE
RRRRRRRRRRRRRRRRRRRRRRRRRRRRRRRRRRRRRRRRRRRRRRRRRRRRRRRRRRRRRRRRRRRRRRRRRRRRRRRRRRRRRRRRRRRRRRRRRRRRRRRRRRRRRRRRRRRRRRRRRRRRRRRRRRRRRRRRRRRRRRRRRRRRRRRRRRRRRRRRRRRRRRRRRRRRRRRRRRRRRRRRRRRRRRRRRRRRR
SSSSSSSSSSSSSSSSSSSSSSSSSSSSSSSSSSSSSSSSSSSSSSSSSSSSSSSSSSSSSSSSSSSSSSSSSSSSSSSSSSSSSSSSSSSSSSSSSSSSSSSSSSSSSSSSSSSSSSSSSSSSSSSSSSSSSSSSSSSSSSSSSSSSSSSSSSSSSSSSSSSSSSSSSSSSSSSSSSSSSSSSSSSSSSSSSSSSS
AAAAAAAAAAAAAAAAAAAAAAAAAAAAAAAAAAAAAAAAAAAAAAAAAAAAAAAAAAAAAAAAAAAAAAAAAAAAAAAAAAAAAAAAAAAAAAAAAAAAAAAAAAAAAAAAAAAAAAAAAAAAAAAAAAAAAAAAAAAAAAAAAAAAAAAAAAAAAAAAAAAAAAAAAAAAAAAAAAAAAAAAAAAAA
FFFFFFFFFFFFFFFFFFFFFFFFFFFFFFFFFFFFFFFFFFFFFFFFFFFFFFFFFFFFFFFFFFFFFFFFFFFFFFFFFFFFFFFFFFFFFFFFFFFFFFFFFFFFFFFFFFFFFFFFFFFFFFFFFFFFFFFFFFFFFFFFFFFFFFFFFFFFFFFFFFFFFFFFFFFFFFFFFFFFFFFFFFFFFFFFFFFFFFFFFFFFFFFFFFFFFFFFFF
EEEEEEEEEEEEEEEEEEEEEEEEEEEEEEEEEEEEEEEEEEEEEEEEEEEEEEEEEEEEEEEEEEEEEEEEEEEEEEEEEEEEEEEEEEEEEEEEEEEEEEEEEEEEEEEEEEEEEEEEEEEEEEEEEEEEEEEEEEEEEEEEEEEEEEEEEEEEEEEEEEEEEEEEEEEEEEEEEEEEEEEEEEEEEEEEEEEEEEEEEEEEEEE
RRRRRRRRRRRRRRRRRRRRRRRRRRRRRRRRRRRRRRRRRRRRRRRRRRRRRRRRRRRRRRRRRRRRRRRRRRRRRRRRRRRRRRRRRRRRRRRRRRRRRRRRRRRRRRRRRRRRRRRRRRRRRRRRRRRRRRRRRRRRRRRRRRRRRRRRRRRRRRRRRRRRRRRRRRRRRRRRRRRRRRRRRRRRRRRRRRRRR
SSSSSSSSSSSSSSSSSSSSSSSSSSSSSSSSSSSSSSSSSSSSSSSSSSSSSSSSSSSSSSSSSSSSSSSSSSSSSSSSSSSSSSSSSSSSSSSSSSSSSSSSSSSSSSSSSSSSSSSSSSSSSSSSSSSSSSSSSSSSSSSSSSSSSSSSSSSSSSSSSSSSSSSSSSSSSSSSSSSSSSSSSSSSSSSSSSSS
AAAAAAAAAAAAAAAAAAAAAAAAAAAAAAAAAAAAAAAAAAAAAAAAAAAAAAAAAAAAAAAAAAAAAAAAAAAAAAAAAAAAAAAAAAAAAAAAAAAAAAAAAAAAAAAAAAAAAAAAAAAAAAAAAAAAAAAAAAAAAAAAAAAAAAAAAAAAAAAAAAAAAAAAAAAAAAAAAAAAAAAAAAAAA
FFFFFFFFFFFFFFFFFFFFFFFFFFFFFFFFFFFFFFFFFFFFFFFFFFFFFFFFFFFFFFFFFFFFFFFFFFFFFFFFFFFFFFFFFFFFFFFFFFFFFFFFFFFFFFFFFFFFFFFFFFFFFFFFFFFFFFFFFFFFFFFFFFFFFFFFFFFFFFFFFFFFFFFFFFFFFFFFFFFFFFFFFFFFFFFFFFFFFFFFFFFFFFFFFFFFFFFFFF
EEEEEEEEEEEEEEEEEEEEEEEEEEEEEEEEEEEEEEEEEEEEEEEEEEEEEEEEEEEEEEEEEEEEEEEEEEEEEEEEEEEEEEEEEEEEEEEEEEEEEEEEEEEEEEEEEEEEEEEEEEEEEEEEEEEEEEEEEEEEEEEEEEEEEEEEEEEEEEEEEEEEEEEEEEEEEEEEEEEEEEEEEEEEEEEEEEEEEEEEEEEEEEEE
RRRRRRRRRRRRRRRRRRRRRRRRRRRRRRRRRRRRRRRRRRRRRRRRRRRRRRRRRRRRRRRRRRRRRRRRRRRRRRRRRRRRRRRRRRRRRRRRRRRRRRRRRRRRRRRRRRRRRRRRRRRRRRRRRRRRRRRRRRRRRRRRRRRRRRRRRRRRRRRRRRRRRRRRRRRRRRRRRRRRRRRRRRRRRRRRRRRRR
SSSSSSSSSSSSSSSSSSSSSSSSSSSSSSSSSSSSSSSSSSSSSSSSSSSSSSSSSSSSSSSSSSSSSSSSSSSSSSSSSSSSSSSSSSSSSSSSSSSSSSSSSSSSSSSSSSSSSSSSSSSSSSSSSSSSSSSSSSSSSSSSSSSSSSSSSSSSSSSSSSSSSSSSSSSSSSSSSSSSSSSSSSSSSSSSSSSSS
AAAAAAAAAAAAAAAAAAAAAAAAAAAAAAAAAAAAAAAAAAAAAAAAAAAAAAAAAAAAAAAAAAAAAAAAAAAAAAAAAAAAAAAAAAAAAAAAAAAAAAAAAAAAAAAAAAAAAAAAAAAAAAAAAAAAAAAAAAAAAAAAAAAAAAAAAAAAAAAAAAAAAAAAAAAAAAAAAAAAAAAAAAAAA
FFFFFFFFFFFFFFFFFFFFFFFFFFFFFFFFFFFFFFFFFFFFFFFFFFFFFFFFFFFFFFFFFFFFFFFFFFFFFFFFFFFFFFFFFFFFFFFFFFFFFFFFFFFFFFFFFFFFFFFFFFFFFFFFFFFFFFFFFFFFFFFFFFFFFFFFFFFFFFFFFFFFFFFFFFFFFFFFFFFFFFFFFFFFFFFFFFFFFFFFFFFFFFFFFFFFFFFFFF
EEEEEEEEEEEEEEEEEEEEEEEEEEEEEEEEEEEEEEEEEEEEEEEEEEEEEEEEEEEEEEEEEEEEEEEEEEEEEEEEEEEEEEEEEEEEEEEEEEEEEEEEEEEEEEEEEEEEEEEEEEEEEEEEEEEEEEEEEEEEEEEEEEEEEEEEEEEEEEEEEEEEEEEEEEEEEEEEEEEEEEEEEEEEEEEEEEEEEEEEEEEEEEEE
RRRRRRRRRRRRRRRRRRRRRRRRRRRRRRRRRRRRRRRRRRRRRRRRRRRRRRRRRRRRRRRRRRRRRRRRRRRRRRRRRRRRRRRRRRRRRRRRRRRRRRRRRRRRRRRRRRRRRRRRRRRRRRRRRRRRRRRRRRRRRRRRRRRRRRRRRRRRRRRRRRRRRRRRRRRRRRRRRRRRRRRRRRRRRRRRRRRRR
SSSSSSSSSSSSSSSSSSSSSSSSSSSSSSSSSSSSSSSSSSSSSSSSSSSSSSSSSSSSSSSSSSSSSSSSSSSSSSSSSSSSSSSSSSSSSSSSSSSSSSSSSSSSSSSSSSSSSSSSSSSSSSSSSSSSSSSSSSSSSSSSSSSSSSSSSSSSSSSSSSSSSSSSSSSSSSSSSSSSSSSSSSSSSSSSSSSSS
AAAAAAAAAAAAAAAAAAAAAAAAAAAAAAAAAAAAAAAAAAAAAAAAAAAAAAAAAAAAAAAAAAAAAAAAAAAAAAAAAAAAAAAAAAAAAAAAAAAAAAAAAAAAAAAAAAAAAAAAAAAAAAAAAAAAAAAAAAAAAAAAAAAAAAAAAAAAAAAAAAAAAAAAAAAAAAAAAAAAAAAAAAAAA
FFFFFFFFFFFFFFFFFFFFFFFFFFFFFFFFFFFFFFFFFFFFFFFFFFFFFFFFFFFFFFFFFFFFFFFFFFFFFFFFFFFFFFFFFFFFFFFFFFFFFFFFFFFFFFFFFFFFFFFFFFFFFFFFFFFFFFFFFFFFFFFFFFFFFFFFFFFFFFFFFFFFFFFFFFFFFFFFFFFFFFFFFFFFFFFFFFFFFFFFFFFFFFFFFFFFFFFFFF
EEEEEEEEEEEEEEEEEEEEEEEEEEEEEEEEEEEEEEEEEEEEEEEEEEEEEEEEEEEEEEEEEEEEEEEEEEEEEEEEEEEEEEEEEEEEEEEEEEEEEEEEEEEEEEEEEEEEEEEEEEEEEEEEEEEEEEEEEEEEEEEEEEEEEEEEEEEEEEEEEEEEEEEEEEEEEEEEEEEEEEEEEEEEEEEEEEEEEEEEEEEEEEE
RRRRRRRRRRRRRRRRRRRRRRRRRRRRRRRRRRRRRRRRRRRRRRRRRRRRRRRRRRRRRRRRRRRRRRRRRRRRRRRRRRRRRRRRRRRRRRRRRRRRRRRRRRRRRRRRRRRRRRRRRRRRRRRRRRRRRRRRRRRRRRRRRRRRRRRRRRRRRRRRRRRRRRRRRRRRRRRRRRRRRRRRRRRRRRRRRRRRR
SSSSSSSSSSSSSSSSSSSSSSSSSSSSSSSSSSSSSSSSSSSSSSSSSSSSSSSSSSSSSSSSSSSSSSSSSSSSSSSSSSSSSSSSSSSSSSSSSSSSSSSSSSSSSSSSSSSSSSSSSSSSSSSSSSSSSSSSSSSSSSSSSSSSSSSSSSSSSSSSSSSSSSSSSSSSSSSSSSSSSSSSSSSSSSSSSSSS
AAAAAAAAAAAAAAAAAAAAAAAAAAAAAAAAAAAAAAAAAAAAAAAAAAAAAAAAAAAAAAAAAAAAAAAAAAAAAAAAAAAAAAAAAAAAAAAAAAAAAAAAAAAAAAAAAAAAAAAAAAAAAAAAAAAAAAAAAAAAAAAAAAAAAAAAAAAAAAAAAAAAAAAAAAAAAAAAAAAAAAAAAAAAA
FFFFFFFFFFFFFFFFFFFFFFFFFFFFFFFFFFFFFFFFFFFFFFFFFFFFFFFFFFFFFFFFFFFFFFFFFFFFFFFFFFFFFFFFFFFFFFFFFFFFFFFFFFFFFFFFFFFFFFFFFFFFFFFFFFFFFFFFFFFFFFFFFFFFFFFFFFFFFFFFFFFFFFFFFFFFFFFFFFFFFFFFFFFFFFFFFFFFFFFFFFFFFFFFFFFFFFFFFF
EEEEEEEEEEEEEEEEEEEEEEEEEEEEEEEEEEEEEEEEEEEEEEEEEEEEEEEEEEEEEEEEEEEEEEEEEEEEEEEEEEEEEEEEEEEEEEEEEEEEEEEEEEEEEEEEEEEEEEEEEEEEEEEEEEEEEEEEEEEEEEEEEEEEEEEEEEEEEEEEEEEEEEEEEEEEEEEEEEEEEEEEEEEEEEEEEEEEEEEEEEEEEEEE
RRRRRRRRRRRRRRRRRRRRRRRRRRRRRRRRRRRRRRRRRRRRRRRRRRRRRRRRRRRRRRRRRRRRRRRRRRRRRRRRRRRRRRRRRRRRRRRRRRRRRRRRRRRRRRRRRRRRRRRRRRRRRRRRRRRRRRRRRRRRRRRRRRRRRRRRRRRRRRRRRRRRRRRRRRRRRRRRRRRRRRRRRRRRRRRRRRRRR
SSSSSSSSSSSSSSSSSSSSSSSSSSSSSSSSSSSSSSSSSSSSSSSSSSSSSSSSSSSSSSSSSSSSSSSSSSSSSSSSSSSSSSSSSSSSSSSSSSSSSSSSSSSSSSSSSSSSSSSSSSSSSSSSSSSSSSSSSSSSSSSSSSSSSSSSSSSSSSSSSSSSSSSSSSSSSSSSSSSSSSSSSSSSSSSSSSSSS
AAAAAAAAAAAAAAAAAAAAAAAAAAAAAAAAAAAAAAAAAAAAAAAAAAAAAAAAAAAAAAAAAAAAAAAAAAAAAAAAAAAAAAAAAAAAAAAAAAAAAAAAAAAAAAAAAAAAAAAAAAAAAAAAAAAAAAAAAAAAAAAAAAAAAAAAAAAAAAAAAAAAAAAAAAAAAAAAAAAAAAAAAAAAA
FFFFFFFFFFFFFFFFFFFFFFFFFFFFFFFFFFFFFFFFFFFFFFFFFFFFFFFFFFFFFFFFFFFFFFFFFFFFFFFFFFFFFFFFFFFFFFFFFFFFFFFFFFFFFFFFFFFFFFFFFFFFFFFFFFFFFFFFFFFFFFFFFFFFFFFFFFFFFFFFFFFFFFFFFFFFFFFFFFFFFFFFFFFFFFFFFFFFFFFFFFFFFFFFFFFFFFFFF
EEEEEEEEEEEEEEEEEEEEEEEEEEEEEEEEEEEEEEEEEEEEEEEEEEEEEEEEEEEEEEEEEEEEEEEEEEEEEEEEEEEEEEEEEEEEEEEEEEEEEEEEEEEEEEEEEEEEEEEEEEEEEEEEEEEEEEEEEEEEEEEEEEEEEEEEEEEEEEEEEEEEEEEEEEEEEEEEEEEEEEEEEEEEEEEEEEEEEEEEEEEEEEEE
RRRRRRRRRRRRRRRRRRRRRRRRRRRRRRRRRRRRRRRRRRRRRRRRRRRRRRRRRRRRRRRRRRRRRRRRRRRRRRRRRRRRRRRRRRRRRRRRRRRRRRRRRRRRRRRRRRRRRRRRRRRRRRRRRRRRRRRRRRRRRRRRRRRRRRRRRRRRRRRRRRRRRRRRRRRRRRRRRRRRRRRRRRRRRRRRRRRRR
SSSSSSSSSSSSSSSSSSSSSSSSSSSSSSSSSSSSSSSSSSSSSSSSSSSSSSSSSSSSSSSSSSSSSSSSSSSSSSSSSSSSSSSSSSSSSSSSSSSSSSSSSSSSSSSSSSSSSSSSSSSSSSSSSSSSSSSSSSSSSSSSSSSSSSSSSSSSSSSSSSSSSSSSSSSSSSSSSSSSSSSSSSSSSSSSSSSS
AAAAAAAAAAAAAAAAAAAAAAAAAAAAAAAAAAAAAAAAAAAAAAAAAAAAAAAAAAAAAAAAAAAAAAAAAAAAAAAAAAAAAAAAAAAAAAAAAAAAAAAAAAAAAAAAAAAAAAAAAAAAAAAAAAAAAAAAAAAAAAAAAAAAAAAAAAAAAAAAAAAAAAAAAAAAAAAAAAAAAAAAAAAAA
FFFFFFFFFFFFFFFFFFFFFFFFFFFFFFFFFFFFFFFFFFFFFFFFFFFFFFFFFFFFFFFFFFFFFFFFFFFFFFFFFFFFFFFFFFFFFFFFFFFFFFFFFFFFFFFFFFFFFFFFFFFFFFFFFFFFFFFFFFFFFFFFFFFFFFFFFFFFFFFFFFFFFFFFFFFFFFFFFFFFFFFFFFFFFFFFFFFFFFFFFFFFFFFFFFFFFFFFFF
EEEEEEEEEEEEEEEEEEEEEEEEEEEEEEEEEEEEEEEEEEEEEEEEEEEEEEEEEEEEEEEEEEEEEEEEEEEEEEEEEEEEEEEEEEEEEEEEEEEEEEEEEEEEEEEEEEEEEEEEEEEEEEEEEEEEEEEEEEEEEEEEEEEEEEEEEEEEEEEEEEEEEEEEEEEEEEEEEEEEEEEEEEEEEEEEEEEEEEEEEEEEEEE
RRRRRRRRRRRRRRRRRRRRRRRRRRRRRRRRRRRRRRRRRRRRRRRRRRRRRRRRRRRRRRRRRRRRRRRRRRRRRRRRRRRRRRRRRRRRRRRRRRRRRRRRRRRRRRRRRRRRRRRRRRRRRRRRRRRRRRRRRRRRRRRRRRRRRRRRRRRRRRRRRRRRRRRRRRRRRRRRRRRRRRRRRRRRRRRRRRRR
SSSSSSSSSSSSSSSSSSSSSSSSSSSSSSSSSSSSSSSSSSSSSSSSSSSSSSSSSSSSSSSSSSSSSSSSSSSSSSSSSSSSSSSSSSSSSSSSSSSSSSSSSSSSSSSSSSSSSSSSSSSSSSSSSSSSSSSSSSSSSSSSSSSSSSSSSSSSSSSSSSSSSSSSSSSSSSSSSSSSSSSSSSSSSSSSSSSS
AAAAAAAAAAAAAAAAAAAAAAAAAAAAAAAAAAAAAAAAAAAAAAAAAAAAAAAAAAAAAAAAAAAAAAAAAAAAAAAAAAAAAAAAAAAAAAAAAAAAAAAAAAAAAAAAAAAAAAAAAAAAAAAAAAAAAAAAAAAAAAAAAAAAAAAAAAAAAAAAAAAAAAAAAAAAAAAAAAAAAAAAAAAAA
FFFFFFFFFFFFFFFFFFFFFFFFFFFFFFFFFFFFFFFFFFFFFFFFFFFFFFFFFFFFFFFFFFFFFFFFFFFFFFFFFFFFFFFFFFFFFFFFFFFFFFFFFFFFFFFFFFFFFFFFFFFFFFFFFFFFFFFFFFFFFFFFFFFFFFFFFFFFFFFFFFFFFFFFFFFFFFFFFFFFFFFFFFFFFFFFFFFFFFFFFFFFFFFFFFFFFFFFF
EEEEEEEEEEEEEEEEEEEEEEEEEEEEEEEEEEEEEEEEEEEEEEEEEEEEEEEEEEEEEEEEEEEEEEEEEEEEEEEEEEEEEEEEEEEEEEEEEEEEEEEEEEEEEEEEEEEEEEEEEEEEEEEEEEEEEEEEEEEEEEEEEEEEEEEEEEEEEEEEEEEEEEEEEEEEEEEEEEEEEEEEEEEEEEEEEEEEEEEEEEEEEEE
RRRRRRRRRRRRRRRRRRRRRRRRRRRRRRRRRRRRRRRRRRRRRRRRRRRRRRRRRRRRRRRRRRRRRRRRRRRRRRRRRRRRRRRRRRRRRRRRRRRRRRRRRRRRRRRRRRRRRRRRRRRRRRRRRRRRRRRRRRRRRRRRRRRRRRRRRRRRRRRRRRRRRRRRRRRRRRRRRRRRRRRRRRRRRRRRRRRR
SSSSSSSSSSSSSSSSSSSSSSSSSSSSSSSSSSSSSSSSSSSSSSSSSSSSSSSSSSSSSSSSSSSSSSSSSSSSSSSSSSSSSSSSSSSSSSSSSSSSSSSSSSSSSSSSSSSSSSSSSSSSSSSSSSSSSSSSSSSSSSSSSSSSSSSSSSSSSSSSSSSSSSSSSSSSSSSSSSSSSSSSSSSSSSSSSSS
AAAAAAAAAAAAAAAAAAAAAAAAAAAAAAAAAAAAAAAAAAAAAAAAAAAAAAAAAAAAAAAAAAAAAAAAAAAAAAAAAAAAAAAAAAAAAAAAAAAAAAAAAAAAAAAAAAAAAAAAAAAAAAAAAAAAAAAAAAAAAAAAAAAAAAAAAAAAAAAAAAAAAAAAAAAAAAAAAAAAAAAAAAAAA
FFFFFFFFFFFFFFFFFFFFFFFFFFFFFFFFFFFFFFFFFFFFFFFFFFFFFFFFFFFFFFFFFFFFFFFFFFFFFFFFFFFFFFFFFFFFFFFFFFFFFFFFFFFFFFFFFFFFFFFFFFFFFFFFFFFFFFFFFFFFFFFFFFFFFFFFFFFFFFFFFFFFFFFFFFFFFFFFFFFFFFFFFFFFFFFFFFFFFFFFFFFFFFFFFFFFFFFFF
EEEEEEEEEEEEEEEEEEEEEEEEEEEEEEEEEEEEEEEEEEEEEEEEEEEEEEEEEEEEEEEEEEEEEEEEEEEEEEEEEEEEEEEEEEEEEEEEEEEEEEEEEEEEEEEEEEEEEEEEEEEEEEEEEEEEEEEEEEEEEEEEEEEEEEEEEEEEEEEEEEEEEEEEEEEEEEEEEEEEEEEEEEEEEEEEEEEEEEEEEEEEEEE
RRRRRRRRRRRRRRRRRRRRRRRRRRRRRRRRRRRRRRRRRRRRRRRRRRRRRRRRRRRRRRRRRRRRRRRRRRRRRRRRRRRRRRRRRRRRRRRRRRRRRRRRRRRRRRRRRRRRRRRRRRRRRRRRRRRRRRRRRRRRRRRRRRRRRRRRRRRRRRRRRRRRRRRRRRRRRRRRRRRRRRRRRRRRRRRRRRRR
SSSSSSSSSSSSSSSSSSSSSSSSSSSSSSSSSSSSSSSSSSSSSSSSSSSSSSSSSSSSSSSSSSSSSSSSSSSSSSSSSSSSSSSSSSSSSSSSSSSSSSSSSSSSSSSSSSSSSSSSSSSSSSSSSSSSSSSSSSSSSSSSSSSSSSSSSSSSSSSSSSSSSSSSSSSSSSSSSSSSSSSSSSSSSSSSSSSS
AAAAAAAAAAAAAAAAAAAAAAAAAAAAAAAAAAAAAAAAAAAAAAAAAAAAAAAAAAAAAAAAAAAAAAAAAAAAAAAAAAAAAAAAAAAAAAAAAAAAAAAAAAAAAAAAAAAAAAAAAAAAAAAAAAAAAAAAAAAAAAAAAAAAAAAAAAAAAAAAAAAAAAAAAAAAAAAAAAAAAAAAAAAAA
FFFFFFFFFFFFFFFFFFFFFFFFFFFFFFFFFFFFFFFFFFFFFFFFFFFFFFFFFFFFFFFFFFFFFFFFFFFFFFFFFFFFFFFFFFFFFFFFFFFFFFFFFFFFFFFFFFFFFFFFFFFFFFFFFFFFFFFFFFFFFFFFFFFFFFFFFFFFFFFFFFFFFFFFFFFFFFFFFFFFFFFFFFFFFFFFFFFFFFFFFFFFFFFFFFFFFFFFFF
EEEEEEEEEEEEEEEEEEEEEEEEEEEEEEEEEEEEEEEEEEEEEEEEEEEEEEEEEEEEEEEEEEEEEEEEEEEEEEEEEEEEEEEEEEEEEEEEEEEEEEEEEEEEEEEEEEEEEEEEEEEEEEEEEEEEEEEEEEEEEEEEEEEEEEEEEEEEEEEEEEEEEEEEEEEEEEEEEEEEEEEEEEEEEEEEEEEEEEEEEEEEEEE
RRRRRRRRRRRRRRRRRRRRRRRRRRRRRRRRRRRRRRRRRRRRRRRRRRRRRRRRRRRRRRRRRRRRRRRRRRRRRRRRRRRRRRRRRRRRRRRRRRRRRRRRRRRRRRRRRRRRRRRRRRRRRRRRRRRRRRRRRRRRRRRRRRRRRRRRRRRRRRRRRRRRRRRRRRRRRRRRRRRRRRRRRRRRRRRRRRRRR
SSSSSSSSSSSSSSSSSSSSSSSSSSSSSSSSSSSSSSSSSSSSSSSSSSSSSSSSSSSSSSSSSSSSSSSSSSSSSSSSSSSSSSSSSSSSSSSSSSSSSSSSSSSSSSSSSSSSSSSSSSSSSSSSSSSSSSSSSSSSSSSSSSSSSSSSSSSSSSSSSSSSSSSSSSSSSSSSSSSSSSSSSSSSSSSSSSSS
AAAAAAAAAAAAAAAAAAAAAAAAAAAAAAAAAAAAAAAAAAAAAAAAAAAAAAAAAAAAAAAAAAAAAAAAAAAAAAAAAAAAAAAAAAAAAAAAAAAAAAAAAAAAAAAAAAAAAAAAAAAAAAAAAAAAAAAAAAAAAAAAAAAAAAAAAAAAAAAAAAAAAAAAAAAAAAAAAAAAAAAAAAAAA
FFFFFFFFFFFFFFFFFFFFFFFFFFFFFFFFFFFFFFFFFFFFFFFFFFFFFFFFFFFFFFFFFFFFFFFFFFFFFFFFFFFFFFFFFFFFFFFFFFFFFFFFFFFFFFFFFFFFFFFFFFFFFFFFFFFFFFFFFFFFFFFFFFFFFFFFFFFFFFFFFFFFFFFFFFFFFFFFFFFFFFFFFFFFFFFFFFFFFFFFFFFFFFFFFFFFFFFFF
EEEEEEEEEEEEEEEEEEEEEEEEEEEEEEEEEEEEEEEEEEEEEEEEEEEEEEEEEEEEEEEEEEEEEEEEEEEEEEEEEEEEEEEEEEEEEEEEEEEEEEEEEEEEEEEEEEEEEEEEEEEEEEEEEEEEEEEEEEEEEEEEEEEEEEEEEEEEEEEEEEEEEEEEEEEEEEEEEEEEEEEEEEEEEEEEEEEEEEEEEEEEEEE
RRRRRRRRRRRRRRRRRRRRRRRRRRRRRRRRRRRRRRRRRRRRRRRRRRRRRRRRRRRRRRRRRRRRRRRRRRRRRRRRRRRRRRRRRRRRRRRRRRRRRRRRRRRRRRRRRRRRRRRRRRRRRRRRRRRRRRRRRRRRRRRRRRRRRRRRRRRRRRRRRRRRRRRRRRRRRRRRRRRRRRRRRRRRRRRRRRRR
SSSSSSSSSSSSSSSSSSSSSSSSSSSSSSSSSSSSSSSSSSSSSSSSSSSSSSSSSSSSSSSSSSSSSSSSSSSSSSSSSSSSSSSSSSSSSSSSSSSSSSSSSSSSSSSSSSSSSSSSSSSSSSSSSSSSSSSSSSSSSSSSSSSSSSSSSSSSSSSSSSSSSSSSSSSSSSSSSSSSSSSSSSSSSSSSSSSS
AAAAAAAAAAAAAAAAAAAAAAAAAAAAAAAAAAAAAAAAAAAAAAAAAAAAAAAAAAAAAAAAAAAAAAAAAAAAAAAAAAAAAAAAAAAAAAAAAAAAAAAAAAAAAAAAAAAAAAAAAAAAAAAAAAAAAAAAAAAAAAAAAAAAAAAAAAAAAAAAAAAAAAAAAAAAAAAAAAAAAAAAAAAAA
FFFFFFFFFFFFFFFFFFFFFFFFFFFFFFFFFFFFFFFFFFFFFFFFFFFFFFFFFFFFFFFFFFFFFFFFFFFFFFFFFFFFFFFFFFFFFFFFFFFFFFFFFFFFFFFFFFFFFFFFFFFFFFFFFFFFFFFFFFFFFFFFFFFFFFFFFFFFFFFFFFFFFFFFFFFFFFFFFFFFFFFFFFFFFFFFFFFFFFFFFFFFFFFFFFFFFFFFF
EEEEEEEEEEEEEEEEEEEEEEEEEEEEEEEEEEEEEEEEEEEEEEEEEEEEEEEEEEEEEEEEEEEEEEEEEEEEEEEEEEEEEEEEEEEEEEEEEEEEEEEEEEEEEEEEEEEEEEEEEEEEEEEEEEEEEEEEEEEEEEEEEEEEEEEEEEEEEEEEEEEEEEEEEEEEEEEEEEEEEEEEEEEEEEEEEEEEEEEEEEEEEEE
RRRRRRRRRRRRRRRRRRRRRRRRRRRRRRRRRRRRRRRRRRRRRRRRRRRRRRRRRRRRRRRRRRRRRRRRRRRRRRRRRRRRRRRRRRRRRRRRRRRRRRRRRRRRRRRRRRRRRRRRRRRRRRRRRRRRRRRRRRRRRRRRRRRRRRRRRRRRRRRRRRRRRRRRRRRRRRRRRRRRRRRRRRRRRRRRRRRR
SSSSSSSSSSSSSSSSSSSSSSSSSSSSSSSSSSSSSSSSSSSSSSSSSSSSSSSSSSSSSSSSSSSSSSSSSSSSSSSSSSSSSSSSSSSSSSSSSSSSSSSSSSSSSSSSSSSSSSSSSSSSSSSSSSSSSSSSSSSSSSSSSSSSSSSSSSSSSSSSSSSSSSSSSSSSSSSSSSSSSSSSSSSSSSSSSSSS
AAAAAAAAAAAAAAAAAAAAAAAAAAAAAAAAAAAAAAAAAAAAAAAAAAAAAAAAAAAAAAAAAAAAAAAAAAAAAAAAAAAAAAAAAAAAAAAAAAAAAAAAAAAAAAAAAAAAAAAAAAAAAAAAAAAAAAAAAAAAAAAAAAAAAAAAAAAAAAAAAAAAAAAAAAAAAAAAAAAAAAAAAAAAA
FFFFFFFFFFFFFFFFFFFFFFFFFFFFFFFFFFFFFFFFFFFFFFFFFFFFFFFFFFFFFFFFFFFFFFFFFFFFFFFFFFFFFFFFFFFFFFFFFFFFFFFFFFFFFFFFFFFFFFFFFFFFFFFFFFFFFFFFFFFFFFFFFFFFFFFFFFFFFFFFFFFFFFFFFFFFFFFFFFFFFFFFFFFFFFFFFFFFFFFFFFFFFFFFFFFFFFFFF
EEEEEEEEEEEEEEEEEEEEEEEEEEEEEEEEEEEEEEEEEEEEEEEEEEEEEEEEEEEEEEEEEEEEEEEEEEEEEEEEEEEEEEEEEEEEEEEEEEEEEEEEEEEEEEEEEEEEEEEEEEEEEEEEEEEEEEEEEEEEEEEEEEEEEEEEEEEEEEEEEEEEEEEEEEEEEEEEEEEEEEEEEEEEEEEEEEEEEEEEEEEEEEE
RRRRRRRRRRRRRRRRRRRRRRRRRRRRRRRRRRRRRRRRRRRRRRRRRRRRRRRRRRRRRRRRRRRRRRRRRRRRRRRRRRRRRRRRRRRRRRRRRRRRRRRRRRRRRRRRRRRRRRRRRRRRRRRRRRRRRRRRRRRRRRRRRRRRRRRRRRRRRRRRRRRRRRRRRRRRRRRRRRRRRRRRRRRRRRRRRRRR
SSSSSSSSSSSSSSSSSSSSSSSSSSSSSSSSSSSSSSSSSSSSSSSSSSSSSSSSSSSSSSSSSSSSSSSSSSSSSSSSSSSSSSSSSSSSSSSSSSSSSSSSSSSSSSSSSSSSSSSSSSSSSSSSSSSSSSSSSSSSSSSSSSSSSSSSSSSSSSSSSSSSSSSSSSSSSSSSSSSSSSSSSSSSSSSSSSSS
AAAAAAAAAAAAAAAAAAAAAAAAAAAAAAAAAAAAAAAAAAAAAAAAAAAAAAAAAAAAAAAAAAAAAAAAAAAAAAAAAAAAAAAAAAAAAAAAAAAAAAAAAAAAAAAAAAAAAAAAAAAAAAAAAAAAAAAAAAAAAAAAAAAAAAAAAAAAAAAAAAAAAAAAAAAAAAAAAAAAAAAAAAAAA
FFFFFFFFFFFFFFFFFFFFFFFFFFFFFFFFFFFFFFFFFFFFFFFFFFFFFFFFFFFFFFFFFFFFFFFFFFFFFFFFFFFFFFFFFFFFFFFFFFFFFFFFFFFFFFFFFFFFFFFFFFFFFFFFFFFFFFFFFFFFFFFFFFFFFFFFFFFFFFFFFFFFFFFFFFFFFFFFFFFFFFFFFFFFFFFFFFFFFFFFFFFFFFFFFFFFFFFFF
EEEEEEEEEEEEEEEEEEEEEEEEEEEEEEEEEEEEEEEEEEEEEEEEEEEEEEEEEEEEEEEEEEEEEEEEEEEEEEEEEEEEEEEEEEEEEEEEEEEEEEEEEEEEEEEEEEEEEEEEEEEEEEEEEEEEEEEEEEEEEEEEEEEEEEEEEEEEEEEEEEEEEEEEEEEEEEEEEEEEEEEEEEEEEEEEEEEEEEEEEEEEEEE
RRRRRRRRRRRRRRRRRRRRRRRRRRRRRRRRRRRRRRRRRRRRRRRRRRRRRRRRRRRRRRRRRRRRRRRRRRRRRRRRRRRRRRRRRRRRRRRRRRRRRRRRRRRRRRRRRRRRRRRRRRRRRRRRRRRRRRRRRRRRRRRRRRRRRRRRRRRRRRRRRRRRRRRRRRRRRRRRRRRRRRRRRRRRRRRRRRRR
SSSSSSSSSSSSSSSSSSSSSSSSSSSSSSSSSSSSSSSSSSSSSSSSSSSSSSSSSSSSSSSSSSSSSSSSSSSSSSSSSSSSSSSSSSSSSSSSSSSSSSSSSSSSSSSSSSSSSSSSSSSSSSSSSSSSSSSSSSSSSSSSSSSSSSSSSSSSSSSSSSSSSSSSSSSSSSSSSSSSSSSSSSSSSSSSSSSS
AAAAAAAAAAAAAAAAAAAAAAAAAAAAAAAAAAAAAAAAAAAAAAAAAAAAAAAAAAAAAAAAAAAAAAAAAAAAAAAAAAAAAAAAAAAAAAAAAAAAAAAAAAAAAAAAAAAAAAAAAAAAAAAAAAAAAAAAAAAAAAAAAAAAAAAAAAAAAAAAAAAAAAAAAAAAAAAAAAAAAAAAAAAAA
FFFFFFFFFFFFFFFFFFFFFFFFFFFFFFFFFFFFFFFFFFFFFFFFFFFFFFFFFFFFFFFFFFFFFFFFFFFFFFFFFFFFFFFFFFFFFFFFFFFFFFFFFFFFFFFFFFFFFFFFFFFFFFFFFFFFFFFFFFFFFFFFFFFFFFFFFFFFFFFFFFFFFFFFFFFFFFFFFFFFFFFFFFFFFFFFFFFFFFFFFFFFFFFFFFFFFFFFF
EEEEEEEEEEEEEEEEEEEEEEEEEEEEEEEEEEEEEEEEEEEEEEEEEEEEEEEEEEEEEEEEEEEEEEEEEEEEEEEEEEEEEEEEEEEEEEEEEEEEEEEEEEEEEEEEEEEEEEEEEEEEEEEEEEEEEEEEEEEEEEEEEEEEEEEEEEEEEEEEEEEEEEEEEEEEEEEEEEEEEEEEEEEEEEEEEEEEEEEEEEEEEEE
RRRRRRRRRRRRRRRRRRRRRRRRRRRRRRRRRRRRRRRRRRRRRRRRRRRRRRRRRRRRRRRRRRRRRRRRRRRRRRRRRRRRRRRRRRRRRRRRRRRRRRRRRRRRRRRRRRRRRRRRRRRRRRRRRRRRRRRRRRRRRRRRRRRRRRRRRRRRRRRRRRRRRRRRRRRRRRRRRRRRRRRRRRRRRRRRRRRR
SSSSSSSSSSSSSSSSSSSSSSSSSSSSSSSSSSSSSSSSSSSSSSSSSSSSSSSSSSSSSSSSSSSSSSSSSSSSSSSSSSSSSSSSSSSSSSSSSSSSSSSSSSSSSSSSSSSSSSSSSSSSSSSSSSSSSSSSSSSSSSSSSSSSSSSSSSSSSSSSSSSSSSSSSSSSSSSSSSSSSSSSSSSSSSSSSSS
AAAAAAAAAAAAAAAAAAAAAAAAAAAAAAAAAAAAAAAAAAAAAAAAAAAAAAAAAAAAAAAAAAAAAAAAAAAAAAAAAAAAAAAAAAAAAAAAAAAAAAAAAAAAAAAAAAAAAAAAAAAAAAAAAAAAAAAAAAAAAAAAAAAAAAAAAAAAAAAAAAAAAAAAAAAAAAAAAAAAAAAAAAAAA
FFFFFFFFFFFFFFFFFFFFFFFFFFFFFFFFFFFFFFFFFFFFFFFFFFFFFFFFFFFFFFFFFFFFFFFFFFFFFFFFFFFFFFFFFFFFFFFFFFFFFFFFFFFFFFFFFFFFFFFFFFFFFFFFFFFFFFFFFFFFFFFFFFFFFFFFFFFFFFFFFFFFFFFFFFFFFFFFFFFFFFFFFFFFFFFFFFFFFFFFFFFFFFFFFFFFFFFFF
EEEEEEEEEEEEEEEEEEEEEEEEEEEEEEEEEEEEEEEEEEEEEEEEEEEEEEEEEEEEEEEEEEEEEEEEEEEEEEEEEEEEEEEEEEEEEEEEEEEEEEEEEEEEEEEEEEEEEEEEEEEEEEEEEEEEEEEEEEEEEEEEEEEEEEEEEEEEEEEEEEEEEEEEEEEEEEEEEEEEEEEEEEEEEEEEEEEEEEEEEEEEEEE
RRRRRRRRRRRRRRRRRRRRRRRRRRRRRRRRRRRRRRRRRRRRRRRRRRRRRRRRRRRRRRRRRRRRRRRRRRRRRRRRRRRRRRRRRRRRRRRRRRRRRRRRRRRRRRRRRRRRRRRRRRRRRRRRRRRRRRRRRRRRRRRRRRRRRRRRRRRRRRRRRRRRRRRRRRRRRRRRRRRRRRRRRRRRRRRRRRRR
SSSSSSSSSSSSSSSSSSSSSSSSSSSSSSSSSSSSSSSSSSSSSSSSSSSSSSSSSSSSSSSSSSSSSSSSSSSSSSSSSSSSSSSSSSSSSSSSSSSSSSSSSSSSSSSSSSSSSSSSSSSSSSSSSSSSSSSSSSSSSSSSSSSSSSSSSSSSSSSSSSSSSSSSSSSSSSSSSSSSSSSSSSSSSSSSSSSS
AAAAAAAAAAAAAAAAAAAAAAAAAAAAAAAAAAAAAAAAAAAAAAAAAAAAAAAAAAAAAAAAAAAAAAAAAAAAAAAAAAAAAAAAAAAAAAAAAAAAAAAAAAAAAAAAAAAAAAAAAAAAAAAAAAAAAAAAAAAAAAAAAAAAAAAAAAAAAAAAAAAAAAAAAAAAAAAAAAAAAAAAAAAAA
FFFFFFFFFFFFFFFFFFFFFFFFFFFFFFFFFFFFFFFFFFFFFFFFFFFFFFFFFFFFFFFFFFFFFFFFFFFFFFFFFFFFFFFFFFFFFFFFFFFFFFFFFFFFFFFFFFFFFFFFFFFFFFFFFFFFFFFFFFFFFFFFFFFFFFFFFFFFFFFFFFFFFFFFFFFFFFFFFFFFFFFFFFFFFFFFFFFFFFFFFFFFFFFFFFFFFFFFF
EEEEEEEEEEEEEEEEEEEEEEEEEEEEEEEEEEEEEEEEEEEEEEEEEEEEEEEEEEEEEEEEEEEEEEEEEEEEEEEEEEEEEEEEEEEEEEEEEEEEEEEEEEEEEEEEEEEEEEEEEEEEEEEEEEEEEEEEEEEEEEEEEEEEEEEEEEEEEEEEEEEEEEEEEEEEEEEEEEEEEEEEEEEEEEEEEEEEEEEEEEEEEEE
RRRRRRRRRRRRRRRRRRRRRRRRRRRRRRRRRRRRRRRRRRRRRRRRRRRRRRRRRRRRRRRRRRRRRRRRRRRRRRRRRRRRRRRRRRRRRRRRRRRRRRRRRRRRRRRRRRRRRRRRRRRRRRRRRRRRRRRRRRRRRRRRRRRRRRRRRRRRRRRRRRRRRRRRRRRRRRRRRRRRRRRRRRRRRRRRRRRRR
SSSSSSSSSSSSSSSSSSSSSSSSSSSSSSSSSSSSSSSSSSSSSSSSSSSSSSSSSSSSSSSSSSSSSSSSSSSSSSSSSSSSSSSSSSSSSSSSSSSSSSSSSSSSSSSSSSSSSSSSSSSSSSSSSSSSSSSSSSSSSSSSSSSSSSSSSSSSSSSSSSSSSSSSSSSSSSSSSSSSSSSSSSSSSSSSSSS
AAAAAAAAAAAAAAAAAAAAAAAAAAAAAAAAAAAAAAAAAAAAAAAAAAAAAAAAAAAAAAAAAAAAAAAAAAAAAAAAAAAAAAAAAAAAAAAAAAAAAAAAAAAAAAAAAAAAAAAAAAAAAAAAAAAAAAAAAAAAAAAAAAAAAAAAAAAAAAAAAAAAAAAAAAAAAAAAAAAAAAAAAAAAA
FFFFFFFFFFFFFFFFFFFFFFFFFFFFFFFFFFFFFFFFFFFFFFFFFFFFFFFFFFFFFFFFFFFFFFFFFFFFFFFFFFFFFFFFFFFFFFFFFFFFFFFFFFFFFFFFFFFFFFFFFFFFFFFFFFFFFFFFFFFFFFFFFFFFFFFFFFFFFFFFFFFFFFFFFFFFFFFFFFFFFFFFFFFFFFFFFFFFFFFFFFFFFFFFFFFFFFFFF
EEEEEEEEEEEEEEEEEEEEEEEEEEEEEEEEEEEEEEEEEEEEEEEEEEEEEEEEEEEEEEEEEEEEEEEEEEEEEEEEEEEEEEEEEEEEEEEEEEEEEEEEEEEEEEEEEEEEEEEEEEEEEEEEEEEEEEEEEEEEEEEEEEEEEEEEEEEEEEEEEEEEEEEEEEEEEEEEEEEEEEEEEEEEEEEEEEEEEEEEEEEEEEE
RRRRRRRRRRRRRRRRRRRRRRRRRRRRRRRRRRRRRRRRRRRRRRRRRRRRRRRRRRRRRRRRRRRRRRRRRRRRRRRRRRRRRRRRRRRRRRRRRRRRRRRRRRRRRRRRRRRRRRRRRRRRRRRRRRRRRRRRRRRRRRRRRRRRRRRRRRRRRRRRRRRRRRRRRRRRRRRRRRRRRRRRRRRRRRRRRRRR
SSSSSSSSSSSSSSSSSSSSSSSSSSSSSSSSSSSSSSSSSSSSSSSSSSSSSSSSSSSSSSSSSSSSSSSSSSSSSSSSSSSSSSSSSSSSSSSSSSSSSSSSSSSSSSSSSSSSSSSSSSSSSSSSSSSSSSSSSSSSSSSSSSSSSSSSSSSSSSSSSSSSSSSSSSSSSSSSSSSSSSSSSSSSSSSSSSS
AAAAAAAAAAAAAAAAAAAAAAAAAAAAAAAAAAAAAAAAAAAAAAAAAAAAAAAAAAAAAAAAAAAAAAAAAAAAAAAAAAAAAAAAAAAAAAAAAAAAAAAAAAAAAAAAAAAAAAAAAAAAAAAAAAAAAAAAAAAAAAAAAAAAAAAAAAAAAAAAAAAAAAAAAAAAAAAAAAAAAAAAAAAAA
FFFFFFFFFFFFFFFFFFFFFFFFFFFFFFFFFFFFFFFFFFFFFFFFFFFFFFFFFFFFFFFFFFFFFFFFFFFFFFFFFFFFFFFFFFFFFFFFFFFFFFFFFFFFFFFFFFFFFFFFFFFFFFFFFFFFFFFFFFFFFFFFFFFFFFFFFFFFFFFFFFFFFFFFFFFFFFFFFFFFFFFFFFFFFFFFFFFFFFFFFFFFFFFFFFFFFFFFF
EEEEEEEEEEEEEEEEEEEEEEEEEEEEEEEEEEEEEEEEEEEEEEEEEEEEEEEEEEEEEEEEEEEEEEEEEEEEEEEEEEEEEEEEEEEEEEEEEEEEEEEEEEEEEEEEEEEEEEEEEEEEEEEEEEEEEEEEEEEEEEEEEEEEEEEEEEEEEEEEEEEEEEEEEEEEEEEEEEEEEEEEEEEEEEEEEEEEEEEEEEEEEEE
RRRRRRRRRRRRRRRRRRRRRRRRRRRRRRRRRRRRRRRRRRRRRRRRRRRRRRRRRRRRRRRRRRRRRRRRRRRRRRRRRRRRRRRRRRRRRRRRRRRRRRRRRRRRRRRRRRRRRRRRRRRRRRRRRRRRRRRRRRRRRRRRRRRRRRRRRRRRRRRRRRRRRRRRRRRRRRRRRRRRRRRRRRRRRRRRRRRR
SSSSSSSSSSSSSSSSSSSSSSSSSSSSSSSSSSSSSSSSSSSSSSSSSSSSSSSSSSSSSSSSSSSSSSSSSSSSSSSSSSSSSSSSSSSSSSSSSSSSSSSSSSSSSSSSSSSSSSSSSSSSSSSSSSSSSSSSSSSSSSSSSSSSSSSSSSSSSSSSSSSSSSSSSSSSSSSSSSSSSSSSSSSSSSSSSSS

SSSSSSSSSSSSSSSSSSSSSSSSSSSSSSSSSSSSSSSSSSSSSSSSSSSSSSSSSSSSSSSSSSSSSSSSSSSSSSSSSSSSSSSSSSSSSSSSSSSSSSSSSSSSSSSSSSSSSSSS
AAAAAAAAAAAAAAAAAAAAAAAAAAAAAAAAAAAAAAAAAAAAAAAAAAAAAAAAAAAAAAAAAAAAAAAAAAAAAAAAAAAAAAAAAAAAAAAAAAAAAAAAAAAAAAAAAAAAAAAA
FFFFFFFFFFFFFFFFFFFFFFFFFFFFFFFFFFFFFFFFFFFFFFFFFFFFFFFFFFFFFFFFFFFFFFFFFFFFFFFFFFFFFFFFFFFFFFFFFFFFFFFFFFFFFFFFFFFFFFFF
EEEEEEEEEEEEEEEEEEEEEEEEEEEEEEEEEEEEEEEEEEEEEEEEEEEEEEEEEEEEEEEEEEEEEEEEEEEEEEEEEEEEEEEEEEEEEEEEEEEEEEEEEEEEEEEEEEEEEEEE
RRRRRRRRRRRRRRRRRRRRRRRRRRRRRRRRRRRRRRRRRRRRRRRRRRRRRRRRRRRRRRRRRRRRRRRRRRRRRRRRRRRRRRRRRRRRRRRRRRRRRRRRRRRRRRRRRRRRRRRR
SSSSSSSSSSSSSSSSSSSSSSSSSSSSSSSSSSSSSSSSSSSSSSSSSSSSSSSSSSSSSSSSSSSSSSSSSSSSSSSSSSSSSSSSSSSSSSSSSSSSSSSSSSSSSSSSSSSSSSSS
AAAAAAAAAAAAAAAAAAAAAAAAAAAAAAAAAAAAAAAAAAAAAAAAAAAAAAAAAAAAAAAAAAAAAAAAAAAAAAAAAAAAAAAAAAAAAAAAAAAAAAAAAAAAAAAAAAAAAAAA
FFFFFFFFFFFFFFFFFFFFFFFFFFFFFFFFFFFFFFFFFFFFFFFFFFFFFFFFFFFFFFFFFFFFFFFFFFFFFFFFFFFFFFFFFFFFFFFFFFFFFFFFFFFFFFFFFFFFFFFF
EEEEEEEEEEEEEEEEEEEEEEEEEEEEEEEEEEEEEEEEEEEEEEEEEEEEEEEEEEEEEEEEEEEEEEEEEEEEEEEEEEEEEEEEEEEEEEEEEEEEEEEEEEEEEEEEEEEEEEEE
RRRRRRRRRRRRRRRRRRRRRRRRRRRRRRRRRRRRRRRRRRRRRRRRRRRRRRRRRRRRRRRRRRRRRRRRRRRRRRRRRRRRRRRRRRRRRRRRRRRRRRRRRRRRRRRRRRRRRRRR
SSSSSSSSSSSSSSSSSSSSSSSSSSSSSSSSSSSSSSSSSSSSSSSSSSSSSSSSSSSSSSSSSSSSSSSSSSSSSSSSSSSSSSSSSSSSSSSSSSSSSSSSSSSSSSSSSSSSSSSS
AAAAAAAAAAAAAAAAAAAAAAAAAAAAAAAAAAAAAAAAAAAAAAAAAAAAAAAAAAAAAAAAAAAAAAAAAAAAAAAAAAAAAAAAAAAAAAAAAAAAAAAAAAAAAAAAAAAAAAAA
FFFFFFFFFFFFFFFFFFFFFFFFFFFFFFFFFFFFFFFFFFFFFFFFFFFFFFFFFFFFFFFFFFFFFFFFFFFFFFFFFFFFFFFFFFFFFFFFFFFFFFFFFFFFFFFFFFFFFFFF
EEEEEEEEEEEEEEEEEEEEEEEEEEEEEEEEEEEEEEEEEEEEEEEEEEEEEEEEEEEEEEEEEEEEEEEEEEEEEEEEEEEEEEEEEEEEEEEEEEEEEEEEEEEEEEEEEEEEEEEE
RRRRRRRRRRRRRRRRRRRRRRRRRRRRRRRRRRRRRRRRRRRRRRRRRRRRRRRRRRRRRRRRRRRRRRRRRRRRRRRRRRRRRRRRRRRRRRRRRRRRRRRRRRRRRRRRRRRRRRRR
SSSSSSSSSSSSSSSSSSSSSSSSSSSSSSSSSSSSSSSSSSSSSSSSSSSSSSSSSSSSSSSSSSSSSSSSSSSSSSSSSSSSSSSSSSSSSSSSSSSSSSSSSSSSSSSSSSSSSSSS
AAAAAAAAAAAAAAAAAAAAAAAAAAAAAAAAAAAAAAAAAAAAAAAAAAAAAAAAAAAAAAAAAAAAAAAAAAAAAAAAAAAAAAAAAAAAAAAAAAAAAAAAAAAAAAAAAAAAAAAA
FFFFFFFFFFFFFFFFFFFFFFFFFFFFFFFFFFFFFFFFFFFFFFFFFFFFFFFFFFFFFFFFFFFFFFFFFFFFFFFFFFFFFFFFFFFFFFFFFFFFFFFFFFFFFFFFFFFFFFFF
EEEEEEEEEEEEEEEEEEEEEEEEEEEEEEEEEEEEEEEEEEEEEEEEEEEEEEEEEEEEEEEEEEEEEEEEEEEEEEEEEEEEEEEEEEEEEEEEEEEEEEEEEEEEEEEEEEEEEEEE
RRRRRRRRRRRRRRRRRRRRRRRRRRRRRRRRRRRRRRRRRRRRRRRRRRRRRRRRRRRRRRRRRRRRRRRRRRRRRRRRRRRRRRRRRRRRRRRRRRRRRRRRRRRRRRRRRRRRRRRR
SSSSSSSSSSSSSSSSSSSSSSSSSSSSSSSSSSSSSSSSSSSSSSSSSSSSSSSSSSSSSSSSSSSSSSSSSSSSSSSSSSSSSSSSSSSSSSSSSSSSSSSSSSSSSSSSSSSSSSSS
AAAAAAAAAAAAAAAAAAAAAAAAAAAAAAAAAAAAAAAAAAAAAAAAAAAAAAAAAAAAAAAAAAAAAAAAAAAAAAAAAAAAAAAAAAAAAAAAAAAAAAAAAAAAAAAAAAAAAAAA
FFFFFFFFFFFFFFFFFFFFFFFFFFFFFFFFFFFFFFFFFFFFFFFFFFFFFFFFFFFFFFFFFFFFFFFFFFFFFFFFFFFFFFFFFFFFFFFFFFFFFFFFFFFFFFFFFFFFFFFF
EEEEEEEEEEEEEEEEEEEEEEEEEEEEEEEEEEEEEEEEEEEEEEEEEEEEEEEEEEEEEEEEEEEEEEEEEEEEEEEEEEEEEEEEEEEEEEEEEEEEEEEEEEEEEEEEEEEEEEEE
RRRRRRRRRRRRRRRRRRRRRRRRRRRRRRRRRRRRRRRRRRRRRRRRRRRRRRRRRRRRRRRRRRRRRRRRRRRRRRRRRRRRRRRRRRRRRRRRRRRRRRRRRRRRRRRRRRRRRRRR
SSSSSSSSSSSSSSSSSSSSSSSSSSSSSSSSSSSSSSSSSSSSSSSSSSSSSSSSSSSSSSSSSSSSSSSSSSSSSSSSSSSSSSSSSSSSSSSSSSSSSSSSSSSSSSSSSSSSSSSS
AAAAAAAAAAAAAAAAAAAAAAAAAAAAAAAAAAAAAAAAAAAAAAAAAAAAAAAAAAAAAAAAAAAAAAAAAAAAAAAAAAAAAAAAAAAAAAAAAAAAAAAAAAAAAAAAAAAAAAAA
FFFFFFFFFFFFFFFFFFFFFFFFFFFFFFFFFFFFFFFFFFFFFFFFFFFFFFFFFFFFFFFFFFFFFFFFFFFFFFFFFFFFFFFFFFFFFFFFFFFFFFFFFFFFFFFFFFFFFFFF
EEEEEEEEEEEEEEEEEEEEEEEEEEEEEEEEEEEEEEEEEEEEEEEEEEEEEEEEEEEEEEEEEEEEEEEEEEEEEEEEEEEEEEEEEEEEEEEEEEEEEEEEEEEEEEEEEEEEEEEE
RRRRRRRRRRRRRRRRRRRRRRRRRRRRRRRRRRRRRRRRRRRRRRRRRRRRRRRRRRRRRRRRRRRRRRRRRRRRRRRRRRRRRRRRRRRRRRRRRRRRRRRRRRRRRRRRRRRRRRRR
SSSSSSSSSSSSSSSSSSSSSSSSSSSSSSSSSSSSSSSSSSSSSSSSSSSSSSSSSSSSSSSSSSSSSSSSSSSSSSSSSSSSSSSSSSSSSSSSSSSSSSSSSSSSSSSSSSSSSSSS
AAAAAAAAAAAAAAAAAAAAAAAAAAAAAAAAAAAAAAAAAAAAAAAAAAAAAAAAAAAAAAAAAAAAAAAAAAAAAAAAAAAAAAAAAAAAAAAAAAAAAAAAAAAAAAAAAAAAAAAA
FFFFFFFFFFFFFFFFFFFFFFFFFFFFFFFFFFFFFFFFFFFFFFFFFFFFFFFFFFFFFFFFFFFFFFFFFFFFFFFFFFFFFFFFFFFFFFFFFFFFFFFFFFFFFFFFFFFFFFFF
EEEEEEEEEEEEEEEEEEEEEEEEEEEEEEEEEEEEEEEEEEEEEEEEEEEEEEEEEEEEEEEEEEEEEEEEEEEEEEEEEEEEEEEEEEEEEEEEEEEEEEEEEEEEEEEEEEEEEEEE
RRRRRRRRRRRRRRRRRRRRRRRRRRRRRRRRRRRRRRRRRRRRRRRRRRRRRRRRRRRRRRRRRRRRRRRRRRRRRRRRRRRRRRRRRRRRRRRRRRRRRRRRRRRRRRRRRRRRRRRR
SSSSSSSSSSSSSSSSSSSSSSSSSSSSSSSSSSSSSSSSSSSSSSSSSSSSSSSSSSSSSSSSSSSSSSSSSSSSSSSSSSSSSSSSSSSSSSSSSSSSSSSSSSSSSSSSSSSSSSSS
AAAAAAAAAAAAAAAAAAAAAAAAAAAAAAAAAAAAAAAAAAAAAAAAAAAAAAAAAAAAAAAAAAAAAAAAAAAAAAAAAAAAAAAAAAAAAAAAAAAAAAAAAAAAAAAAAAAAAAAA
FFFFFFFFFFFFFFFFFFFFFFFFFFFFFFFFFFFFFFFFFFFFFFFFFFFFFFFFFFFFFFFFFFFFFFFFFFFFFFFFFFFFFFFFFFFFFFFFFFFFFFFFFFFFFFFFFFFFFFFF
EEEEEEEEEEEEEEEEEEEEEEEEEEEEEEEEEEEEEEEEEEEEEEEEEEEEEEEEEEEEEEEEEEEEEEEEEEEEEEEEEEEEEEEEEEEEEEEEEEEEEEEEEEEEEEEEEEEEEEEE
RRRRRRRRRRRRRRRRRRRRRRRRRRRRRRRRRRRRRRRRRRRRRRRRRRRRRRRRRRRRRRRRRRRRRRRRRRRRRRRRRRRRRRRRRRRRRRRRRRRRRRRRRRRRRRRRRRRRRRRR
SSSSSSSSSSSSSSSSSSSSSSSSSSSSSSSSSSSSSSSSSSSSSSSSSSSSSSSSSSSSSSSSSSSSSSSSSSSSSSSSSSSSSSSSSSSSSSSSSSSSSSSSSSSSSSSSSSSSSSSS
AAAAAAAAAAAAAAAAAAAAAAAAAAAAAAAAAAAAAAAAAAAAAAAAAAAAAAAAAAAAAAAAAAAAAAAAAAAAAAAAAAAAAAAAAAAAAAAAAAAAAAAAAAAAAAAAAAAAAAAA
FFFFFFFFFFFFFFFFFFFFFFFFFFFFFFFFFFFFFFFFFFFFFFFFFFFFFFFFFFFFFFFFFFFFFFFFFFFFFFFFFFFFFFFFFFFFFFFFFFFFFFFFFFFFFFFFFFFFFFFF
EEEEEEEEEEEEEEEEEEEEEEEEEEEEEEEEEEEEEEEEEEEEEEEEEEEEEEEEEEEEEEEEEEEEEEEEEEEEEEEEEEEEEEEEEEEEEEEEEEEEEEEEEEEEEEEEEEEEEEEE
RRRRRRRRRRRRRRRRRRRRRRRRRRRRRRRRRRRRRRRRRRRRRRRRRRRRRRRRRRRRRRRRRRRRRRRRRRRRRRRRRRRRRRRRRRRRRRRRRRRRRRRRRRRRRRRRRRRRRRRR
SSSSSSSSSSSSSSSSSSSSSSSSSSSSSSSSSSSSSSSSSSSSSSSSSSSSSSSSSSSSSSSSSSSSSSSSSSSSSSSSSSSSSSSSSSSSSSSSSSSSSSSSSSSSSSSSSSSSSSSS
AAAAAAAAAAAAAAAAAAAAAAAAAAAAAAAAAAAAAAAAAAAAAAAAAAAAAAAAAAAAAAAAAAAAAAAAAAAAAAAAAAAAAAAAAAAAAAAAAAAAAAAAAAAAAAAAAAAAAAAA
FFFFFFFFFFFFFFFFFFFFFFFFFFFFFFFFFFFFFFFFFFFFFFFFFFFFFFFFFFFFFFFFFFFFFFFFFFFFFFFFFFFFFFFFFFFFFFFFFFFFFFFFFFFFFFFFFFFFFFFF
EEEEEEEEEEEEEEEEEEEEEEEEEEEEEEEEEEEEEEEEEEEEEEEEEEEEEEEEEEEEEEEEEEEEEEEEEEEEEEEEEEEEEEEEEEEEEEEEEEEEEEEEEEEEEEEEEEEEEEEE
RRRRRRRRRRRRRRRRRRRRRRRRRRRRRRRRRRRRRRRRRRRRRRRRRRRRRRRRRRRRRRRRRRRRRRRRRRRRRRRRRRRRRRRRRRRRRRRRRRRRRRRRRRRRRRRRRRRRRRRR
SSSSSSSSSSSSSSSSSSSSSSSSSSSSSSSSSSSSSSSSSSSSSSSSSSSSSSSSSSSSSSSSSSSSSSSSSSSSSSSSSSSSSSSSSSSSSSSSSSSSSSSSSSSSSSSSSSSSSSSS
AAAAAAAAAAAAAAAAAAAAAAAAAAAAAAAAAAAAAAAAAAAAAAAAAAAAAAAAAAAAAAAAAAAAAAAAAAAAAAAAAAAAAAAAAAAAAAAAAAAAAAAAAAAAAAAAAAAAAAAA
FFFFFFFFFFFFFFFFFFFFFFFFFFFFFFFFFFFFFFFFFFFFFFFFFFFFFFFFFFFFFFFFFFFFFFFFFFFFFFFFFFFFFFFFFFFFFFFFFFFFFFFFFFFFFFFFFFFFFFFF
EEEEEEEEEEEEEEEEEEEEEEEEEEEEEEEEEEEEEEEEEEEEEEEEEEEEEEEEEEEEEEEEEEEEEEEEEEEEEEEEEEEEEEEEEEEEEEEEEEEEEEEEEEEEEEEEEEEEEEEE
RRRRRRRRRRRRRRRRRRRRRRRRRRRRRRRRRRRRRRRRRRRRRRRRRRRRRRRRRRRRRRRRRRRRRRRRRRRRRRRRRRRRRRRRRRRRRRRRRRRRRRRRRRRRRRRRRRRRRRRR
SSSSSSSSSSSSSSSSSSSSSSSSSSSSSSSSSSSSSSSSSSSSSSSSSSSSSSSSSSSSSSSSSSSSSSSSSSSSSSSSSSSSSSSSSSSSSSSSSSSSSSSSSSSSSSSSSSSSSSSS
AAAAAAAAAAAAAAAAAAAAAAAAAAAAAAAAAAAAAAAAAAAAAAAAAAAAAAAAAAAAAAAAAAAAAAAAAAAAAAAAAAAAAAAAAAAAAAAAAAAAAAAAAAAAAAAAAAAAAAAA
FFFFFFFFFFFFFFFFFFFFFFFFFFFFFFFFFFFFFFFFFFFFFFFFFFFFFFFFFFFFFFFFFFFFFFFFFFFFFFFFFFFFFFFFFFFFFFFFFFFFFFFFFFFFFFFFFFFFFFFF
EEEEEEEEEEEEEEEEEEEEEEEEEEEEEEEEEEEEEEEEEEEEEEEEEEEEEEEEEEEEEEEEEEEEEEEEEEEEEEEEEEEEEEEEEEEEEEEEEEEEEEEEEEEEEEEEEEEEEEEE
RRRRRRRRRRRRRRRRRRRRRRRRRRRRRRRRRRRRRRRRRRRRRRRRRRRRRRRRRRRRRRRRRRRRRRRRRRRRRRRRRRRRRRRRRRRRRRRRRRRRRRRRRRRRRRRRRRRRRRRR
SSSSSSSSSSSSSSSSSSSSSSSSSSSSSSSSSSSSSSSSSSSSSSSSSSSSSSSSSSSSSSSSSSSSSSSSSSSSSSSSSSSSSSSSSSSSSSSSSSSSSSSSSSSSSSSSSSSSSSSS
AAAAAAAAAAAAAAAAAAAAAAAAAAAAAAAAAAAAAAAAAAAAAAAAAAAAAAAAAAAAAAAAAAAAAAAAAAAAAAAAAAAAAAAAAAAAAAAAAAAAAAAAAAAAAAAAAAAAAAAA
FFFFFFFFFFFFFFFFFFFFFFFFFFFFFFFFFFFFFFFFFFFFFFFFFFFFFFFFFFFFFFFFFFFFFFFFFFFFFFFFFFFFFFFFFFFFFFFFFFFFFFFFFFFFFFFFFFFFFFFF
EEEEEEEEEEEEEEEEEEEEEEEEEEEEEEEEEEEEEEEEEEEEEEEEEEEEEEEEEEEEEEEEEEEEEEEEEEEEEEEEEEEEEEEEEEEEEEEEEEEEEEEEEEEEEEEEEEEEEEEE
RRRRRRRRRRRRRRRRRRRRRRRRRRRRRRRRRRRRRRRRRRRRRRRRRRRRRRRRRRRRRRRRRRRRRRRRRRRRRRRRRRRRRRRRRRRRRRRRRRRRRRRRRRRRRRRRRRRRRRRR
SSSSSSSSSSSSSSSSSSSSSSSSSSSSSSSSSSSSSSSSSSSSSSSSSSSSSSSSSSSSSSSSSSSSSSSSSSSSSSSSSSSSSSSSSSSSSSSSSSSSSSSSSSSSSSSSSSSSSSSS
AAAAAAAAAAAAAAAAAAAAAAAAAAAAAAAAAAAAAAAAAAAAAAAAAAAAAAAAAAAAAAAAAAAAAAAAAAAAAAAAAAAAAAAAAAAAAAAAAAAAAAAAAAAAAAAAAAAAAAAA
FFFFFFFFFFFFFFFFFFFFFFFFFFFFFFFFFFFFFFFFFFFFFFFFFFFFFFFFFFFFFFFFFFFFFFFFFFFFFFFFFFFFFFFFFFFFFFFFFFFFFFFFFFFFFFFFFFFFFFFF
EEEEEEEEEEEEEEEEEEEEEEEEEEEEEEEEEEEEEEEEEEEEEEEEEEEEEEEEEEEEEEEEEEEEEEEEEEEEEEEEEEEEEEEEEEEEEEEEEEEEEEEEEEEEEEEEEEEEEEEE
RRRRRRRRRRRRRRRRRRRRRRRRRRRRRRRRRRRRRRRRRRRRRRRRRRRRRRRRRRRRRRRRRRRRRRRRRRRRRRRRRRRRRRRRRRRRRRRRRRRRRRRRRRRRRRRRRRRRRRRR
SSSSSSSSSSSSSSSSSSSSSSSSSSSSSSSSSSSSSSSSSSSSSSSSSSSSSSSSSSSSSSSSSSSSSSSSSSSSSSSSSSSSSSSSSSSSSSSSSSSSSSSSSSSSSSSSSSSSSSSS
AAAAAAAAAAAAAAAAAAAAAAAAAAAAAAAAAAAAAAAAAAAAAAAAAAAAAAAAAAAAAAAAAAAAAAAAAAAAAAAAAAAAAAAAAAAAAAAAAAAAAAAAAAAAAAAAAAAAAAAA
FFFFFFFFFFFFFFFFFFFFFFFFFFFFFFFFFFFFFFFFFFFFFFFFFFFFFFFFFFFFFFFFFFFFFFFFFFFFFFFFFFFFFFFFFFFFFFFFFFFFFFFFFFFFFFFFFFFFFFFF
EEEEEEEEEEEEEEEEEEEEEEEEEEEEEEEEEEEEEEEEEEEEEEEEEEEEEEEEEEEEEEEEEEEEEEEEEEEEEEEEEEEEEEEEEEEEEEEEEEEEEEEEEEEEEEEEEEEEEEEE
RRRRRRRRRRRRRRRRRRRRRRRRRRRRRRRRRRRRRRRRRRRRRRRRRRRRRRRRRRRRRRRRRRRRRRRRRRRRRRRRRRRRRRRRRRRRRRRRRRRRRRRRRRRRRRRRRRRRRRRR
SSSSSSSSSSSSSSSSSSSSSSSSSSSSSSSSSSSSSSSSSSSSSSSSSSSSSSSSSSSSSSSSSSSSSSSSSSSSSSSSSSSSSSSSSSSSSSSSSSSSSSSSSSSSSSSSSSSSSSSS
AAAAAAAAAAAAAAAAAAAAAAAAAAAAAAAAAAAAAAAAAAAAAAAAAAAAAAAAAAAAAAAAAAAAAAAAAAAAAAAAAAAAAAAAAAAAAAAAAAAAAAAAAAAAAAAAAAAAAAAA
FFFFFFFFFFFFFFFFFFFFFFFFFFFFFFFFFFFFFFFFFFFFFFFFFFFFFFFFFFFFFFFFFFFFFFFFFFFFFFFFFFFFFFFFFFFFFFFFFFFFFFFFFFFFFFFFFFFFFFFF
EEEEEEEEEEEEEEEEEEEEEEEEEEEEEEEEEEEEEEEEEEEEEEEEEEEEEEEEEEEEEEEEEEEEEEEEEEEEEEEEEEEEEEEEEEEEEEEEEEEEEEEEEEEEEEEEEEEEEEEE
RRRRRRRRRRRRRRRRRRRRRRRRRRRRRRRRRRRRRRRRRRRRRRRRRRRRRRRRRRRRRRRRRRRRRRRRRRRRRRRRRRRRRRRRRRRRRRRRRRRRRRRRRRRRRRRRRRRRRRRR
SSSSSSSSSSSSSSSSSSSSSSSSSSSSSSSSSSSSSSSSSSSSSSSSSSSSSSSSSSSSSSSSSSSSSSSSSSSSSSSSSSSSSSSSSSSSSSSSSSSSSSSSSSSSSSSSSSSSSSSS
AAAAAAAAAAAAAAAAAAAAAAAAAAAAAAAAAAAAAAAAAAAAAAAAAAAAAAAAAAAAAAAAAAAAAAAAAAAAAAAAAAAAAAAAAAAAAAAAAAAAAAAAAAAAAAAAAAAAAAAA
FFFFFFFFFFFFFFFFFFFFFFFFFFFFFFFFFFFFFFFFFFFFFFFFFFFFFFFFFFFFFFFFFFFFFFFFFFFFFFFFFFFFFFFFFFFFFFFFFFFFFFFFFFFFFFFFFFFFFFFF
EEEEEEEEEEEEEEEEEEEEEEEEEEEEEEEEEEEEEEEEEEEEEEEEEEEEEEEEEEEEEEEEEEEEEEEEEEEEEEEEEEEEEEEEEEEEEEEEEEEEEEEEEEEEEEEEEEEEEEEE
RRRRRRRRRRRRRRRRRRRRRRRRRRRRRRRRRRRRRRRRRRRRRRRRRRRRRRRRRRRRRRRRRRRRRRRRRRRRRRRRRRRRRRRRRRRRRRRRRRRRRRRRRRRRRRRRRRRRRRRR
SSSSSSSSSSSSSSSSSSSSSSSSSSSSSSSSSSSSSSSSSSSSSSSSSSSSSSSSSSSSSSSSSSSSSSSSSSSSSSSSSSSSSSSSSSSSSSSSSSSSSSSSSSSSSSSSSSSSSSSS
AAAAAAAAAAAAAAAAAAAAAAAAAAAAAAAAAAAAAAAAAAAAAAAAAAAAAAAAAAAAAAAAAAAAAAAAAAAAAAAAAAAAAAAAAAAAAAAAAAAAAAAAAAAAAAAAAAAAAAAA
FFFFFFFFFFFFFFFFFFFFFFFFFFFFFFFFFFFFFFFFFFFFFFFFFFFFFFFFFFFFFFFFFFFFFFFFFFFFFFFFFFFFFFFFFFFFFFFFFFFFFFFFFFFFFFFFFFFFFFFF
EEEEEEEEEEEEEEEEEEEEEEEEEEEEEEEEEEEEEEEEEEEEEEEEEEEEEEEEEEEEEEEEEEEEEEEEEEEEEEEEEEEEEEEEEEEEEEEEEEEEEEEEEEEEEEEEEEEEEEEE
RRRRRRRRRRRRRRRRRRRRRRRRRRRRRRRRRRRRRRRRRRRRRRRRRRRRRRRRRRRRRRRRRRRRRRRRRRRRRRRRRRRRRRRRRRRRRRRRRRRRRRRRRRRRRRRRRRRRRRRR
SSSSSSSSSSSSSSSSSSSSSSSSSSSSSSSSSSSSSSSSSSSSSSSSSSSSSSSSSSSSSSSSSSSSSSSSSSSSSSSSSSSSSSSSSSSSSSSSSSSSSSSSSSSSSSSSSSSSSSSS
AAAAAAAAAAAAAAAAAAAAAAAAAAAAAAAAAAAAAAAAAAAAAAAAAAAAAAAAAAAAAAAAAAAAAAAAAAAAAAAAAAAAAAAAAAAAAAAAAAAAAAAAAAAAAAAAAAAAAAAA
FFFFFFFFFFFFFFFFFFFFFFFFFFFFFFFFFFFFFFFFFFFFFFFFFFFFFFFFFFFFFFFFFFFFFFFFFFFFFFFFFFFFFFFFFFFFFFFFFFFFFFFFFFFFFFFFFFFFFFFF
EEEEEEEEEEEEEEEEEEEEEEEEEEEEEEEEEEEEEEEEEEEEEEEEEEEEEEEEEEEEEEEEEEEEEEEEEEEEEEEEEEEEEEEEEEEEEEEEEEEEEEEEEEEEEEEEEEEEEEEE
RRRRRRRRRRRRRRRRRRRRRRRRRRRRRRRRRRRRRRRRRRRRRRRRRRRRRRRRRRRRRRRRRRRRRRRRRRRRRRRRRRRRRRRRRRRRRRRRRRRRRRRRRRRRRRRRRRRRRRRR
SSSSSSSSSSSSSSSSSSSSSSSSSSSSSSSSSSSSSSSSSSSSSSSSSSSSSSSSSSSSSSSSSSSSSSSSSSSSSSSSSSSSSSSSSSSSSSSSSSSSSSSSSSSSSSSSSSSSSSSS
AAAAAAAAAAAAAAAAAAAAAAAAAAAAAAAAAAAAAAAAAAAAAAAAAAAAAAAAAAAAAAAAAAAAAAAAAAAAAAAAAAAAAAAAAAAAAAAAAAAAAAAAAAAAAAAAAAAAAAAA
FFFFFFFFFFFFFFFFFFFFFFFFFFFFFFFFFFFFFFFFFFFFFFFFFFFFFFFFFFFFFFFFFFFFFFFFFFFFFFFFFFFFFFFFFFFFFFFFFFFFFFFFFFFFFFFFFFFFFFFF
EEEEEEEEEEEEEEEEEEEEEEEEEEEEEEEEEEEEEEEEEEEEEEEEEEEEEEEEEEEEEEEEEEEEEEEEEEEEEEEEEEEEEEEEEEEEEEEEEEEEEEEEEEEEEEEEEEEEEEEE
RRRRRRRRRRRRRRRRRRRRRRRRRRRRRRRRRRRRRRRRRRRRRRRRRRRRRRRRRRRRRRRRRRRRRRRRRRRRRRRRRRRRRRRRRRRRRRRRRRRRRRRRRRRRRRRRRRRRRRRR
SSSSSSSSSSSSSSSSSSSSSSSSSSSSSSSSSSSSSSSSSSSSSSSSSSSSSSSSSSSSSSSSSSSSSSSSSSSSSSSSSSSSSSSSSSSSSSSSSSSSSSSSSSSSSSSSSSSSSSSS
AAAAAAAAAAAAAAAAAAAAAAAAAAAAAAAAAAAAAAAAAAAAAAAAAAAAAAAAAAAAAAAAAAAAAAAAAAAAAAAAAAAAAAAAAAAAAAAAAAAAAAAAAAAAAAAAAAAAAAAA
FFFFFFFFFFFFFFFFFFFFFFFFFFFFFFFFFFFFFFFFFFFFFFFFFFFFFFFFFFFFFFFFFFFFFFFFFFFFFFFFFFFFFFFFFFFFFFFFFFFFFFFFFFFFFFFFFFFFFFFF
EEEEEEEEEEEEEEEEEEEEEEEEEEEEEEEEEEEEEEEEEEEEEEEEEEEEEEEEEEEEEEEEEEEEEEEEEEEEEEEEEEEEEEEEEEEEEEEEEEEEEEEEEEEEEEEEEEEEEEEE
RRRRRRRRRRRRRRRRRRRRRRRRRRRRRRRRRRRRRRRRRRRRRRRRRRRRRRRRRRRRRRRRRRRRRRRRRRRRRRRRRRRRRRRRRRRRRRRRRRRRRRRRRRRRRRRRRRRRRRRR
SSSSSSSSSSSSSSSSSSSSSSSSSSSSSSSSSSSSSSSSSSSSSSSSSSSSSSSSSSSSSSSSSSSSSSSSSSSSSSSSSSSSSSSSSSSSSSSSSSSSSSSSSSSSSSSSSSSSSSSS
AAAAAAAAAAAAAAAAAAAAAAAAAAAAAAAAAAAAAAAAAAAAAAAAAAAAAAAAAAAAAAAAAAAAAAAAAAAAAAAAAAAAAAAAAAAAAAAAAAAAAAAAAAAAAAAAAAAAAAAA
FFFFFFFFFFFFFFFFFFFFFFFFFFFFFFFFFFFFFFFFFFFFFFFFFFFFFFFFFFFFFFFFFFFFFFFFFFFFFFFFFFFFFFFFFFFFFFFFFFFFFFFFFFFFFFFFFFFFFFFF
EEEEEEEEEEEEEEEEEEEEEEEEEEEEEEEEEEEEEEEEEEEEEEEEEEEEEEEEEEEEEEEEEEEEEEEEEEEEEEEEEEEEEEEEEEEEEEEEEEEEEEEEEEEEEEEEEEEEEEEE
RRRRRRRRRRRRRRRRRRRRRRRRRRRRRRRRRRRRRRRRRRRRRRRRRRRRRRRRRRRRRRRRRRRRRRRRRRRRRRRRRRRRRRRRRRRRRRRRRRRRRRRRRRRRRRRRRRRRRRRR
SSSSSSSSSSSSSSSSSSSSSSSSSSSSSSSSSSSSSSSSSSSSSSSSSSSSSSSSSSSSSSSSSSSSSSSSSSSSSSSSSSSSSSSSSSSSSSSSSSSSSSSSSSSSSSSSSSSSSSSS
AAAAAAAAAAAAAAAAAAAAAAAAAAAAAAAAAAAAAAAAAAAAAAAAAAAAAAAAAAAAAAAAAAAAAAAAAAAAAAAAAAAAAAAAAAAAAAAAAAAAAAAAAAAAAAAAAAAAAAAA
FFFFFFFFFFFFFFFFFFFFFFFFFFFFFFFFFFFFFFFFFFFFFFFFFFFFFFFFFFFFFFFFFFFFFFFFFFFFFFFFFFFFFFFFFFFFFFFFFFFFFFFFFFFFFFFFFFFFFFFF
EEEEEEEEEEEEEEEEEEEEEEEEEEEEEEEEEEEEEEEEEEEEEEEEEEEEEEEEEEEEEEEEEEEEEEEEEEEEEEEEEEEEEEEEEEEEEEEEEEEEEEEEEEEEEEEEEEEEEEEE
RRRRRRRRRRRRRRRRRRRRRRRRRRRRRRRRRRRRRRRRRRRRRRRRRRRRRRRRRRRRRRRRRRRRRRRRRRRRRRRRRRRRRRRRRRRRRRRRRRRRRRRRRRRRRRRRRRRRRRRR
SSSSSSSSSSSSSSSSSSSSSSSSSSSSSSSSSSSSSSSSSSSSSSSSSSSSSSSSSSSSSSSSSSSSSSSSSSSSSSSSSSSSSSSSSSSSSSSSSSSSSSSSSSSSSSSSSSSSSSSS
AAAAAAAAAAAAAAAAAAAAAAAAAAAAAAAAAAAAAAAAAAAAAAAAAAAAAAAAAAAAAAAAAAAAAAAAAAAAAAAAAAAAAAAAAAAAAAAAAAAAAAAAAAAAAAAAAAAAAAAA
FFFFFFFFFFFFFFFFFFFFFFFFFFFFFFFFFFFFFFFFFFFFFFFFFFFFFFFFFFFFFFFFFFFFFFFFFFFFFFFFFFFFFFFFFFFFFFFFFFFFFFFFFFFFFFFFFFFFFFFF
EEEEEEEEEEEEEEEEEEEEEEEEEEEEEEEEEEEEEEEEEEEEEEEEEEEEEEEEEEEEEEEEEEEEEEEEEEEEEEEEEEEEEEEEEEEEEEEEEEEEEEEEEEEEEEEEEEEEEEEE
RRRRRRRRRRRRRRRRRRRRRRRRRRRRRRRRRRRRRRRRRRRRRRRRRRRRRRRRRRRRRRRRRRRRRRRRRRRRRRRRRRRRRRRRRRRRRRRRRRRRRRRRRRRRRRRRRRRRRRRR
SSSSSSSSSSSSSSSSSSSSSSSSSSSSSSSSSSSSSSSSSSSSSSSSSSSSSSSSSSSSSSSSSSSSSSSSSSSSSSSSSSSSSSSSSSSSSSSSSSSSSSSSSSSSSSSSSSSSSSSS
AAAAAAAAAAAAAAAAAAAAAAAAAAAAAAAAAAAAAAAAAAAAAAAAAAAAAAAAAAAAAAAAAAAAAAAAAAAAAAAAAAAAAAAAAAAAAAAAAAAAAAAAAAAAAAAAAAAAAAAA
FFFFFFFFFFFFFFFFFFFFFFFFFFFFFFFFFFFFFFFFFFFFFFFFFFFFFFFFFFFFFFFFFFFFFFFFFFFFFFFFFFFFFFFFFFFFFFFFFFFFFFFFFFFFFFFFFFFFFFFF
EEEEEEEEEEEEEEEEEEEEEEEEEEEEEEEEEEEEEEEEEEEEEEEEEEEEEEEEEEEEEEEEEEEEEEEEEEEEEEEEEEEEEEEEEEEEEEEEEEEEEEEEEEEEEEEEEEEEEEEE
RRRRRRRRRRRRRRRRRRRRRRRRRRRRRRRRRRRRRRRRRRRRRRRRRRRRRRRRRRRRRRRRRRRRRRRRRRRRRRRRRRRRRRRRRRRRRRRRRRRRRRRRRRRRRRRRRRRRRRRR
SSSSSSSSSSSSSSSSSSSSSSSSSSSSSSSSSSSSSSSSSSSSSSSSSSSSSSSSSSSSSSSSSSSSSSSSSSSSSSSSSSSSSSSSSSSSSSSSSSSSSSSSSSSSSSSSSSSSSSSS
AAAAAAAAAAAAAAAAAAAAAAAAAAAAAAAAAAAAAAAAAAAAAAAAAAAAAAAAAAAAAAAAAAAAAAAAAAAAAAAAAAAAAAAAAAAAAAAAAAAAAAAAAAAAAAAAAAAAAAAA

fig. 1
Testing of the *Tumbleweed* inflatable rover in Antarctica, January 24, 2004

Cara McCarty

# NASA: ADVANCING ULTRA-PERFORMANCE

Since the advent of modernity in the nineteenth century, we have come to rely on the inventions made by science and technology not only to define these times but to assert an optimism and a belief in constant progress. This was particularly true after the Second World War, when governments emerged as major actors in the production of new science, often in competition with other economically advanced nations. Sometimes this science produced competing technologies—the making of the nuclear bomb, for instance, which threatened international disaster but managed to be curtailed through agreements to restrain its proliferation. By and large, however, science, and its application to technology, has come to be seen as socially beneficial, economically essential, and a vital ingredient of a modern society.

Perhaps the most significant race after World War II was between the Soviet Union and the United States for the conquest of space. The competition was about prestige and national security; it was also about influencing the allegiances of the nonaligned world. In response to the orbiting of the Soviet Union's *Sputnik 1* in October 1957, NASA was formed a year later in a climate of American doubt about the prowess of its scientific education and research. Creating NASA and the project to put a man on the moon also suited our insatiable quest to explore new frontiers: in his "Urgent National Needs" speech to Congress in 1961, President John F. Kennedy argued that the nation "needed to respond extraordinarily" and "no single space project will be more impressive to mankind."[1]

Never before in American history had the government, with wide citizen approval, mandated the funds, and the collaboration between its own research capacity and that of universities and private research laboratories, to propel science and technology to such a focused and lofty goal. The lunar landing was the centerpiece of NASA's work, but it was accompanied by myriad advancements in, among others, aeronautics, satellites for weather observation and global communication, space science, and high-performance materials, including textiles. The birth of NASA and the dawn of the space age were pivotal moments in our history, which depended on concurrent breakthroughs in miniaturization and semiconductor electronics, and on our ability to invent and manipulate materials on a molecular level.

Two themes from NASA's civil space program have been essential in the search for innovation in American science and technology: a partnership between government, university, and industry research and development, and an environment of competition for advancement in ideas and products. In the case of textile innovation, for instance, the sophistication of the resulting materials could only have been achieved by teams of highly qualified scientists with wide-ranging expertise working together.

The NASA space program provided a catalyst for accelerating the revolution in materials and textiles that occurred during the second half of the twentieth century. In venturing beyond the earth, design for travel and exploration had to contend with contrasts of climate, temperature extremes, antigravity conditions, solar and cosmic radiation, and flying micrometeoroids. In addition, there were the physical and psychological issues that arise when one leaves the earth: everything in the body changes when in space. As a consequence, the textiles and materials required for protection in such hostile environments were about high performance, about keeping the wearer alive and the equipment safe. There was little place for design considerations such as styling and fashion.

Many of the traditional demands of textiles—function, comfort, protection, and insulation—apply to space textiles, but to a vastly higher degree. Textiles are made of fibers, either woven or nonwoven. However, more aggressive focused research looked to textile composites as a breakthrough technology to address the rigorous demands of outer space. Analyzing fibers at a molecular level and manipulating their structural and chemical properties under laboratory conditions have created new composites and hybrid textiles valued for their light weight, superior strength, durability, and ability to perform in extreme conditions. These highly engineered textiles have promoted reductions in the weight and size of spacecraft and equipment while maintaining protection and function.

The Space Act of 1958 required NASA to make its discoveries and inventions available to private industry. Later legislation made all federal scientists and engineers responsible for technology transfer, gathering over seven hundred laboratories under one umbrella organization, the National Technology Transfer Center.[2] This movement of processes and materials from one industry to another was believed to have many benefits: stimulating the economy, increasing competitiveness, enlarging visibility within the technical community, and enabling commercial companies to license NASA-developed technologies for use in everyday products including automobiles, airplanes, medical devices, and sporting equipment, such as tennis rackets, skis, fishing rods, and bicycle frames.

Perhaps no image is more emblematic of the optimism, collective ambitions, and remarkable achievements of the space age than the spacesuit. It is among the most palpable examples of NASA's technology transfer program, and its most tangible design. Just as protective armor characterized cultures of older times, so does the spacesuit embody ours. It allows survival in life-threatening situations and exploration where no humans have ever ventured. Through our travels and work in space we have penetrated the realm of the nonhuman, and have been able to survive for extended periods.

Spacesuit technology has been central in the development of new textiles. Spacesuits are complex assemblies in which the collective layers of textiles and materials work together to ensure ultimate safety and insulation. Among the deadly hazards astronauts encounter are extreme temperatures. The side of a suit facing the sun may be heated to a temperature as high as 121°C (250°F); the other side, exposed to darkness, may get as cold as –156°C (–250°F) (fig. 2).

Occasionally, NASA adopts existing materials for space use and inspires further development, which ultimately expands the range of applications. Among the most familiar materials that have followed NASA research are Kevlar, Teflon, and Gore-Tex. Kevlar, referred to as the "muscle fiber," was first marketed in 1971 by DuPont. Best known as the material from which bullet-resistant vests are made, it has found many applications in space, most recently in ropes that secure the airbags in the landing apparatus of the Mars *Pathfinder*. Because of the tenacity of its yarns, Kevlar offers great strength at high temperatures. The ultra-performance, heat- and flame-resistant suits of firefighters and industrial workers exposed to heat, steam, and hot liquids derive from advanced thermal technology developed for the Apollo spacesuits. An example of such a suit is the Jamesville Aero-Commando coat and pants, an extremely lightweight proximity system manufactured by Lion Apparel that enables firefighters to walk into fire (figs. 3, 4). It is made of a Kevlar thermal liner and moisture barrier, and Nomex—nonwoven fibers developed by scientists at DuPont, and noted for their advanced thermal qualities. The outer shell has the most demanding role, to resist ignition from direct flames and to protect the inner layers from rips, slashes,

fig. 2
Pressure suit for *Skylab 2*
Made by ILC Dover Inc.
U.S.A., 1972
Teflon® V fabric, Beta cloth, aluminum, nylon, Velcro®, polyester
Courtesy of Smithsonian Institution, National Air and Space Museum
1976-1192-000

This suit was worn by astronaut Charles "Pete" Conrad on the first manned Skylab mission in May 1973.

fig. 3 (left)
Fire Proximity System
Designed by Lion Apparel Inc., textiles produced by W. L. Gore & Associates Inc., Safety Components, and Gentex U.S.A., coat and pant designed 1990–91, gloves designed 2002, helmet designed 2003, system manufactured 2005
Coat and pant outer shell: knitted PBI®, Kevlar® laminated with reflective material, thermal liner of nonwoven E-89™ Dri and AraFlo® Dri quilted to Glide™ II woven filament face cloth, moisture barrier of breathable woven Crosstech® 2C (PTFE) membrane laminated to woven Nomex®; glove outer shell: PBI®, Kevlar®, moisture barrier Crosstech® with thermal liner Direct Grip®, wristlet Nomex®; proximity helmet ensemble: outer shell of PBI®, Kevlar®, moisture barrier of Gore RT7100™ laminated to nonwoven Nomex®, shroud laminated with reflective material

fig. 4 (above)
Many firefighters' tools, such as protective clothing, are spinoffs from technologies first developed for astronauts.

fig. 6 (left)
Cooling vest worn by Dale Earnhardt, 1986 NASCAR Winston Cup Champion

fig. 5 (above left)
Fire- and heat-resistant webbing holding breathing apparatus
Designed by Karen Donnelly, manufactured by Offray Specialty Narrow Fabrics Inc.
U.S.A., designed 1996, manufactured 2004
Woven Kevlar®

and abrasions. Even the webbing of the straps that hold the breathing apparatus for firefighters has been designed to meet the high standards for flame and heat resistance (fig. 5). This particular strap, manufactured by Offray Specialty Narrow Fabrics, has the capacity to experience a fully involved flashover (1,093°C or 2,000°F) and still function. The fire-blocking fibers of these suits and accessories have found civil application in, among others, airline seat cushions and auto-racing uniforms.

Another spinoff from spacesuit technology is a UV-blocking suit and cooling vest. To stabilize an astronaut's body temperature, spacesuits are equipped with an undergarment containing a network of small tubes that are held against the body (fig. 30, p.172). Water, chilled by a cooling reservoir outside the suit, is pumped through the tubes. Heat transfer between the skin and the cold water in the tubes removes the body's metabolic heat. These cooling suits developed by NASA in the 1960s have since been used in industrial settings such as nuclear power plants and steel mills as well as in military applications. The same technology has now replaced icepacks and elastic bandages for sports injuries. Cool vests have become part of a personal cooling system many race-car drivers use to combat cockpit temperatures that often reach 54°C (130°F) (fig. 6).

The liquid cooling and ventilation garment is also finding applications in the field of health care. It is used as a microclimatic environment for patients whose bodies do not cool properly: for people with multiple sclerosis, these garments can eliminate 40%–60% of stored body heat. Similarly, the garment helps cool down burn victims who have lost the ability to do it themselves. Microclimate cooling suits with an antifreeze solution chilled by a battery-powered refrigeration unit were developed for the Hypohidrotic Ectodermal (HED) Foundation to protect HED, Sun and Light Reaction Syndrome, and Xeroderma Pigmentosum patients and others with related disorders that affect the body's ability to cool itself. The UV-blocking suits serve as a protective coating, enabling those afflicted to go outside unharmed by light.

fig. 7
Denver International Airport
Curtiss Worth Fentress and James H. Bradburn, Fentress Bradburn Architects
U.S.A., 1991–94

Teflon is the miracle material we have come to associate with slipperiness. First used to line kitchen pans in the 1960s, Teflon was discovered by DuPont chemist Roy J. Plunkett while studying refrigerants in 1938. The original form of Teflon is polytetrafluorethylene, or PTFE, popularly known by its DuPont trademark, Teflon. It is the "superhero of polymers: neither heat nor cold can hurt it and no acid rain can attack it."[3] Beginning with the Apollo lunar missions, Teflon-coated glass-fiber Beta cloth has been used for the protective outer layer of spacesuits; the fiberglass fabric provides strength, while the Teflon coating gives additional durability, and fire and water resistance. Teflon has since found many uses in medicine: as a paste most commonly used to correct defects of the vocal cords, as sheets, strands, or tubes in implants, or as a mesh in the repair of hernias.

Architects using a skin technology derived from astronauts' spacesuits are also playing a role in the way buildings are now designed and built. Demands for lower construction costs and for large structural composites with complex shapes have increased the prominence of tensioned membrane structures in architecture, both for temporary situations as well as for structures designed to last for several decades. The architectural opportunities lie in the making of very large, light-filled interiors and in the potential to shape the roof in various complex curves. Such lightweight, floating roof constructions have become a preferred solution for the roofing of large, spatially continuous volumes such as airports, exhibition halls, and stadiums. The conical forms of the Denver International Airport, for example, create an artificial landscape that contrasts severely with the airport's flat and endless surroundings (fig. 7). The yarn for the Denver airport's architectural fabric is fiberglass

figs. 8 (a, b), 9
Gore-Tex® apparel
Manufactured by W. L. Gore & Associates Inc.
U.S.A., 1989
Gore-Tex® fabric

Worn by the International Trans-Antarctica Expedition Team (left) and project leaders Will Steger of the United States and Dr. Jean-Louis Etienne of France (above), 1989-90.

and glass filament woven with uniform tension and then coated with Teflon. Manufactured under the trade name Sheerfill by Saint-Gobain Performance Plastics (formerly Chemfab Corporation), the membrane was fabricated, assembled, and erected by Taiyo Birdair Corporation. It successfully repels water, resists soiling and airborne chemicals, and maintains its color and translucency. The translucent roof reduces lighting needs, and its reflectivity, as high as 75%, lowers cooling costs. The fabric also absorbs sound.

Teflon was later extended into Gore-Tex, the revolutionary water-resistant, windproof, breathable material developed by engineer Bob Gore. Gore-Tex is a membrane with microscopic holes—a single square inch contains nine billion pores, each twenty thousand times smaller than a raindrop, but seven hundred times larger than a molecule of water vapor. So, while water cannot pass through the fabric, perspiration can. In 1976 Gore-Tex became commercially available under the slogan, "Guaranteed to Keep You Dry." It is essentially a climate-management system keeping thermal properties intact by shedding external water and allowing the interior to breathe. Although Gore-Tex was not invented for NASA, in 1981 the astronauts of NASA's first space shuttle mission, Columbia, wore spacesuits made with Gore-Tex. Today, W. L. Gore & Associates produces a range of protective apparel from rain gear to outfits created for an international expedition team in Antarctica (figs. 8, 9). In fact, in 1990, an explorer in Antarctica credited Gore-Tex with saving his life while he traversed the continent. Most commonly used for a wide variety of sports clothing, it is also the preferred material for surgeons' garments since its breathable surface also blocks bacteria. In 2003, Gore-Tex was the chosen material for a sixty-eight-foot-diameter L-3 Communications ESSCO radome on Kwajalein, an atoll in the Marshall Islands. This metal space frame dome was constructed with a Gore-Tex panel membrane, designed to shelter and protect a Millimeter Wave radar system, a precision instrument requiring far more stringent tolerances than those of typical radar systems (fig. 10).

Over the years, a number of other special insulation textiles have been developed under the aegis of the NASA programs. The former Chemfab Corporation, cooperating with NASA and Rockwell International, developed a woven glass fiber with a Teflon coating to provide a static-free thermal liner for the space shuttle's cargo bay, and later for the Discovery's Hubble Space Telescope mission. Ski gloves and apparel made by Gateway Technologies in Colorado are commercial adaptations of astronauts' insulated gloves. By integrating microencapsulated phase-change materials (microPCMs) into manmade fiber before extrusion, heat is held onto more efficiently at the molecular level and disbursed evenly within the fabric, yielding a thinner and more dynamic thermal barrier.

Many of the ultra-performance insulating textiles are being created by combining pliable fibers with materials we typically associate with rigidity, such as glass and ceramics. Since 1963, NASA has funded major research

fig. 10
L-3 Communications ESSCO radome
with Gore-Tex® panels
Kwajalein, 2003

fig. 11
*Delta II* rocket
U.S.A., 1998

Nextel™ ceramic fabrics sewn into thermal blankets to protect the liquid engine from the plume of the solid boosters.

efforts in the field of ceramics. Thermal blankets made of layers of woven ceramic fabric and insulating batting protect the space shuttle orbiter from intense heat, especially during reentry into Earth's atmosphere. Such blankets, called "durable advanced flexible reusable surface insulation," have been used in NASA's *Delta II*, an expendable rocket for launching spacecraft or satellites into orbit (fig. 11). In cooperation with the 3M company, NASA has employed their Nextel™ ceramic technology to manufacture blankets as heat shields, some custom-fitted to and sewn onto the orbiter cap's conical support structure, an improvement over conventional aluminum shielding. Normally, we think of blankets as protecting us from cold temperatures, but for these missions, blankets offer protection from intense heat.

Nextel fibers are also the key components of Stuffed Whipple shields, produced at NASA centers in Huntsville, Alabama, and Houston, Texas. These shields take up less room and offer spacecrafts great protection from flying debris. Unlike metals, ceramics cannot bend to absorb impact. But by using ceramic fibers, brittleness can be controlled, and the material can be easily bent and braided. Ceramic fibers are especially valued for their ability to withstand extremely high temperatures without suffering damage; under intense heat, the fibers move but do not weaken the material. Nextel aerospace fabrics can withstand flame-penetration tests of up to 1,093°C (2,000°F), and retain strength and flexibility with little shrinkage up to 1,100°C (2,012°F). One transfer of this technology is to NASCAR racing, where the ceramics, thin as paper and strong as steel, can decrease the heat experienced by drivers, who would face extreme ambient temperatures and floorboards hot enough to boil water.

As space travel takes us closer to the sun, the challenges of insulation increase. The sun appears eleven times brighter at Mercury than on Earth, causing the designers of the Mercury *Messenger* to provide a heat-resistant and highly reflective 8-by-6-foot sunshade that will safeguard the spacecraft and its instruments. This high-tech parasol, fabricated from Nextel ceramic cloth and Kapton™ insulation, permits *Messenger*'s sun-facing side to heat to above 310°C (590°F) while maintaining internal temperatures at about 20°C (68°F), or room temperature (fig. 12).

In 1997, *Cassini-Huygens*, the most sophisticated U.S. planetary spacecraft ever built, was launched for a seven-year trip to Saturn, more than six hundred million miles from Earth. The mission is named for the seventeenth-century Italian-French astronomer Jean-Dominique Cassini and the Dutch scientist Christiaan Huygens, and is a joint project of NASA, the European Space Agency, and the Italian Space Agency. As with previous space missions, the project involves thousands of scientists at university, government, and business research laboratories. To counteract the extreme climates encountered en route to Saturn, the spacecraft's temperature is controlled by shades, shields, louvers, heaters, and thermal blankets. Made using industrial sewing machines and brown butcher-paper patterns, the blankets have

fig. 12
Rendering of the *Messenger* spacecraft in orbit at Mercury with heat-resistant, reflective sunshade constructed of Nextel™ ceramic cloth, 2004
Illustrated by Steven Gribben

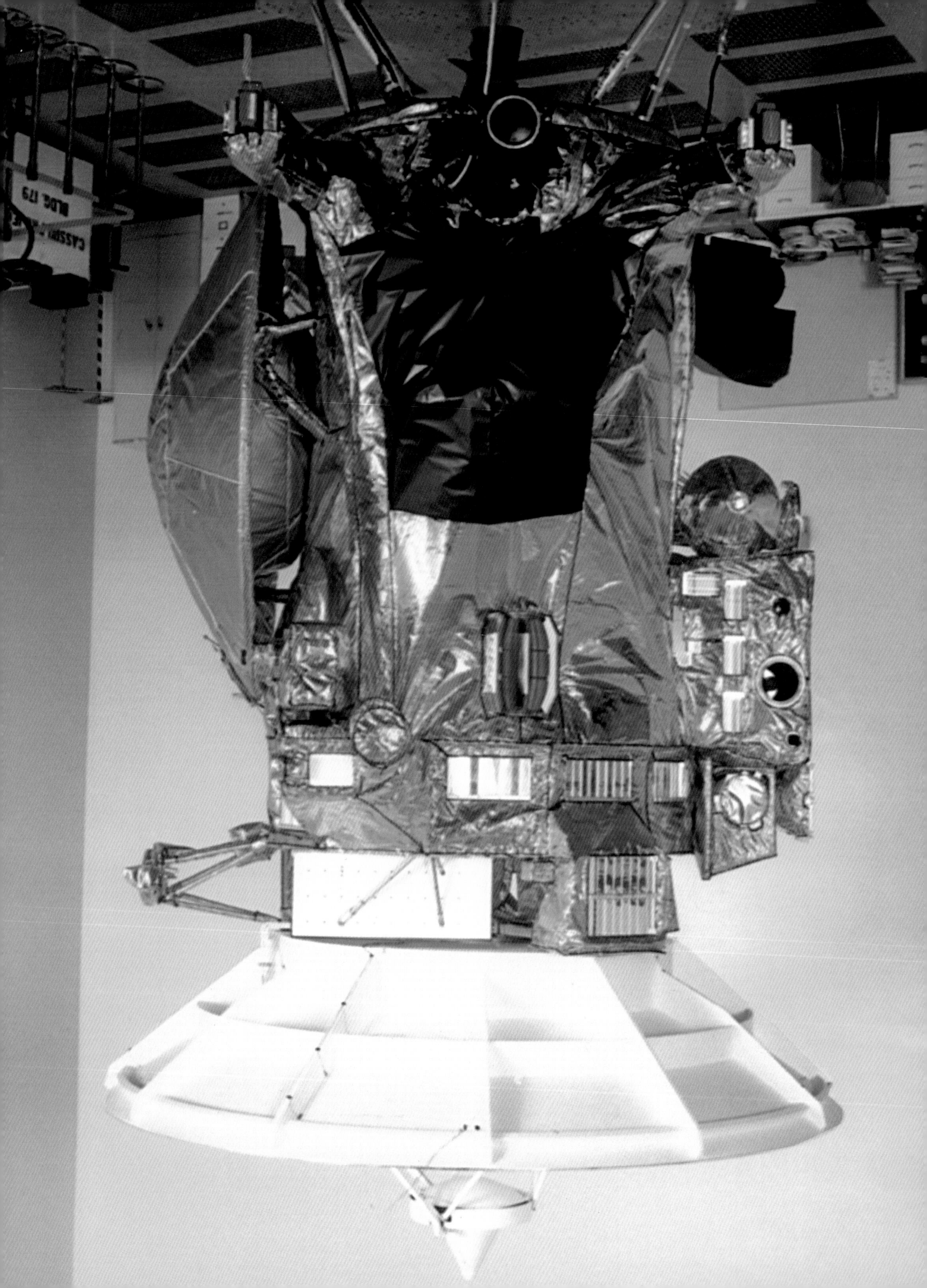

twenty-four layers of different fabrics, among which are Kapton, Mylar, and Dacron®, with a reflective gold-color finish to protect the Cassini's systems (fig. 13). Because *Cassini-Huygens* is traveling through air full of charged particles, each blanket is electrically grounded with thin strips of aluminum.

The 1997 Mars *Pathfinder*'s inflatable airbag landing system, which was launched from Earth deflated, was designed to inflate upon arrival and to strike the rocky surface of Mars and bounce several times before coming to a rest. The airbags were made of Vectran, developed by Hoechst Celanese scientists initially for military purposes. This lightweight, vibration-damping, low–moisture absorption fiber is ideal for use in such demanding applications as aerospace, ocean exploration, electronic support structures, and for recreation and leisure activities (sailcloth, golf clubs, skis, snowboards, and paragliders) and protective garments.

Both Mars missions (1997, 2003) relied on parachutes deployed about six miles above the Martian surface to help slow the descent, ensuring a safe, decelerated landing speed. With each Mars mission, parachute technology has continued to develop. The parachutes are made of nylon and polyester, two durable textiles that are light enough to be pressure-packed into a small compartment onboard the spacecraft. The airbag is tethered to the end of the parachutes with a long bridle made of braided Zylon, an advanced fiber material similar to Kevlar and often used in lines for sailing (fig. 14).

Ultra-light, space-inflatable vehicles are currently an important area of investigation at the Jet Propulsion Laboratory, NASA's primary center for the robotic exploration of the solar system. ILC Dover has helped develop and fabricate a number of airbag and ballute (decelerator) systems for NASA's spacecraft and launch-vehicle components for their safe, cushioned landing on the surface of Mars or Earth. Textiles are now being made into structures that perform tasks. Presently under development at the Jet Propulsion Laboratory is the *Tumbleweed* inflatable rover, a robotic explorer (fig. 15). It is a large beach ball–shaped nylon airbag that can be used as a parachute for descent, cushion for landing, and vehicle for mobility during unmanned exploration on Mars, Earth, Venus, Titan, and Neptune's moon, Triton. The ball could travel along tough terrains, propelled by wind, rolling over rocks instead of around them, greatly increasing the rover's versatility, speed, and range. Movement is controlled through remote control—partially deflating the bags in order to stop for scientific investigation, and re-inflating for mobility with a built-in pump. In early 2004, *Tumbleweed*'s final test deployment was in Antarctica, an environment comparable to the Martian polar caps. That venture proved the *Tumbleweed* an effective and relatively simple means for gathering data in remote regions (fig. 1).

The *Tumbleweed* rover is a good example of biomimicry, or biologically inspired engineering. The wind-driven design of these spherical vehicles eliminates the need for motors or fuel systems, thereby decreasing the overall weight and number of moving parts that are susceptible to breakage.

fig. 13
*Cassini-Huygens* spacecraft wrapped in gold thermal insulation blanket, 1997

fig. 14
Testing Mars Exploration Rover (MER) parachute-deployment in the world's largest wind tunnel at NASA's Ames Research Center, Moffet Field, California, 2003

As technology takes on more characteristics of our natural environment, nature becomes more a subject of study.

Centuries ago, explorers relied on sails to carry them across oceans to new frontiers. Now we are again about to use sails, this time to probe deeper into the solar system. Solar sails, effectively giant mirrors that reflect photons, require no fuel; instead, they depend on energy from the sun to propel the spacecraft between the planets. These space-deployed structures would unfurl like a fan, be as large as possible, extremely light (under one gram per square meter) and durable, and be able to fly within eighteen million miles of the sun. Theoretically, the sails could enable travel at speeds much higher than NASA's *Voyagers*, which reach thirty-eight thousand miles per hour. Much of this technology is in the conceptual research stage, and various prototypes are being developed. While most are of Mylar film, NASA's Jet

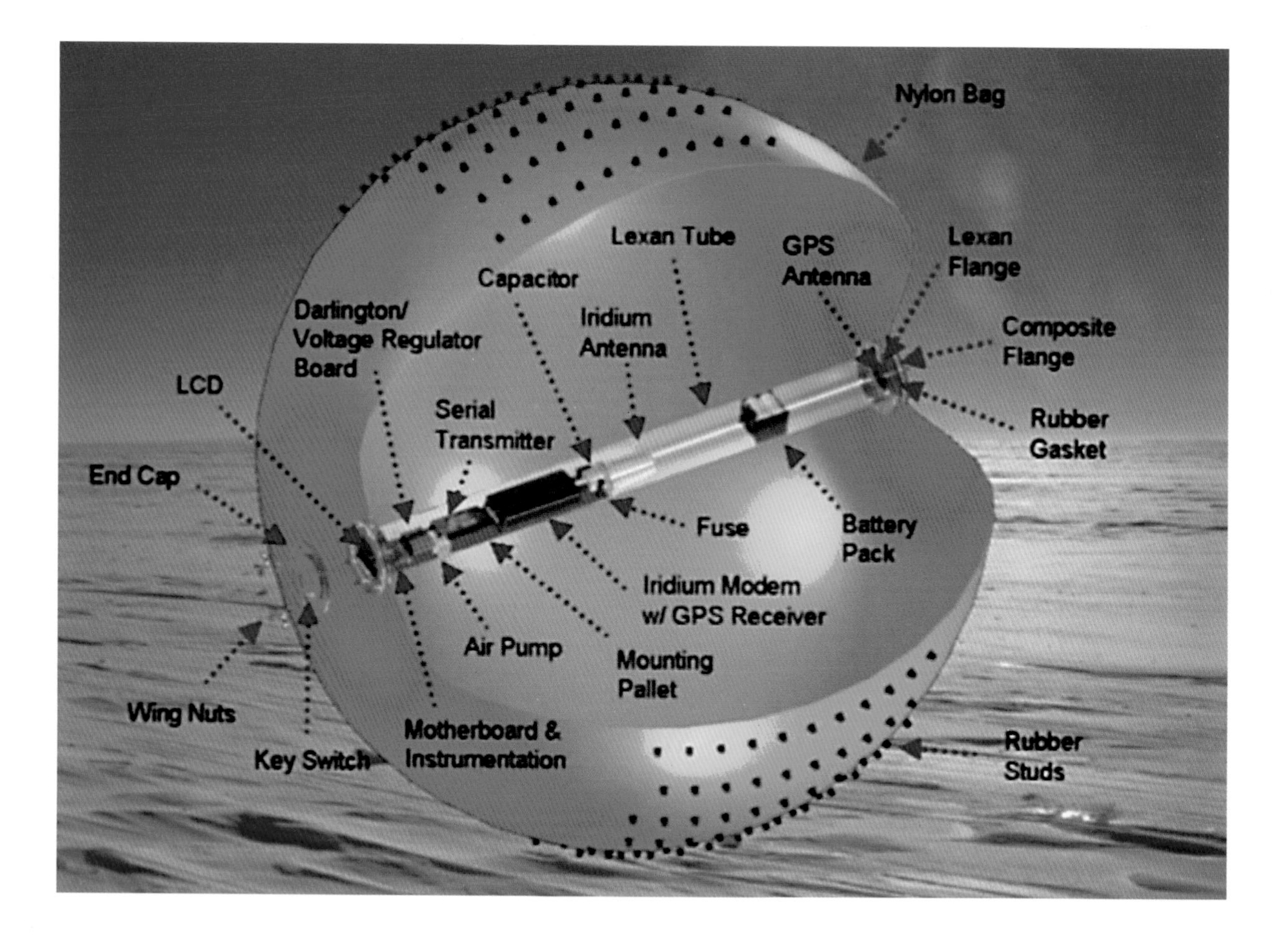


fig. 15
Diagram of *Tumbleweed* inflatable rover system
Designed by NASA Jet Propulsion Laboratory
U.S.A., 2004

Propulsion Laboratory and Marshall Space Flight Center have invented an ultra-lightweight yet extremely strong solar sailcloth made from nets of carbon fibers (fig. 16). Carbon is preferable to polymeric film because it thrives on heat, getting stronger as the heat increases, and carbon nets can carry greater tensile loads. Already, solar-sail technology promises other applications: the National Oceanic and Atmospheric Administration wants these products to create new kinds of weather-monitoring stations that survey the earth more effectively and provide advanced warning of solar storms.

Yet perhaps the greatest future for revolutionizing textiles lies in nanotechnology—the ability to miniaturize things to an incredibly small scale and to introduce intelligence into them. Nanotechnology allows the manipulation of a material's structure on the molecular scale; the diameter of the head of a pin is equivalent to one million nanometers. NASA is one of the agencies involved with the National Nanotechnology Initiative (NNI), whose research strives for major improvements in medicine, manufacturing, materials, and information and environmental technology. Space exploration increasingly demands smaller, lighter, more durable, smarter, and more powerful components, which nanotechnology will be able to provide. Advanced technology that incorporates electronics in clothing itself can amplify the notion of portable environments, such as those already utilized by the astronauts who walked on the moon and worked in space. Intelligent textiles can persuade

fig. 16
Rendering of solar sail made from carbon-fiber nets, 2004

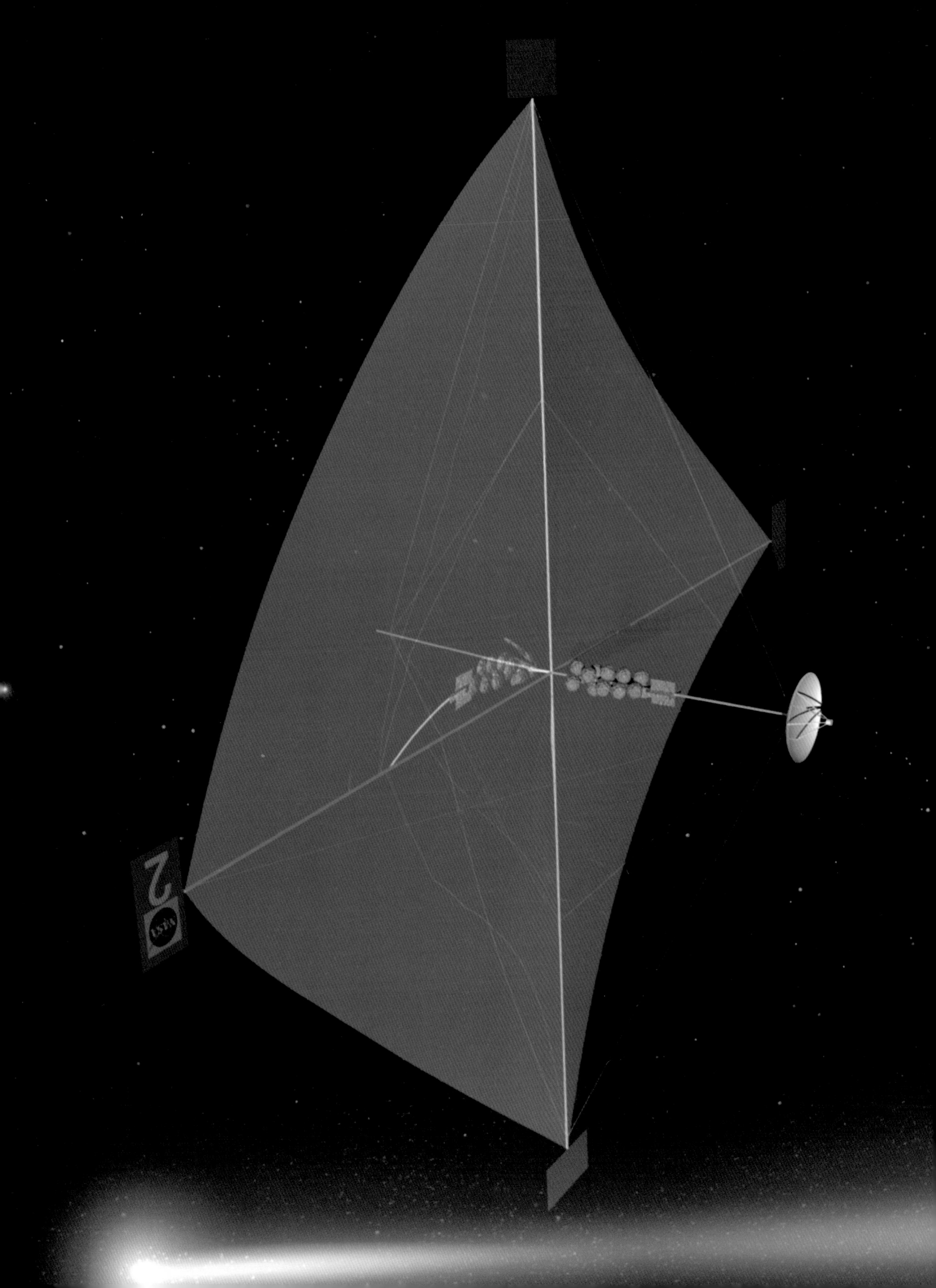

our bodies to act as information systems. Some believe that the combination of miniaturization, artificial intelligence, and our ability to control matter will change the way almost everything is designed and made today, including cars, electronics, machinery, computing, transportation, medicine, and clothing. According to Senator Ron Wyden of Oregon, the author of the bill that introduced the Twenty-first Century Nanotechnology Research and Development Act, "Nanotechnology has the potential to change America on a scale that equals or is greater than the computer revolution."[4]

Designing for space has pushed science and technology to solve problems that have had many repercussions on our lives. Although NASA's space flights are often one-of-a-kind missions, the harsh conditions of outer space have proven to be an excellent environment in which to pioneer new materials and textiles. These textile marvels are the creations of scientists and engineers working together in laboratories in government, universities, and commercial industry. We are now in the fifth decade of the U.S. civil space program. As we venture toward the dark outer reaches of the solar system, our successes will depend on our ability to negotiate the challenges. Interestingly, many of the unmanned spacecraft that are being sent to collect information about the universe, atmosphere, sun, and space environment are being reduced to delicate kite-like structures that consist primarily of textiles. In the past, high strength and light weight were considered mutually exclusive, and metals symbolized the ultimate in strength. But increasingly stringent mass limitations and the mechanical performance properties required at elevated temperatures are driving the need for ultra-light, structurally efficient, and more resistant textile composites.

Another outcome of these engineered textiles is the influence of space-travel aesthetics on contemporary, everyday designs. High-tech, high-performance products with lightweight, reflective, metallic, and durable properties are now common to us. NASA's inventions will continue to inspire new applications, affecting the way we use products and what we expect from them. NASA will promote these ventures, and we will find that the rate of change in textile design, with all of its consequences, may not only be maintained but accelerated. In NASA's own words: "NASA will increasingly look to fields such as biotechnology, information technology, and nanotechnology for the ability to create new structures by building them at the molecular level, atom by atom—enabling advanced performance attributes such as self-corrective maintenance, system compensation in emergencies, or even capabilities such as changing shape."[5]

fig. 17 (facing page)
Copper-wire cloth shielding fabric
Manufactured by Sefar America Inc.
U.S.A., 2003
Plain-woven copper wire

figs. 18, 19 (left)
Slickstop deep-sea containment boom
Manufactured by Slickbar
Products Corporation
U.S.A., designed 1996–97,
manufactured 2000
Woven polyester, urethane coating
Inflated: 152.4 x 190.5 x 40.6 cm
(60 x 75 x 16 in.) per section

This rapid-deployment system for oil-spill containment can be stored, rolled onboard a vessel, and inflated and deployed at a rate of 2,000 feet per hour.

---

fig. 20 (facing page)
Storm™ zipper
Designed by Livio Cossutti, manufactured by Riri Inc.
Switzerland, 2004
Polyester tape covered with a thermoplastic film, teeth, and slider of molded, engineered thermoplastics

Extremes textiles require notions—sewing threads, hardware, and fasteners—which match the level of performance. The Storm zipper is resistant to water, wind, salt corrosion, and UV, making it an ideal zipper to pair with high-performance textiles for outdoor industrial and recreational uses. Much attention was given to refining the zipper's basic functions. The molded thermoplastic teeth are slanted to provide self-adjustment. The woven polyester textile is coated with a thermoplastic film to strengthen the weave and provide an impenetrable surface. Contact between the two edges of the coated tape inside the zipper teeth forms a tight, waterproof seal. Storm is being used for military shelters, flood barriers, wetsuits, life jackets, and the nautical gear of Team Alinghi, winners of the 2003 America's Cup. The same technology is being further refined to create gas- and chemical-resistant zippers for biochemical hazard suits and containment shelters.

fig. 21 (above)
Palmhive® Bobble camouflage net
Developed by Ministry of Defense, United Kingdom; manufactured by Palmhive® Technical Textiles Ltd.
England, designed 1998, manufactured 2004
Warp-knit polyester and Lycra®

fig. 22 (right)
Webshot
Designed and manufactured by Foster-Miller Inc.
U.S.A., 1997
Net construction of Vectran®, Kevlar®, and polyethylene

Line strength, knotting, and spacing were optimized for maximum entanglement with the minimum of human hazard during capture. The cut-resistant fibers prevent escape.

fig. 23
X-Eleven motorcycle helmet
Manufactured by Shoei Co. Ltd.
Japan, designed 1999–2002,
manufactured 2004
Composite of glass, organic, and high-strength polyethylene fibers in plastic resin matrix; polyurethane paint

fig. 24
Wiley Post, suit 3
Made by Wiley Post and Russell Colley
with B. F. Goodrich Company
U.S.A., 1934
Cotton, rubber, leather, aluminum
Courtesy of Smithsonian Institution,
National Air and Space Museum,
1936-0036-000

Wiley Post developed three suits with B. F. Goodrich. The third, from 1934, was the first full-pressure suit to successfully protect a pilot's life at extreme altitudes.

Amanda Young

# THE SPACESUIT

A spacesuit is, for all intents and purposes, a miniature spacecraft designed to keep an astronaut alive and well in the most hostile environments. The Mercury (1961–63), Gemini (1965–66), and Apollo (1968–72) years were a magical period in history, and among the most incredible images of the twentieth century are those of the white-suited astronauts walking on the moon. Those white suits, however, were the culmination of research programs that used textiles, glues, plastics, and metals in ways that had never been done before. In the course of developing these suits, advances were made in materials, and space-program spinoffs became a part of many other aspects of daily life.

One of the earliest pressure/protective suits of the high-altitude/space era was developed in England in the 1930s. Designed by Professor John Haldane and Dr. Robert Davis of Siebe, Gorman & Company, and worn by Mark Ridge in a low-pressure chamber, it was based on a deep-sea diving suit. Ridge had previously tested a suit of his own design, in a steel tank packed with dry ice, to prove that an armor of clothes could protect a person from the cold found at extremely high altitudes. This suit was most likely made of wool and cotton, with an aluminum foil lining. The thermal properties of wool, coupled with those of the quilted cotton layer, would have added a known level of protection, and the interior aluminum foil layer would have provided reflection to help magnify these qualities.

Around the same time, Wiley Post, a licensed transport pilot, won the Los Angeles to Chicago Air Derby in 1930. He was given the airplane *Winnie Mae*, and set off on a trip around the world. During this journey he discovered that the higher he went, the faster he could fly. It was also apparent that at these higher altitudes, along with the lack of oxygen and air pressure, it was very cold. As the technology of the times had not advanced to pressurize the whole airplane or even the cockpit, a practical suit capable of maintaining pressure and keeping the pilot warm was necessary. In 1934 Post worked with Russell Colley and the B. F. Goodrich Company to build a suit for these purposes. Its development required three test versions, which were, again, based in part on the principles of a deep-sea diving suit. The first two suits were made of a single layer of rubber and failed during testing. The third suit consisted of two layers, with a rubber bladder and a quilted cotton exterior, and included leather boots and gloves and an attached helmet (fig. 24). This was the first suit to actually protect a pilot's life at extreme altitudes.

An active developmental period followed. A variety of people, organizations, and countries participated in pressure-suit research, and, throughout the world, record-breaking altitude flights generated improvements in these suits. In the United States, these advancements culminated with the series of "Mark" suits produced by B. F. Goodrich under a U.S. Navy contract. These high-altitude, full-pressure suits were of two-layer construction, with a rubber bladder to contain the oxygen and air pressure, and a green, heavyweight nylon exterior, which protected the bladder. A restraint system, consisting of nonstretching tapes, wires, and zippers, aided by the nylon exterior layer, was designed to prevent the bladder from blowing up like a balloon when pressurized, and was incorporated into the suit. Although some of these suits had attached rubber boots, often the feet were not covered with the nylon layer, and the pilot wore leather boots over the bladder. Gloves, made with a rubber bladder and a nylon and leather cover layer, with a steel and nylon restraint system, attached to the suit with an aluminum locking ring on the wrist, and a zipper that held everything in place.

In April 1959 the first group of astronauts was selected. Known as the "Mercury 7" for the Mercury program, they became the "legendary seven astronauts."[1] The spacesuits they wore were essentially silver versions of the Mark IV high-altitude, full-pressure suits. Because the first astronauts occupied only a small area within the cabin, and were not required to perform activities outside the spacecraft, the suit remained unpressurized during most of the mission. This allowed the suit to act as a back-up to the craft's life-support systems, which, should those have failed, would have kept the astronaut alive until splashdown. The Mercury suits were also made by B. F. Goodrich, and were most likely given the silver color for the thermal qualities of a reflective material, and for its aesthetic value (fig. 25). These seven men were ready to

fig. 25
John Glenn's Mercury suit
Made by B. F. Goodrich Company
U.S.A., 1962
Aluminized nylon, rubber, leather, aluminum
Courtesy of Smithsonian Institution, National Air and Space Museum, 1967-0178-000

John Glenn wore this Mercury suit on *Freedom 7*, the first orbital mission of an American astronaut, in February 1962.

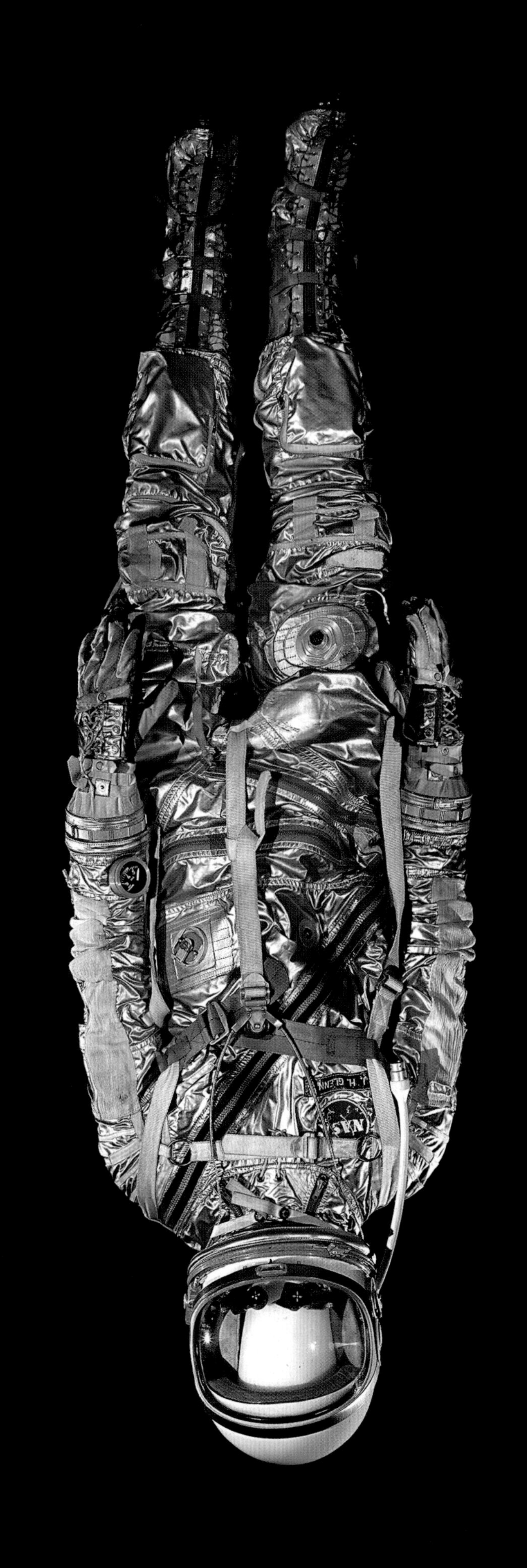

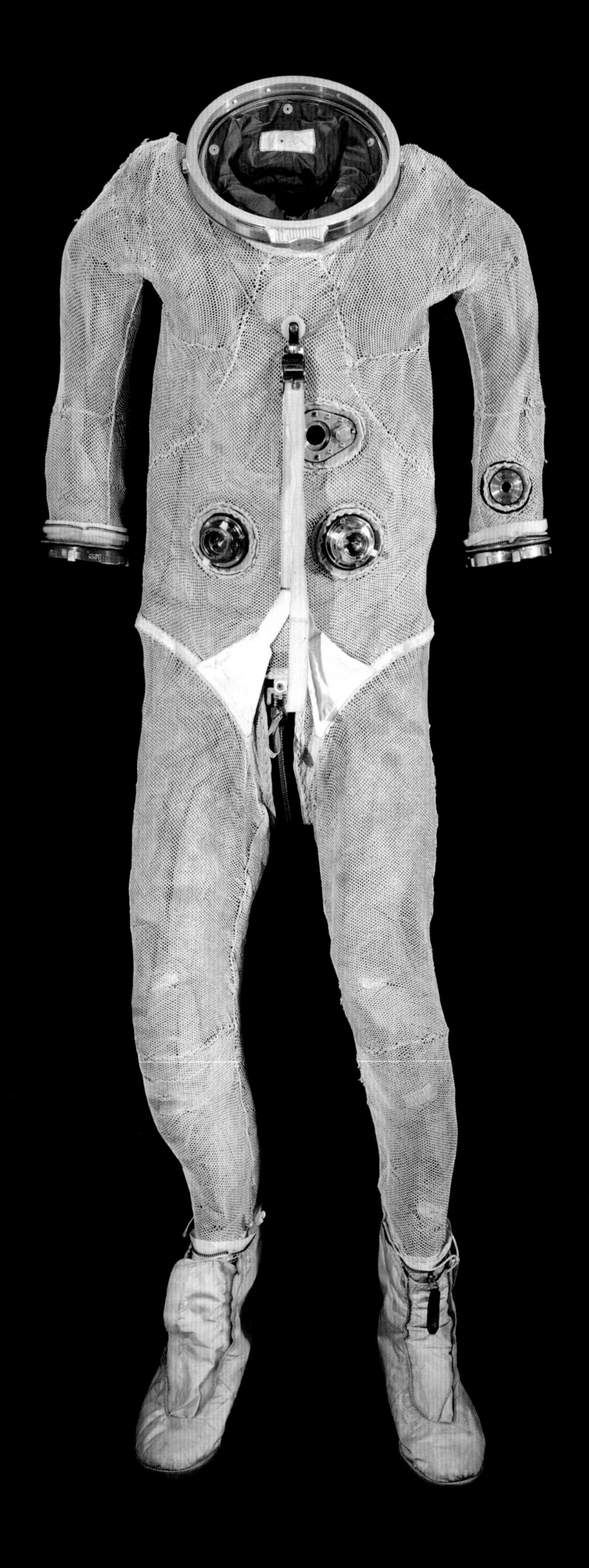

figs. 26, 27
Eugene Cernan's *Gemini 9* suit
Made by David Clark Company
U.S.A., 1966
Interior: HT-Nylon (uncoated Nomex®), rubberized ripstop nylon, aluminum; exterior: HT-Nylon, Chromel-R®
Courtesy of Smithsonian Institution, National Air and Space Museum, 1968-0439-000 (left), 1985-0266-000 (right)

An interior view of Eugene Cernan's G4-C spacesuit, worn on *Gemini 9* in June 1966. The cover layer has been removed to show the Link Net restraint layer (left).

This is the only time a mission suit was constructed with Chromel-R® leg coverings to protect the astronaut's legs from the heat blast generated by the maneuvering unit he was to test (right).

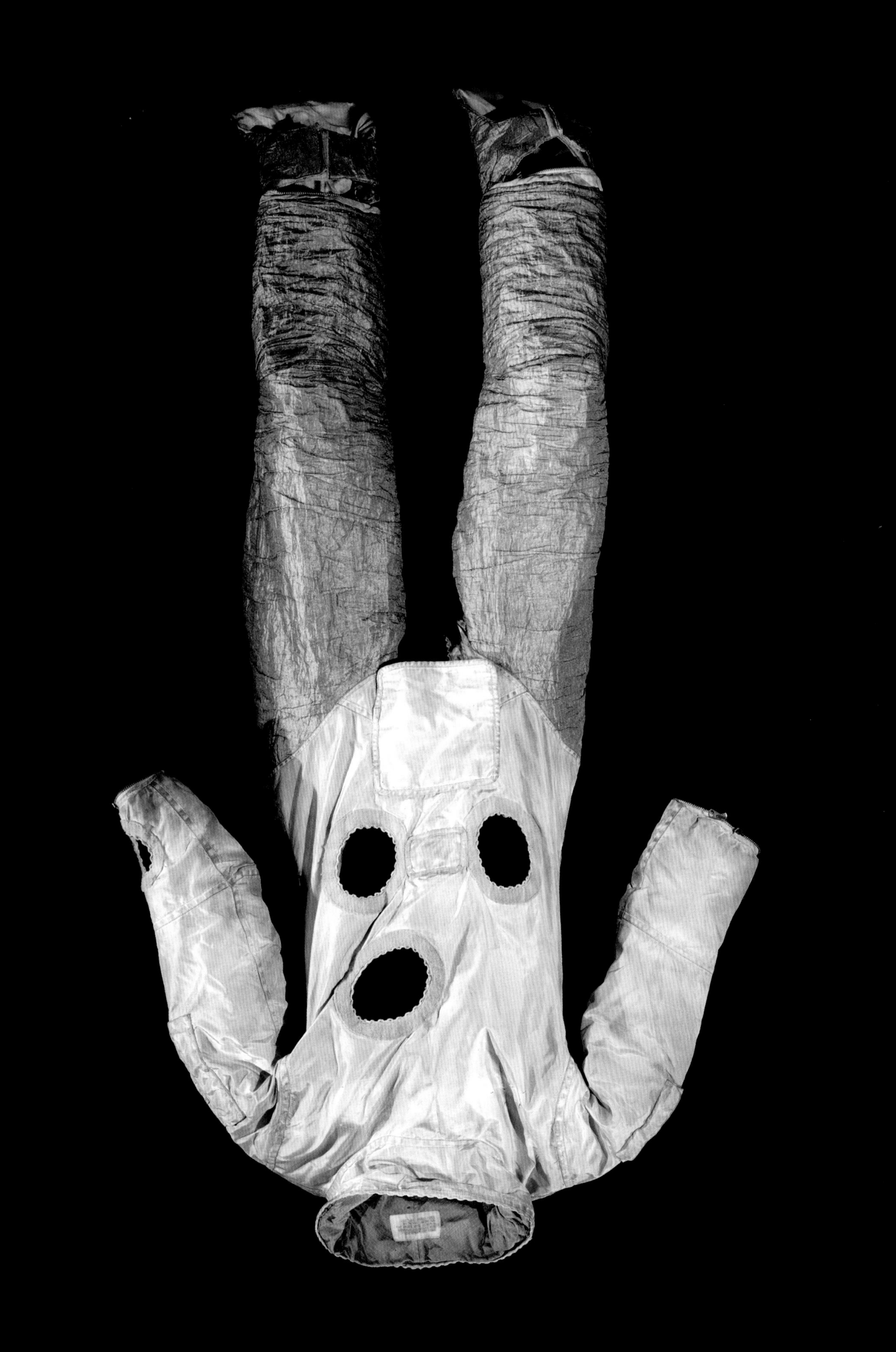

fig. 28 (facing page)
Frank Borman's *Gemini 7* suit
Made by David Clark Company
U.S.A., 1965
HT-Nylon (uncoated Nomex®), polycarbonate, polyester, aluminum
Courtesy of Smithsonian Institution, National Air and Space Museum, 1968-0022-000

The long-duration *Gemini 7* mission of December 1965 required more comfortable suits for the astronauts. This G5-C suit was worn by Frank Borman.

do what no man had done before, and with a wonderful silver suit in which to do it. In actuality, the silver color was from an aluminized powder coating glued to the green nylon exterior layer prior to suit construction. The Mercury suits now have brown and green patches where the aluminized coating has worn away and the glue and nylon have begun to show through.

The second of the manned programs on the way to the moon was Gemini, the first of which was launched in March 1965, and the last in November 1966. Gemini was designed to test spacecraft systems, astronaut stamina and long-term survival, and the ability to perform tasks outside the spacecraft. The new program called for a new spacesuit. Working closely with the NASA Manned Spacecraft Center in Houston, B. F. Goodrich and the David Clark Company, of Worcester, Massachusetts, designed early prototypes of the suits. These first Gemini developmental suits were silver, having the same aluminized coatings as the Mercury suits, and similar in appearance to their predecessor. The David Clark Company was awarded the contract and the second series of Gemini suits was constructed. These were the first of the white suits, although in most cases they still had silver boots.

The ability of the suits to function while the astronauts performed the more difficult tasks became a major focus, and its weight and flexibility were as important as the use of heat-resistant and flame-retardant materials. The Gemini suits, like the Mercury suits, were constructed with two sets of layers attached to each other. They came in two models, the G3-C and later G4-C, and in two configurations: Intra-vehicular (IV) and Extra-vehicular (EV). Both configurations had an interior layer consisting of the rubber bladder with restraint system, and an exterior insulating layer that included aluminized Mylar with Dacron spacers to provide air pockets between the layers. They also contained a cover layer of Nomex—a long-chain molecule developed by DuPont that has extraordinary heat- and flame-resistant properties, and can be used in sheet or honeycomb form, or as a woven fabric. In addition to the cover layer, Nomex was woven into a looped material called Link Net, which acted as part of the restraint system (figs. 26, 27). The IV configuration had fewer layers of insulation, whereas the EV suit had an additional seven layers of aluminized Mylar with Dacron spacers, and a layer of HT-Nylon (uncoated Nomex) to protect the astronaut when he was outside the spacecraft.

Although the Gemini suits were classified as soft suits, they were not very comfortable to wear for extended periods. Thus for the two-week *Gemini 7* mission in December 1965, the spacesuits worn by Frank Borman and Jim Lovell were redesigned soft suits with built-in helmets (fig. 28). In developing this more comfortable spacesuit, NASA and the David Clark Company traded mobility while pressurized for long-term comfort while unpressurized. The G5-C suit was the result, which was affectionately known as the Grasshopper suit and was only used during this mission. The suit weighed twelve pounds and included a large built-in helmet that was worn with a four-pound protective soft helmet underneath. The flexible polycarbonate visor of the attached

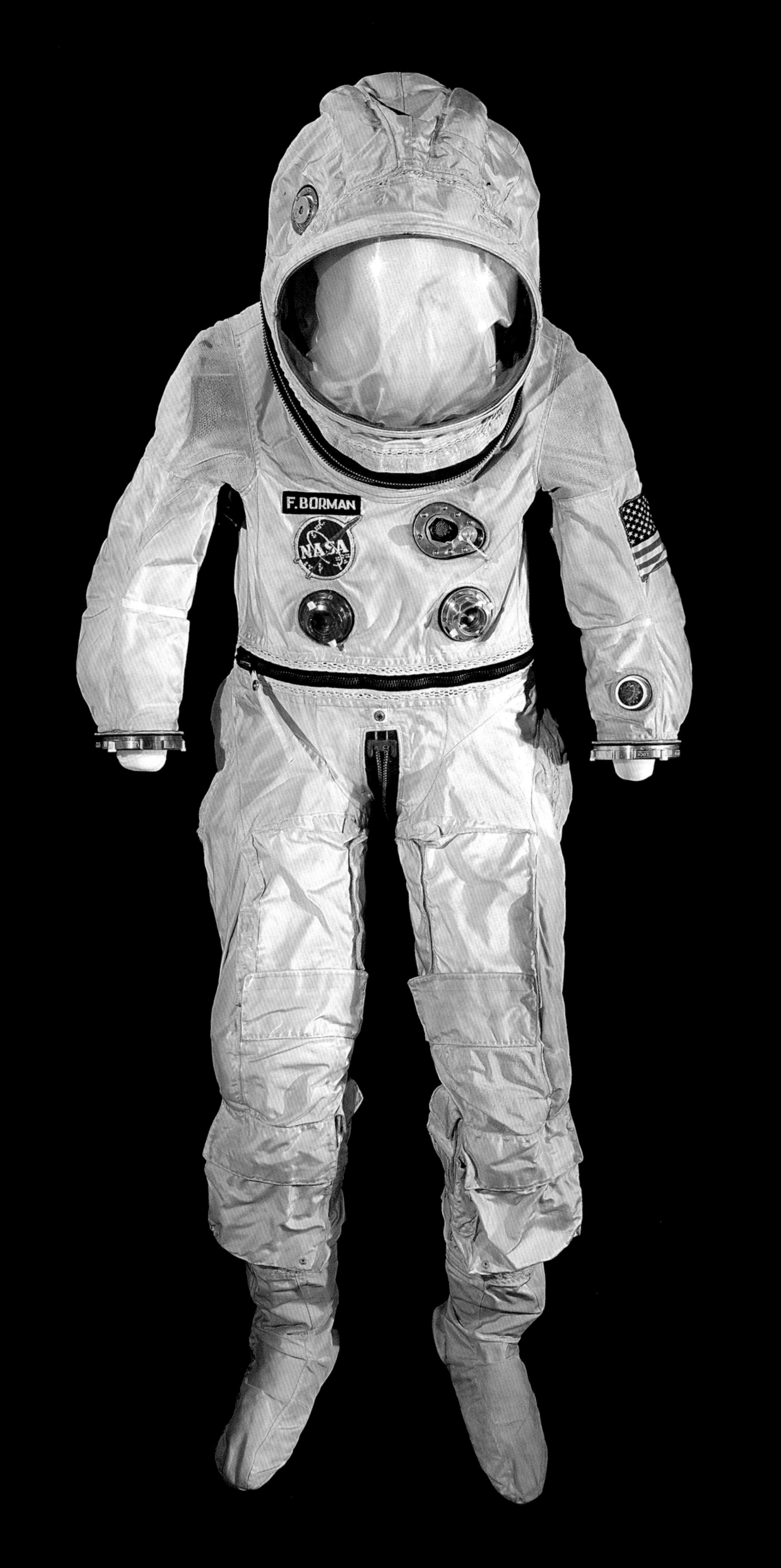
F.BORMAN
NASA

helmet gave the appearance of a large grasshopper eye, hence the name. The suits were fairly comfortable, made of Nomex fabric, with the Link Net restraint layer, and additional small sections of knitted Nomex in the shoulders for improved mobility.

With the successes of the Gemini program, NASA turned its sights on the moon, and the first Apollo mission was launched in October 1968. The Apollo program required the astronauts to perform very different tasks from either the Mercury or Gemini programs, and, consequently, they needed very different suits and life-support equipment. Developments of an Apollo spacesuit began during the Gemini program, and had gone in several directions. For that reason, in 1962, NASA held a competition for an Apollo prototype, choosing the design best suited to their needs. Over the next two years, however, the level of suit mobility did not meet NASA's expectations, and a second competition was announced in June 1965. The three suits submitted in 1965 were made by the David Clark Company (AX1-C), Hamilton Standard and B. F. Goodrich (AX6-H), and ILC Industries (AX5-L), makers of Playtex bras and girdles. ILC Industries had been designing and making the soft parts and restraint systems of spacesuits under contract to Hamilton for years, but this was the first time they were being considered for the role of prime contractor. Their AX5-L soft suit was ultimately selected, and ILC was chosen to head the Apollo suit program. Hamilton continued to make the life-support systems and Air-Lock continued to make the fittings as subcontractors. The first suits worn on the moon were a subsequent version of the competition suit, and had the NASA designation A7-L (A [Apollo], 7 [7th in the series], L [ILC, Prime Contractor]).

However, the competitions and development of the final spacesuit took time, and training for the Apollo missions needed to proceed. Consequently, the suit used during the early Apollo training programs was essentially that used during the Gemini missions, but with a modified helmet. It was given the name A1-C (A [Apollo], 1 [1st in the series], C [David Clark Company, Prime Contractor]). These suits were worn by astronauts Gus Grissom, Ed White, and Roger Chaffee in the tragic *Apollo 1* fire in January 1967.

In October 1968, *Apollo* 7, on a Saturn IB vehicle, became the first of the Apollo missions to be launched. *Apollo* 7, with astronauts Wally Schirra, Donn Eisele, and Walter Cunningham, was an Earth orbital flight lasting over ten days, and was designed to test the spacecraft systems, which had undergone major redesign as a result of the *Apollo 1* fire, and the new A7-L spacesuit, which included fire-retardant Velcro, aluminum, and other fire-retardant measures. The A7-L was a great success. Like the Gemini suits, these soft suits came in both the IV and EV configurations. During the mission, the Apollo astronauts were able to remove their spacesuits, and could either wear their constant-wear garments, which were made of cotton and looked like long underwear, or change into an in-flight coverall, which consisted of a jacket, pair of trousers, and booties, made of a white Teflon fabric that was flame-resistant and had a slippery quality for ease of donning and doffing. The spacesuit was packed in a bag and stowed away.

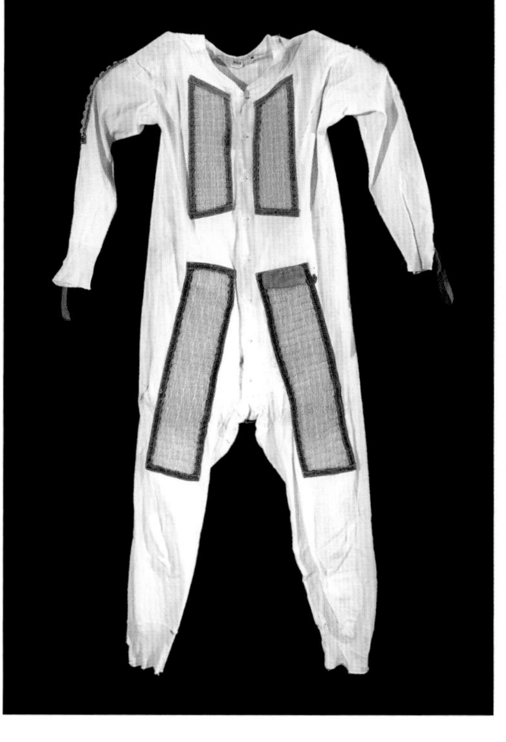

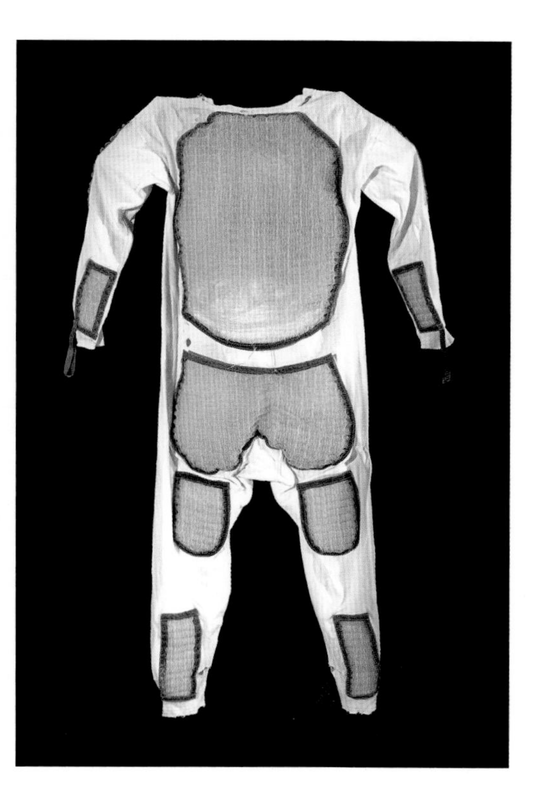

fig. 29
John Glenn's Mercury long underwear
U.S.A., 1962
Cotton knit, Trilok (coiled nylon) spacers
Courtesy of Smithsonian Institution, National Air and Space Museum, 1967-0180-000

John Glenn's Mercury underwear worn on *Freedom 7* in February 1962

The A7-L suits were worn by Neil Armstrong and "Buzz" Aldrin on *Apollo 11*, and by all the astronauts through *Apollo 14*. These spacesuits were made of approximately twenty-six layers, including Teflon-coated Beta cloth, Mylar, Dacron, Kapton, and Chromel-R, a woven stainless steel that provided a high level of protection from micrometeoroids and sharp objects. Like the Gemini suits, they had a pressure-sealing zipper, but they also had a far more complicated inner restraint system, which required a second zipper. In the A7-L suits, this zipper ran from back to front covered with a beavertail flap of insulated Beta cloth.

The last three Apollo missions, *Apollo 15*, *16*, and *17*, were equipped with lunar roving vehicles (LRV) for travel on the moon. The spacesuits worn on these missions were similar in construction to the A7-L, made of the same materials, but modified to enable the astronauts to sit and bend their knees more easily. The most obvious and visible difference between the A7-L and the new suits, called A7-LB, was a wider seat, made so the astronaut could sit without putting undue strain on the zipper. The zipper was therefore repositioned from the back of the suit to the side, which subsequently necessitated moving all the life-support hose connectors. Less visible changes included a convolute of rubber/neoprene and, in the neck, an additional cable that allowed up or down positioning of the helmet with tie-downs to prevent the suit from riding up when the astronaut sat in the rover. These changes made the suit slightly heavier, weighing in at about sixty-five pounds as compared to the fifty-six pounds of the A7-L.

While wearing a spacesuit, maintaining the astronaut's body temperature is of paramount importance. Astronauts during the Mercury program had

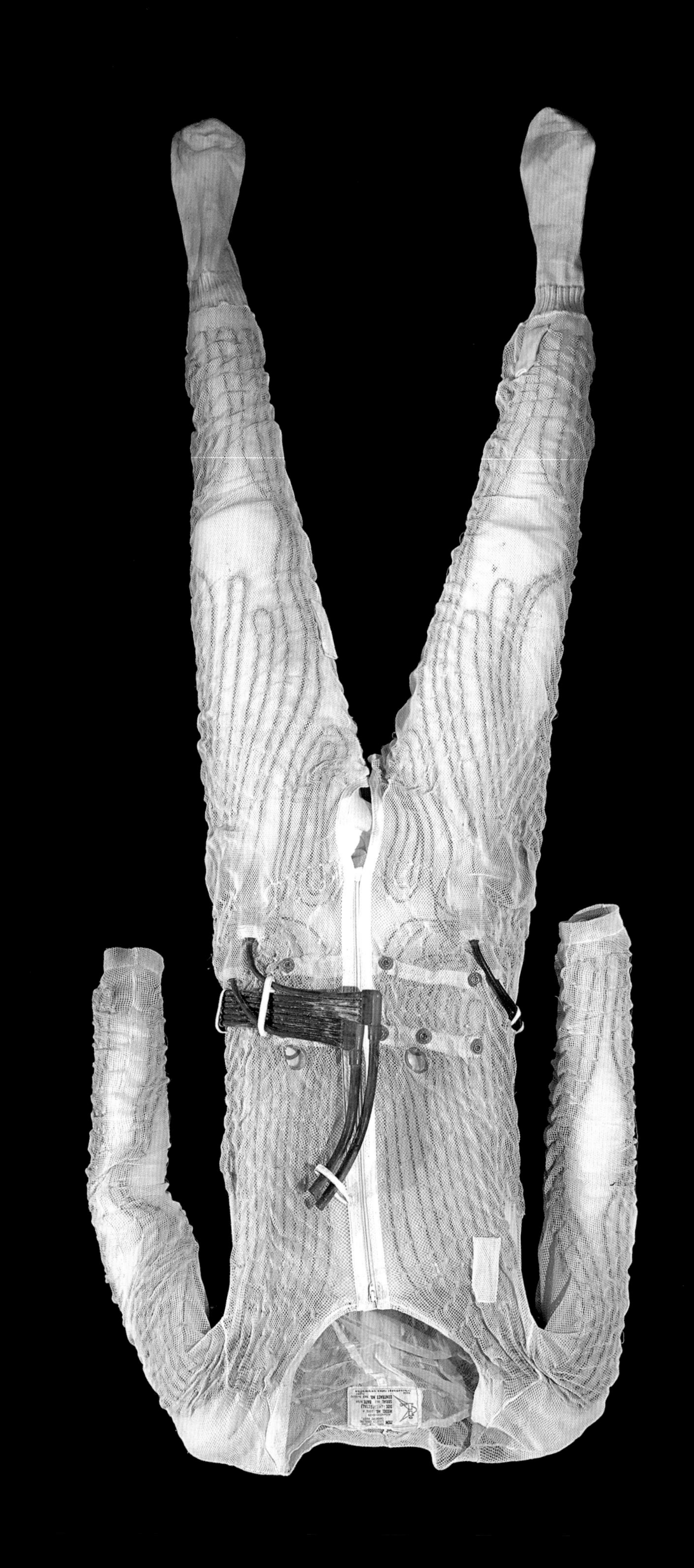

their suits ventilated through an air-hose attachment to the spacecraft. They wore a cotton undergarment with Trilok (coiled nylon fabric) spacers to keep the spacesuit from adhering to the body and preventing air circulation (fig. 29). Gemini astronauts wore a plain, one-piece undergarment made of cotton knit. The spacesuits were ventilated by a series of air channels, the inlet of which was attached to the spacecraft with a hose. The channels were made of stainless-steel springs covered with Lycra mesh and a neoprene cover. With Apollo, the cooling and ventilation system had to be completely self-contained, and the Liquid Cooling Garment (LCG) was developed for this purpose. The LCG is made of nylon spandex mesh with a network of polyvinyl chloride (PVC) tubes running through it at evenly spaced intervals. The lining is of knitted nylon, and the feet of knitted cotton. This garment was powered by the Portable Life Support System (PLSS) while on the lunar surface, or from the spacecraft or lunar module while in those vehicles, and provided a very efficient method of keeping the astronaut cool (fig. 30).

While on the moon, the astronauts were issued specialized equipment to protect them from the sharp rocks, extreme temperature, and UV light. The most memorable of these is the gold-visored helmet seen in all the images of astronauts on the moon. This is actually an over-helmet, weighing about four-and-a-half pounds, and was worn over the pressure bubble helmet. The helmet had two visors — the interior visor was made of UV-stabilized polycarbonate, which provided micrometeoroid and UV protection, and the exterior visor was made of polysulfone with a 24-karat gold coating, that reduced the amount of visible light entering the helmet and limited the heat build-up. The helmets were covered with padded Beta cloth and had a large collar with Velcro closures, also made of Beta cloth. This collar protected the neck-ring disconnect from heating up in the lunar day. Most of these gold-visored helmets were left on the moon, or otherwise expended during the mission, and there are only six in the National Air and Space Museum collection — those worn on *Apollo 11, 15*, and *17*.

The other spacesuit components used on the moon were the EV gloves and boots (fig. 31). While on the lunar surface, the astronauts wore gloves that provided additional insulation and protection. They were known as EV gloves, and were essentially, modified IV gloves, as were worn during launch. The EV gloves weighed less than three pounds, and covered the entire hand and wrist. They were made with an IV glove interior of rubber/neoprene, to which the EV shell of Teflon-coated Beta cloth and Chromel-R, with insulating layers of Mylar, Dacron, and Kapton film, was attached. The blue silicone fingertips enabled the astronaut to have a surprising amount of feeling in his fingers. The gloves were attached to the suit with anodized aluminum disconnect rings — always red for the right glove and blue for the left.

The boots that made the famous footprints on the moon were in fact overshoes worn only while on the lunar surface (fig. 32). There are only two pairs of these boots in the Smithsonian collection that were actually on the moon,

fig. 30
Apollo Liquid Cooling Garment (LCG)
Made by ILC Industries Inc.
U.S.A., 1968
Nylon, polyester, PVC tubing
Courtesy of Smithsonian Institution, National Air and Space Museum, 1973-0120-000

Apollo LCG, the layer worn closest to the skin while outside the spacecraft

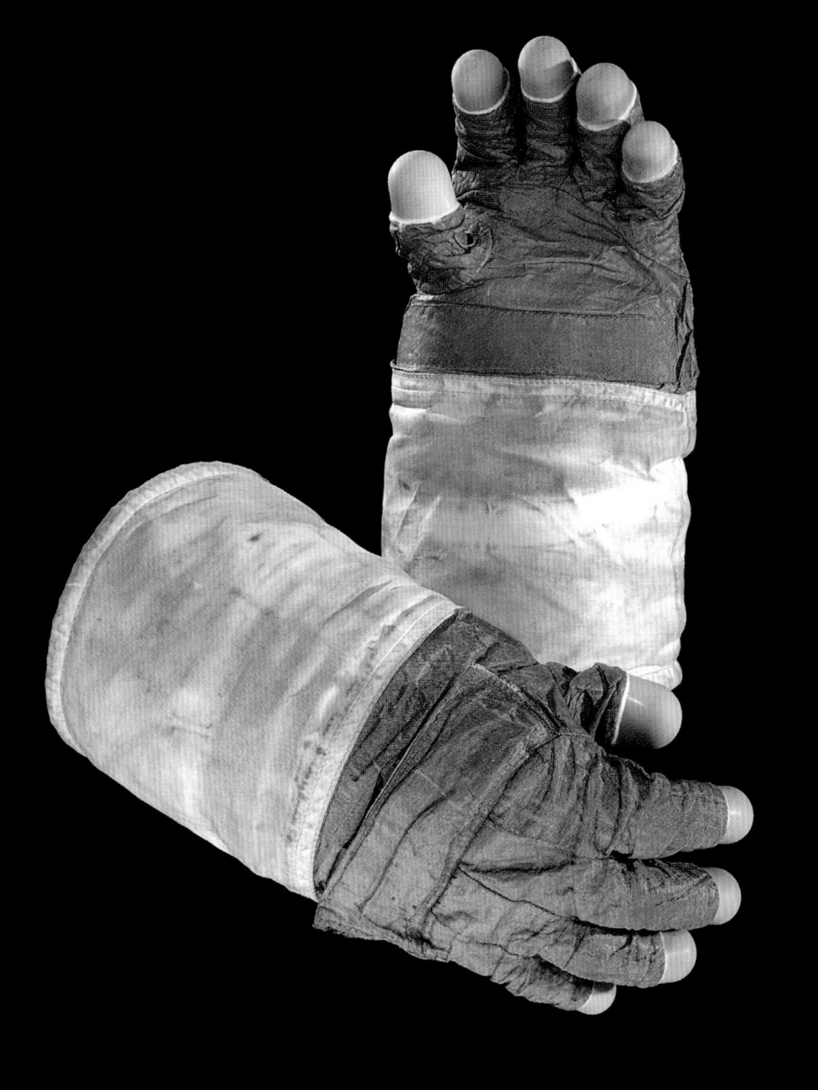

fig. 31 (left)
Charles Duke's lunar gloves
Made by ILC Industries Inc.
U.S.A., 1972
Chromel-R®, Beta cloth, silicone, aluminum
Courtesy of Smithsonian Institution,
National Air and Space Museum,
1974-0150-002/003

Charles Duke was the lunar module pilot of the *Apollo 16* mission in April 1972. His lunar gloves had Chromel-R® hands and Beta cloth gauntlets; the blue fingertips are of silicone.

---

fig. 32 (below)
Harrison Schmitt's lunar boot
Made by ILC Industries Inc.
U.S.A., 1972
Chromel-R®, Beta cloth, silicone
Courtesy of Smithsonian Institution,
National Air and Space Museum,
1974-0183-008

This boot, worn by lunar module pilot Harrison "Jack" Schmitt on *Apollo 17*, is one of two pair of lunar overshoes in the Museum's collection.

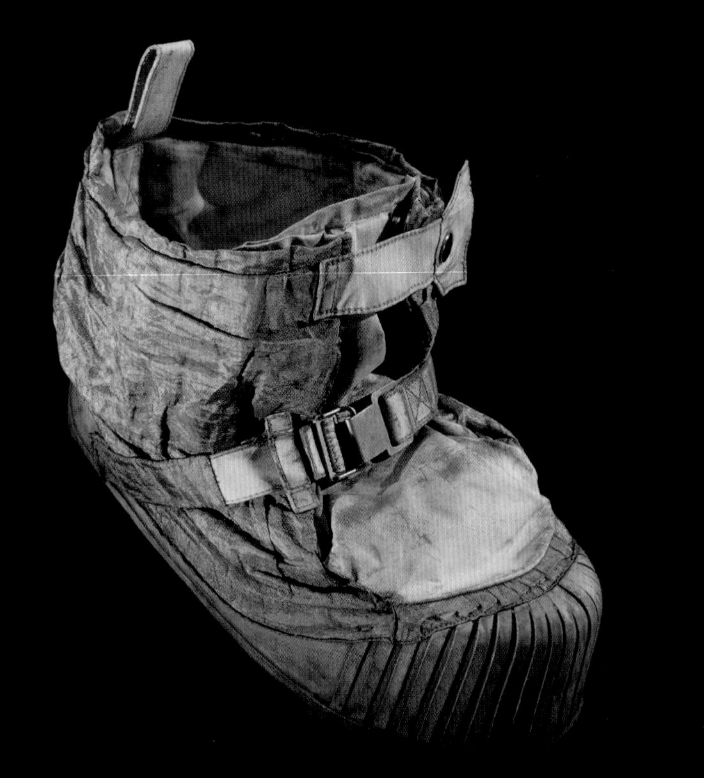

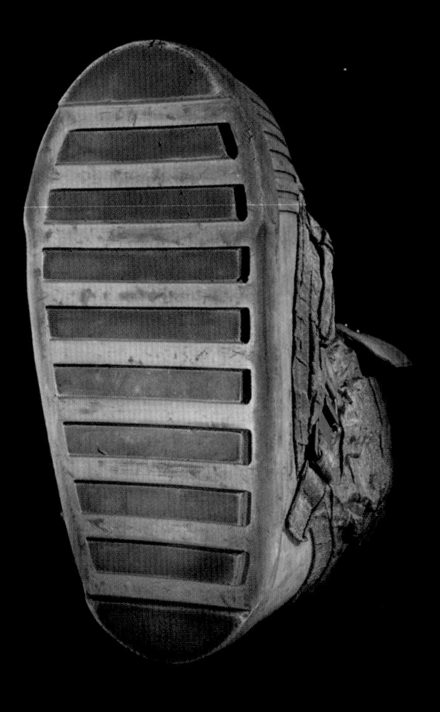

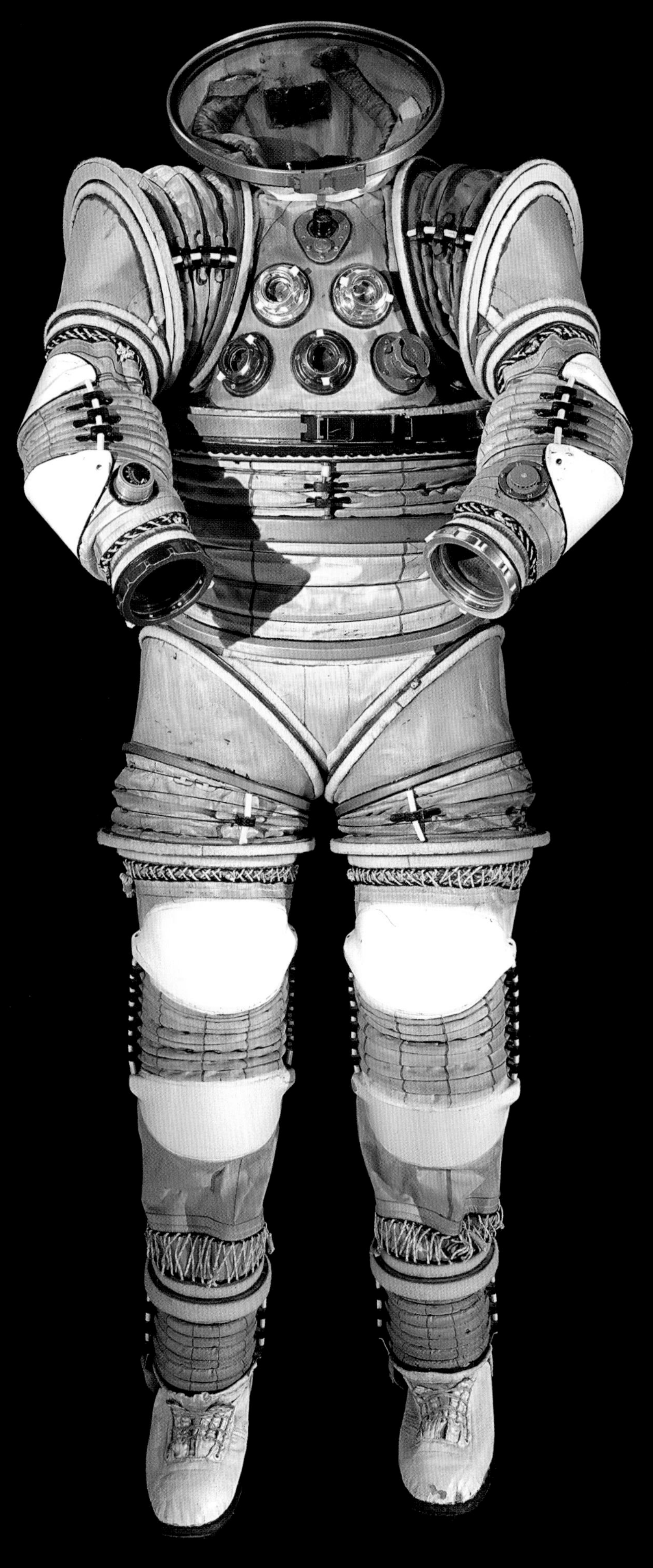

fig. 33
EX1-A laminated fabric suit
Made by the AiResearch Division of
Garret Corporation
(now Honeywell Corporation)
U.S.A., 1968
Nylon, polyester, aluminum, plastic
Courtesy of Smithsonian Institution,
National Air and Space Museum,
1982-0454-000

This Single-Wall Laminate (SWL) suit, designed to be soft while unpressurized and hard when fully pressurized.

those worn by astronauts Eugene Cernan and Harrison "Jack" Schmitt on *Apollo 17*. The others, like the gold-visored helmets, did not return from the mission. The overshoes weighed approximately five pounds and consisted of Chromel-R uppers, with Beta cloth lining, and thirteen layers of aluminized Mylar and Kapton film for insulation and protection. The blue silicone soles had two layers of Beta felt between the layers of Kapton film. These overshoes were designed to protect the spacesuit boots without making it any more difficult to walk or bend the foot.

The last major program of the Apollo era was the Skylab orbital workshop. Originally called the Apollo Applications Program, it was designed to use up leftover pieces of Apollo hardware, test the astronauts' ability to remain in space for extended periods of time, and conduct a series of experiments. Although not the first space station (the Soviets launched *Salyut I* in April 1971), this was the first U.S. space station, and was launched in May 1973—the last time a craft was launched on the *Saturn V* vehicle, which had been used for all the lunar missions. Nine people flew on Skylab, and as these were long-duration missions (the longest was three months), the astronauts needed different garments. They were issued soft in-flight cotton-knit coveralls, to be worn while inside, and a slightly modified A7-LB spacesuit for use outside the workshop.

The Skylab spacesuits were a slightly trimmed down version of the Apollo lunar suits, but otherwise it is difficult to tell them apart (fig. 2, p. 142). The suit had fewer layers, and the gold-visored EV helmet was less bulky and did not have the Beta cloth cover—layer or collar. The astronauts were issued EV gloves, but not EV boots. All nine Skylab suits are in the collection of the National Air and Space Museum.

The final mission of the Apollo era was the highly successful Apollo-Soyuz Test Program (ASTP) that took place in July 1975, in which the U.S. *Apollo* docked with the Soviet *Soyuz 19* spacecraft, and astronaut Tom Stafford and cosmonaut Alexi Leonov shook hands in space. Called the 800 series, the ASTP spacesuits were the final Apollo suits used in space. They were similar to the A7-LB models, but the astronauts were not equipped with gold-visored helmets or EV gloves and boots.

During the course of the space program, research continued in the development of full-pressure suits. The EX1-A, developed for NASA by the AiResearch Corporation in 1968, is a Single-Wall Laminate (SWL) suit, and the only one of this type made (fig. 33). It was constructed of two layers of fabric bonded together to create the single, laminated material, with joints similar to those in other constant volume suits. The EX1-A was designed to be soft while unpressurized and thus easily stowable in the *Apollo* spacecraft. When the suit was pressurized, it became hard and maintained a constant volume. This dual construction was a distinct advantage, as storage space inside the spacecraft was at a minimum. It used Gemini gloves and boots and had a dome-shaped helmet.

## LOSSARY OF AJOR TEXTILES

**ETA CLOTH:** Teflon-coated, small-iameter glass fiber yarn, tightly woven nd used as a flame-resistant layer in pacesuits; developed by the Owens Corning Fiberglass Corporation in Ashton, hode Island, under contract to NASA.

**ETA MARQUISETTE:** nonwoven Teflon-oated textile, used as thermal insulation nd a flame-resistant spacer between pacesuit layers; developed by the Owens Corning Fiberglass Corporation under ontract to NASA.

**HROMEL-R:** single-draw (nonbraided) hromium/steel, woven into fabric, used s a protective layer in selected areas on pacesuits, gloves, and boots; developed y Fabrics Research in Dedham, Massachusetts, under contract to NASA.

**YCOLAC:** thermo-plastic resin known as ABS Resin, used for its high-impact quali-ies and flame-retardant properties; made y GE Plastics, and used by David Clark Company as shields for the A1-C helmets.

**MYLAR:** polymer material used as light-weight, thermal insulation; perforated-luminized Mylar film was developed by the National Research Corporation in Cambridge, Massachusetts.

**NOMEX:** high-performance heat- and flame-resistant textile used for protective clothing, such as the outer layer of spacesuits. In sheet form it is used in high-temperature applications, generators, hoses, etc. In honeycomb form it is used in aircraft; developed by the DuPont Corporation in Wilmington, Delaware.

**TEFLON:** polymer material used for thermal radiative and flame resistance; developed by the DuPont Corporation, fabric manufactured by Stern and Stern Textiles in Hornell, New York.

**VELCRO:** fire-retardant Velcro made with a polyester hook, and Teflon loop. The ground tape to which the hook and loop was attached was made of Beta glass yarns; manufactured by the Velcro Corporation in Manchester, New Hampshire, under contract to NASA and known as Astro Velcro.

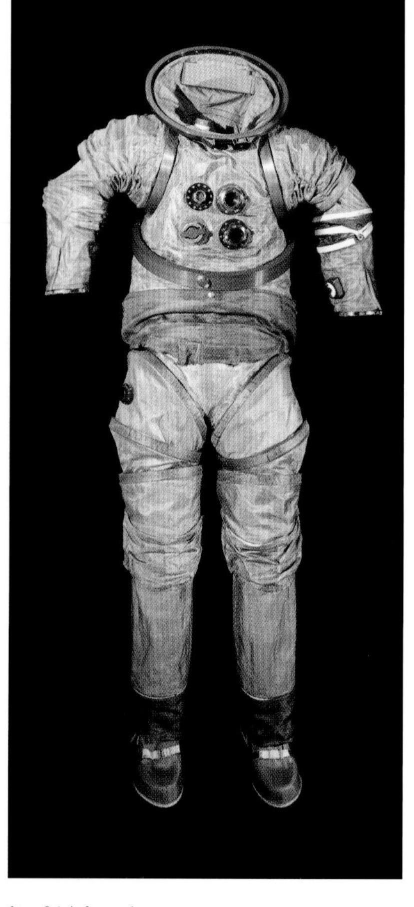

fig. 34 (above)
Advanced Extra-vehicular Suit (AES), Chromel-R® cover
Made by Litton Industries
U.S.A., 1967–68
Chromel-R®, Beta cloth, nylon, aluminum
Courtesy of Smithsonian Institution, National Air and Space Museum, 1982-0455-000

This hybrid AES was made by Litton Industries, and is the only one known to have been configured with a Chromel-R® cover layer.

Another area of research was the Advanced Extra-vehicular Suit (AES) with a Chromel-R exterior layer. Made by Litton Industries in the mid-1960s, it is a hybrid, with a hard upper torso, waist bearing, and partially hard legs (fig. 34). It also experimented with a Chromel-R and fiberglass cover layer. Chromel-R was first used in the Gemini program as a protective textile, however, it was very expensive, difficult to work with, and left no room for error during suit construction.

The spacesuits developed for and used during the U.S. space programs of the 1960s and 1970s were the first of the truly extreme garments, and required the development of extraordinary materials. The suits combined modern materials and natural fibers. The life spans of these compounds were unknown. Spacesuit pressure bladders, for example, were made from a 70% natural rubber and 30% neoprene compound that has deteriorated considerably over time. Current spacesuits use no natural materials at all, but their long-term survival is also unknown. During the development of these spacesuits, other uses were uncovered for the new materials, as discussed in Cara McCarty's essay. The spacesuit research undertaken by NASA continues to influence advanced clothing design—their Ames Research Center has found its way into garments for underwater exploration, and Mars suit studies still build on those used in the very first lunar exploration (fig. 37). The nylon and glues are employed in dentistry and other medical applications, and the science that these suits have enabled the astronauts and specialists to perform in space is just beginning to be appreciated.

The Astronaut Personal Equipment collection of the Department of Space History at the National Air and Space Museum consists of approximately 2,500 objects. It includes examples of the equipment that was assigned to the astronauts during the Mercury, Gemini, and Apollo programs, along with the Skylab missions. This equipment consists of spacesuits and associated components like helmets, boots, gloves, and other clothing, such as the in-flight and constant-wear garments. The museum also holds personal life-support systems and maneuvering units (such as the one used by Ed White on *Gemini 4*), chronographs, and smaller objects, like personal hygiene items, food, and the dollar bills signed by the astronauts and flown on each mission. The collection also includes prototypes and experimental artifacts such as developmental pressure suits, from some of the very earliest high-altitude suits to the constant-volume Advanced Extra-vehicular Suits.

Caring for this collection has its own particular set of problems and rewards. The materials in spacesuits range from natural cotton and woven stainless steel (Chromel-R) to plastics, fiberglass, and aluminum. In the past, it was believed that spacesuits "had withstood the extremes of space, and so would last forever," but spacesuits and the materials of which they are made have become fragile with time, and need to be displayed and stored in specialized temperatures, light, and humidity in order to preserve them for future generations.

fig. 35 (left)
Penguin-3 suit
Manufactured by Zvezda
Russia, 1996
Synthetic fabric with elastic inserts
Courtesy of Smithsonian Institution,
National Air and Space Museum,
1997-0200-000

American astronaut Shannon Lucid wore this suit during her record-breaking six-month stay on board the Russian *Mir* space station from March to September 1996. The suit is designed to ameliorate the debilitating effects of prolonged weightlessness by providing artificial stress on the muscular-skeletal system. The inside of the suit contains a system of elastic bands that can simulate some of the effects of gravity, aiding in the retention of muscle mass and calcium in the bones, when used in conjunction with a rigorous exercise program. The effectiveness of the Penguin-3 suit was demonstrated when Dr. Lucid emerged from the space shuttle *Atlantis* able to walk after six months in orbit. The Russian designers are adapting the suit for therapeutic use by children with cerebral palsy.

fig. 36 (below)
Astronaut Shannon Lucid, Cosmonaut
Yuriy-V Usachov, and Flight
Engineer/Cosmonaut Yuriy-I
Onuufriyenko, 1996

fig. 37
I-Suit Mars- or lunar-surface operations spacesuit prototype
Manufactured by ILC Dover Inc.
U.S.A., 2000
Thermally sealed urethane-coated nylon bladder for gas retention, sewn polyester restraint to support structural loads, stainless-steel rotary bearings

In development at ILC Dover, the I-Suit is a high-mobility suit for future manned missions to Mars, incorporating electronic textiles for enhanced functionality. Surface exploration of Mars is expected to include the use of an assistant robotic rover, and ILC Dover worked with Softswitch Ltd. to incorporate pressure-sensitive textile switches in the gauntlet of the glove for basic rover control, as well as in the shoulder, allowing control of the suit's helmet-mounted lights. Other e-textile-enhanced functions under consideration are embedded communications antennas, physiological monitoring, environmental sensors, and actuators for strength-augmented mobility joints.

MMMMMMMMMMMMMMMMMMMMMMMMMMMMMMMMMMMMMMMMMMMMMMMMMMMMMMMMMMMMMMMMMMMMMMMMMMMMMMMMMMMMMMMMMMMMMMMMMMMM
AAAAAAAAAAAAAAAAAAAAAAAAAAAAAAAAAAAAAAAAAAAAAAAAAAAAAAAAAAAAAAAAAAAAAAAAAAAAAAAAAAAAAAAAAAAAAAAAAAAA
RRRRRRRRRRRRRRRRRRRRRRRRRRRRRRRRRRRRRRRRRRRRRRRRRRRRRRRRRRRRRRRRRRRRRRRRRRRRRRRRRRRRRRRRRRRRRRRRRRRR
TTTTTTTTTTTTTTTTTTTTTTTTTTTTTTTTTTTTTTTTTTTTTTTTTTTTTTTTTTTTTTTTTTTTTTTTTTTTTTTTTTTTTTTTTTTTTTTTTTTT
EEEEEEEEEEEEEEEEEEEEEEEEEEEEEEEEEEEEEEEEEEEEEEEEEEEEEEEEEEEEEEEEEEEEEEEEEEEEEEEEEEEEEEEEEEEEEEEEEEEE
RRRRRRRRRRRRRRRRRRRRRRRRRRRRRRRRRRRRRRRRRRRRRRRRRRRRRRRRRRRRRRRRRRRRRRRRRRRRRRRRRRRRRRRRRRRRRRRRRRRR
SSSSSSSSSSSSSSSSSSSSSSSSSSSSSSSSSSSSSSSSSSSSSSSSSSSSSSSSSSSSSSSSSSSSSSSSSSSSSSSSSSSSSSSSSSSSSSSSSSSS
MMMMMMMMMMMMMMMMMMMMMMMMMMMMMMMMMMMMMMMMMMMMMMMMMMMMMMMMMMMMMMMMMMMMMMMMMMMMMMMMMMMMMMMMMMMMMMMMMMMM
AAAAAAAAAAAAAAAAAAAAAAAAAAAAAAAAAAAAAAAAAAAAAAAAAAAAAAAAAAAAAAAAAAAAAAAAAAAAAAAAAAAAAAAAAAAAAAAAAAAA
RRRRRRRRRRRRRRRRRRRRRRRRRRRRRRRRRRRRRRRRRRRRRRRRRRRRRRRRRRRRRRRRRRRRRRRRRRRRRRRRRRRRRRRRRRRRRRRRRRRR
TTTTTTTTTTTTTTTTTTTTTTTTTTTTTTTTTTTTTTTTTTTTTTTTTTTTTTTTTTTTTTTTTTTTTTTTTTTTTTTTTTTTTTTTTTTTTTTTTTTT
EEEEEEEEEEEEEEEEEEEEEEEEEEEEEEEEEEEEEEEEEEEEEEEEEEEEEEEEEEEEEEEEEEEEEEEEEEEEEEEEEEEEEEEEEEEEEEEEEEEE
RRRRRRRRRRRRRRRRRRRRRRRRRRRRRRRRRRRRRRRRRRRRRRRRRRRRRRRRRRRRRRRRRRRRRRRRRRRRRRRRRRRRRRRRRRRRRRRRRRRR
SSSSSSSSSSSSSSSSSSSSSSSSSSSSSSSSSSSSSSSSSSSSSSSSSSSSSSSSSSSSSSSSSSSSSSSSSSSSSSSSSSSSSSSSSSSSSSSSSSSS
MMMMMMMMMMMMMMMMMMMMMMMMMMMMMMMMMMMMMMMMMMMMMMMMMMMMMMMMMMMMMMMMMMMMMMMMMMMMMMMMMMMMMMMMMMMMMMMMMMMM
AAAAAAAAAAAAAAAAAAAAAAAAAAAAAAAAAAAAAAAAAAAAAAAAAAAAAAAAAAAAAAAAAAAAAAAAAAAAAAAAAAAAAAAAAAAAAAAAAAAA
RRRRRRRRRRRRRRRRRRRRRRRRRRRRRRRRRRRRRRRRRRRRRRRRRRRRRRRRRRRRRRRRRRRRRRRRRRRRRRRRRRRRRRRRRRRRRRRRRRRR
TTTTTTTTTTTTTTTTTTTTTTTTTTTTTTTTTTTTTTTTTTTTTTTTTTTTTTTTTTTTTTTTTTTTTTTTTTTTTTTTTTTTTTTTTTTTTTTTTTTT
EEEEEEEEEEEEEEEEEEEEEEEEEEEEEEEEEEEEEEEEEEEEEEEEEEEEEEEEEEEEEEEEEEEEEEEEEEEEEEEEEEEEEEEEEEEEEEEEEEEE
RRRRRRRRRRRRRRRRRRRRRRRRRRRRRRRRRRRRRRRRRRRRRRRRRRRRRRRRRRRRRRRRRRRRRRRRRRRRRRRRRRRRRRRRRRRRRRRRRRRR
SSSSSSSSSSSSSSSSSSSSSSSSSSSSSSSSSSSSSSSSSSSSSSSSSSSSSSSSSSSSSSSSSSSSSSSSSSSSSSSSSSSSSSSSSSSSSSSSSSSS
MMMMMMMMMMMMMMMMMMMMMMMMMMMMMMMMMMMMMMMMMMMMMMMMMMMMMMMMMMMMMMMMMMMMMMMMMMMMMMMMMMMMMMMMMMMMMMMMMMMM
AAAAAAAAAAAAAAAAAAAAAAAAAAAAAAAAAAAAAAAAAAAAAAAAAAAAAAAAAAAAAAAAAAAAAAAAAAAAAAAAAAAAAAAAAAAAAAAAAAAA
RRRRRRRRRRRRRRRRRRRRRRRRRRRRRRRRRRRRRRRRRRRRRRRRRRRRRRRRRRRRRRRRRRRRRRRRRRRRRRRRRRRRRRRRRRRRRRRRRRRR
TTTTTTTTTTTTTTTTTTTTTTTTTTTTTTTTTTTTTTTTTTTTTTTTTTTTTTTTTTTTTTTTTTTTTTTTTTTTTTTTTTTTTTTTTTTTTTTTTTTT
EEEEEEEEEEEEEEEEEEEEEEEEEEEEEEEEEEEEEEEEEEEEEEEEEEEEEEEEEEEEEEEEEEEEEEEEEEEEEEEEEEEEEEEEEEEEEEEEEEEE
RRRRRRRRRRRRRRRRRRRRRRRRRRRRRRRRRRRRRRRRRRRRRRRRRRRRRRRRRRRRRRRRRRRRRRRRRRRRRRRRRRRRRRRRRRRRRRRRRRR
SSSSSSSSSSSSSSSSSSSSSSSSSSSSSSSSSSSSSSSSSSSSSSSSSSSSSSSSSSSSSSSSSSSSSSSSSSSSSSSSSSSSSSSSSSSSSSSSSSSS
MMMMMMMMMMMMMMMMMMMMMMMMMMMMMMMMMMMMMMMMMMMMMMMMMMMMMMMMMMMMMMMMMMMMMMMMMMMMMMMMMMMMMMMMMMMMMMMMMMMM
AAAAAAAAAAAAAAAAAAAAAAAAAAAAAAAAAAAAAAAAAAAAAAAAAAAAAAAAAAAAAAAAAAAAAAAAAAAAAAAAAAAAAAAAAAAAAAAAAAAA
RRRRRRRRRRRRRRRRRRRRRRRRRRRRRRRRRRRRRRRRRRRRRRRRRRRRRRRRRRRRRRRRRRRRRRRRRRRRRRRRRRRRRRRRRRRRRRRRRRRR
TTTTTTTTTTTTTTTTTTTTTTTTTTTTTTTTTTTTTTTTTTTTTTTTTTTTTTTTTTTTTTTTTTTTTTTTTTTTTTTTTTTTTTTTTTTTTTTTTTTT
EEEEEEEEEEEEEEEEEEEEEEEEEEEEEEEEEEEEEEEEEEEEEEEEEEEEEEEEEEEEEEEEEEEEEEEEEEEEEEEEEEEEEEEEEEEEEEEEEEEE
RRRRRRRRRRRRRRRRRRRRRRRRRRRRRRRRRRRRRRRRRRRRRRRRRRRRRRRRRRRRRRRRRRRRRRRRRRRRRRRRRRRRRRRRRRRRRRRRRRRR
SSSSSSSSSSSSSSSSSSSSSSSSSSSSSSSSSSSSSSSSSSSSSSSSSSSSSSSSSSSSSSSSSSSSSSSSSSSSSSSSSSSSSSSSSSSSSSSSSSSS
MMMMMMMMMMMMMMMMMMMMMMMMMMMMMMMMMMMMMMMMMMMMMMMMMMMMMMMMMMMMMMMMMMMMMMMMMMMMMMMMMMMMMMMMMMMMMMMMMMMM
AAAAAAAAAAAAAAAAAAAAAAAAAAAAAAAAAAAAAAAAAAAAAAAAAAAAAAAAAAAAAAAAAAAAAAAAAAAAAAAAAAAAAAAAAAAAAAAAAAAA
RRRRRRRRRRRRRRRRRRRRRRRRRRRRRRRRRRRRRRRRRRRRRRRRRRRRRRRRRRRRRRRRRRRRRRRRRRRRRRRRRRRRRRRRRRRRRRRRRRR
TTTTTTTTTTTTTTTTTTTTTTTTTTTTTTTTTTTTTTTTTTTTTTTTTTTTTTTTTTTTTTTTTTTTTTTTTTTTTTTTTTTTTTTTTTTTTTTTTTTT
EEEEEEEEEEEEEEEEEEEEEEEEEEEEEEEEEEEEEEEEEEEEEEEEEEEEEEEEEEEEEEEEEEEEEEEEEEEEEEEEEEEEEEEEEEEEEEEEEEEE
RRRRRRRRRRRRRRRRRRRRRRRRRRRRRRRRRRRRRRRRRRRRRRRRRRRRRRRRRRRRRRRRRRRRRRRRRRRRRRRRRRRRRRRRRRRRRRRRRRR
SSSSSSSSSSSSSSSSSSSSSSSSSSSSSSSSSSSSSSSSSSSSSSSSSSSSSSSSSSSSSSSSSSSSSSSSSSSSSSSSSSSSSSSSSSSSSSSSSSSS
MMMMMMMMMMMMMMMMMMMMMMMMMMMMMMMMMMMMMMMMMMMMMMMMMMMMMMMMMMMMMMMMMMMMMMMMMMMMMMMMMMMMMMMMMMMMMMMMMMMM
AAAAAAAAAAAAAAAAAAAAAAAAAAAAAAAAAAAAAAAAAAAAAAAAAAAAAAAAAAAAAAAAAAAAAAAAAAAAAAAAAAAAAAAAAAAAAAAAAAAA
RRRRRRRRRRRRRRRRRRRRRRRRRRRRRRRRRRRRRRRRRRRRRRRRRRRRRRRRRRRRRRRRRRRRRRRRRRRRRRRRRRRRRRRRRRRRRRRRRRR
TTTTTTTTTTTTTTTTTTTTTTTTTTTTTTTTTTTTTTTTTTTTTTTTTTTTTTTTTTTTTTTTTTTTTTTTTTTTTTTTTTTTTTTTTTTTTTTTTTTT
EEEEEEEEEEEEEEEEEEEEEEEEEEEEEEEEEEEEEEEEEEEEEEEEEEEEEEEEEEEEEEEEEEEEEEEEEEEEEEEEEEEEEEEEEEEEEEEEEEEE
RRRRRRRRRRRRRRRRRRRRRRRRRRRRRRRRRRRRRRRRRRRRRRRRRRRRRRRRRRRRRRRRRRRRRRRRRRRRRRRRRRRRRRRRRRRRRRRRRRR
SSSSSSSSSSSSSSSSSSSSSSSSSSSSSSSSSSSSSSSSSSSSSSSSSSSSSSSSSSSSSSSSSSSSSSSSSSSSSSSSSSSSSSSSSSSSSSSSSSSS
MMMMMMMMMMMMMMMMMMMMMMMMMMMMMMMMMMMMMMMMMMMMMMMMMMMMMMMMMMMMMMMMMMMMMMMMMMMMMMMMMMMMMMMMMMMMMMMMMMMM
AAAAAAAAAAAAAAAAAAAAAAAAAAAAAAAAAAAAAAAAAAAAAAAAAAAAAAAAAAAAAAAAAAAAAAAAAAAAAAAAAAAAAAAAAAAAAAAAAAAA
RRRRRRRRRRRRRRRRRRRRRRRRRRRRRRRRRRRRRRRRRRRRRRRRRRRRRRRRRRRRRRRRRRRRRRRRRRRRRRRRRRRRRRRRRRRRRRRRRRR
TTTTTTTTTTTTTTTTTTTTTTTTTTTTTTTTTTTTTTTTTTTTTTTTTTTTTTTTTTTTTTTTTTTTTTTTTTTTTTTTTTTTTTTTTTTTTTTTTTTT
EEEEEEEEEEEEEEEEEEEEEEEEEEEEEEEEEEEEEEEEEEEEEEEEEEEEEEEEEEEEEEEEEEEEEEEEEEEEEEEEEEEEEEEEEEEEEEEEEEEE
RRRRRRRRRRRRRRRRRRRRRRRRRRRRRRRRRRRRRRRRRRRRRRRRRRRRRRRRRRRRRRRRRRRRRRRRRRRRRRRRRRRRRRRRRRRRRRRRRRR
SSSSSSSSSSSSSSSSSSSSSSSSSSSSSSSSSSSSSSSSSSSSSSSSSSSSSSSSSSSSSSSSSSSSSSSSSSSSSSSSSSSSSSSSSSSSSSSSSSSS
MMMMMMMMMMMMMMMMMMMMMMMMMMMMMMMMMMMMMMMMMMMMMMMMMMMMMMMMMMMMMMMMMMMMMMMMMMMMMMMMMMMMMMMMMMMMMMMMMMMM
AAAAAAAAAAAAAAAAAAAAAAAAAAAAAAAAAAAAAAAAAAAAAAAAAAAAAAAAAAAAAAAAAAAAAAAAAAAAAAAAAAAAAAAAAAAAAAAAAAAA
RRRRRRRRRRRRRRRRRRRRRRRRRRRRRRRRRRRRRRRRRRRRRRRRRRRRRRRRRRRRRRRRRRRRRRRRRRRRRRRRRRRRRRRRRRRRRRRRRRR
TTTTTTTTTTTTTTTTTTTTTTTTTTTTTTTTTTTTTTTTTTTTTTTTTTTTTTTTTTTTTTTTTTTTTTTTTTTTTTTTTTTTTTTTTTTTTTTTTTTT
EEEEEEEEEEEEEEEEEEEEEEEEEEEEEEEEEEEEEEEEEEEEEEEEEEEEEEEEEEEEEEEEEEEEEEEEEEEEEEEEEEEEEEEEEEEEEEEEEEEE
RRRRRRRRRRRRRRRRRRRRRRRRRRRRRRRRRRRRRRRRRRRRRRRRRRRRRRRRRRRRRRRRRRRRRRRRRRRRRRRRRRRRRRRRRRRRRRRRRRR
SSSSSSSSSSSSSSSSSSSSSSSSSSSSSSSSSSSSSSSSSSSSSSSSSSSSSSSSSSSSSSSSSSSSSSSSSSSSSSSSSSSSSSSSSSSSSSSSSSSS
MMMMMMMMMMMMMMMMMMMMMMMMMMMMMMMMMMMMMMMMMMMMMMMMMMMMMMMMMMMMMMMMMMMMMMMMMMMMMMMMMMMMMMMMMMMMMMMMMMMM
AAAAAAAAAAAAAAAAAAAAAAAAAAAAAAAAAAAAAAAAAAAAAAAAAAAAAAAAAAAAAAAAAAAAAAAAAAAAAAAAAAAAAAAAAAAAAAAAAAAA
RRRRRRRRRRRRRRRRRRRRRRRRRRRRRRRRRRRRRRRRRRRRRRRRRRRRRRRRRRRRRRRRRRRRRRRRRRRRRRRRRRRRRRRRRRRRRRRRRRR
TTTTTTTTTTTTTTTTTTTTTTTTTTTTTTTTTTTTTTTTTTTTTTTTTTTTTTTTTTTTTTTTTTTTTTTTTTTTTTTTTTTTTTTTTTTTTTTTTTTT
EEEEEEEEEEEEEEEEEEEEEEEEEEEEEEEEEEEEEEEEEEEEEEEEEEEEEEEEEEEEEEEEEEEEEEEEEEEEEEEEEEEEEEEEEEEEEEEEEEEE
RRRRRRRRRRRRRRRRRRRRRRRRRRRRRRRRRRRRRRRRRRRRRRRRRRRRRRRRRRRRRRRRRRRRRRRRRRRRRRRRRRRRRRRRRRRRRRRRRRR
SSSSSSSSSSSSSSSSSSSSSSSSSSSSSSSSSSSSSSSSSSSSSSSSSSSSSSSSSSSSSSSSSSSSSSSSSSSSSSSSSSSSSSSSSSSSSSSSSSSS
MMMMMMMMMMMMMMMMMMMMMMMMMMMMMMMMMMMMMMMMMMMMMMMMMMMMMMMMMMMMMMMMMMMMMMMMMMMMMMMMMMMMMMMMMMMMMMMMMMMM
AAAAAAAAAAAAAAAAAAAAAAAAAAAAAAAAAAAAAAAAAAAAAAAAAAAAAAAAAAAAAAAAAAAAAAAAAAAAAAAAAAAAAAAAAAAAAAAAAAAA
RRRRRRRRRRRRRRRRRRRRRRRRRRRRRRRRRRRRRRRRRRRRRRRRRRRRRRRRRRRRRRRRRRRRRRRRRRRRRRRRRRRRRRRRRRRRRRRRRRR
TTTTTTTTTTTTTTTTTTTTTTTTTTTTTTTTTTTTTTTTTTTTTTTTTTTTTTTTTTTTTTTTTTTTTTTTTTTTTTTTTTTTTTTTTTTTTTTTTTT
EEEEEEEEEEEEEEEEEEEEEEEEEEEEEEEEEEEEEEEEEEEEEEEEEEEEEEEEEEEEEEEEEEEEEEEEEEEEEEEEEEEEEEEEEEEEEEEEEEEE
RRRRRRRRRRRRRRRRRRRRRRRRRRRRRRRRRRRRRRRRRRRRRRRRRRRRRRRRRRRRRRRRRRRRRRRRRRRRRRRRRRRRRRRRRRRRRRRRRRR
SSSSSSSSSSSSSSSSSSSSSSSSSSSSSSSSSSSSSSSSSSSSSSSSSSSSSSSSSSSSSSSSSSSSSSSSSSSSSSSSSSSSSSSSSSSSSSSSSSSS
MMMMMMMMMMMMMMMMMMMMMMMMMMMMMMMMMMMMMMMMMMMMMMMMMMMMMMMMMMMMMMMMMMMMMMMMMMMMMMMMMMMMMMMMMMMMMMMMMMMM
AAAAAAAAAAAAAAAAAAAAAAAAAAAAAAAAAAAAAAAAAAAAAAAAAAAAAAAAAAAAAAAAAAAAAAAAAAAAAAAAAAAAAAAAAAAAAAAAAAAA
RRRRRRRRRRRRRRRRRRRRRRRRRRRRRRRRRRRRRRRRRRRRRRRRRRRRRRRRRRRRRRRRRRRRRRRRRRRRRRRRRRRRRRRRRRRRRRRRRRR
TTTTTTTTTTTTTTTTTTTTTTTTTTTTTTTTTTTTTTTTTTTTTTTTTTTTTTTTTTTTTTTTTTTTTTTTTTTTTTTTTTTTTTTTTTTTTTTTTTTT
EEEEEEEEEEEEEEEEEEEEEEEEEEEEEEEEEEEEEEEEEEEEEEEEEEEEEEEEEEEEEEEEEEEEEEEEEEEEEEEEEEEEEEEEEEEEEEEEEEEE
RRRRRRRRRRRRRRRRRRRRRRRRRRRRRRRRRRRRRRRRRRRRRRRRRRRRRRRRRRRRRRRRRRRRRRRRRRRRRRRRRRRRRRRRRRRRRRRRRRR
SSSSSSSSSSSSSSSSSSSSSSSSSSSSSSSSSSSSSSSSSSSSSSSSSSSSSSSSSSSSSSSSSSSSSSSSSSSSSSSSSSSSSSSSSSSSSSSSSSSS
MMMMMMMMMMMMMMMMMMMMMMMMMMMMMMMMMMMMMMMMMMMMMMMMMMMMMMMMMMMMMMMMMMMMMMMMMMMMMMMMMMMMMMMMMMMMMMMMMMMM
AAAAAAAAAAAAAAAAAAAAAAAAAAAAAAAAAAAAAAAAAAAAAAAAAAAAAAAAAAAAAAAAAAAAAAAAAAAAAAAAAAAAAAAAAAAAAAAAAAAA
RRRRRRRRRRRRRRRRRRRRRRRRRRRRRRRRRRRRRRRRRRRRRRRRRRRRRRRRRRRRRRRRRRRRRRRRRRRRRRRRRRRRRRRRRRRRRRRRRRR
TTTTTTTTTTTTTTTTTTTTTTTTTTTTTTTTTTTTTTTTTTTTTTTTTTTTTTTTTTTTTTTTTTTTTTTTTTTTTTTTTTTTTTTTTTTTTTTTTTTT
EEEEEEEEEEEEEEEEEEEEEEEEEEEEEEEEEEEEEEEEEEEEEEEEEEEEEEEEEEEEEEEEEEEEEEEEEEEEEEEEEEEEEEEEEEEEEEEEEEEE
RRRRRRRRRRRRRRRRRRRRRRRRRRRRRRRRRRRRRRRRRRRRRRRRRRRRRRRRRRRRRRRRRRRRRRRRRRRRRRRRRRRRRRRRRRRRRRRRRRR
SSSSSSSSSSSSSSSSSSSSSSSSSSSSSSSSSSSSSSSSSSSSSSSSSSSSSSSSSSSSSSSSSSSSSSSSSSSSSSSSSSSSSSSSSSSSSSSSSSSS
MMMMMMMMMMMMMMMMMMMMMMMMMMMMMMMMMMMMMMMMMMMMMMMMMMMMMMMMMMMMMMMMMMMMMMMMMMMMMMMMMMMMMMMMMMMMMMMMMMMM
AAAAAAAAAAAAAAAAAAAAAAAAAAAAAAAAAAAAAAAAAAAAAAAAAAAAAAAAAAAAAAAAAAAAAAAAAAAAAAAAAAAAAAAAAAAAAAAAAAAA
RRRRRRRRRRRRRRRRRRRRRRRRRRRRRRRRRRRRRRRRRRRRRRRRRRRRRRRRRRRRRRRRRRRRRRRRRRRRRRRRRRRRRRRRRRRRRRRRRRR
TTTTTTTTTTTTTTTTTTTTTTTTTTTTTTTTTTTTTTTTTTTTTTTTTTTTTTTTTTTTTTTTTTTTTTTTTTTTTTTTTTTTTTTTTTTTTTTTTTTT
EEEEEEEEEEEEEEEEEEEEEEEEEEEEEEEEEEEEEEEEEEEEEEEEEEEEEEEEEEEEEEEEEEEEEEEEEEEEEEEEEEEEEEEEEEEEEEEEEEEE
RRRRRRRRRRRRRRRRRRRRRRRRRRRRRRRRRRRRRRRRRRRRRRRRRRRRRRRRRRRRRRRRRRRRRRRRRRRRRRRRRRRRRRRRRRRRRRRRRRR
SSSSSSSSSSSSSSSSSSSSSSSSSSSSSSSSSSSSSSSSSSSSSSSSSSSSSSSSSSSSSSSSSSSSSSSSSSSSSSSSSSSSSSSSSSSSSSSSSSSS
MMMMMMMMMMMMMMMMMMMMMMMMMMMMMMMMMMMMMMMMMMMMMMMMMMMMMMMMMMMMMMMMMMMMMMMMMMMMMMMMMMMMMMMMMMMMMMMMMMMM
AAAAAAAAAAAAAAAAAAAAAAAAAAAAAAAAAAAAAAAAAAAAAAAAAAAAAAAAAAAAAAAAAAAAAAAAAAAAAAAAAAAAAAAAAAAAAAAAAAAA
RRRRRRRRRRRRRRRRRRRRRRRRRRRRRRRRRRRRRRRRRRRRRRRRRRRRRRRRRRRRRRRRRRRRRRRRRRRRRRRRRRRRRRRRRRRRRRRRRRR
TTTTTTTTTTTTTTTTTTTTTTTTTTTTTTTTTTTTTTTTTTTTTTTTTTTTTTTTTTTTTTTTTTTTTTTTTTTTTTTTTTTTTTTTTTTTTTTTTTT
EEEEEEEEEEEEEEEEEEEEEEEEEEEEEEEEEEEEEEEEEEEEEEEEEEEEEEEEEEEEEEEEEEEEEEEEEEEEEEEEEEEEEEEEEEEEEEEEEEEE
RRRRRRRRRRRRRRRRRRRRRRRRRRRRRRRRRRRRRRRRRRRRRRRRRRRRRRRRRRRRRRRRRRRRRRRRRRRRRRRRRRRRRRRRRRRRRRRRRRR
SSSSSSSSSSSSSSSSSSSSSSSSSSSSSSSSSSSSSSSSSSSSSSSSSSSSSSSSSSSSSSSSSSSSSSSSSSSSSSSSSSSSSSSSSSSSSSSSSSSS
MMMMMMMMMMMMMMMMMMMMMMMMMMMMMMMMMMMMMMMMMMMMMMMMMMMMMMMMMMMMMMMMMMMMMMMMMMMMMMMMMMMMMMMMMMMMMMMMMMMM
AAAAAAAAAAAAAAAAAAAAAAAAAAAAAAAAAAAAAAAAAAAAAAAAAAAAAAAAAAAAAAAAAAAAAAAAAAAAAAAAAAAAAAAAAAAAAAAAAAAA
RRRRRRRRRRRRRRRRRRRRRRRRRRRRRRRRRRRRRRRRRRRRRRRRRRRRRRRRRRRRRRRRRRRRRRRRRRRRRRRRRRRRRRRRRRRRRRRRRRR
TTTTTTTTTTTTTTTTTTTTTTTTTTTTTTTTTTTTTTTTTTTTTTTTTTTTTTTTTTTTTTTTTTTTTTTTTTTTTTTTTTTTTTTTTTTTTTTTTTTT
EEEEEEEEEEEEEEEEEEEEEEEEEEEEEEEEEEEEEEEEEEEEEEEEEEEEEEEEEEEEEEEEEEEEEEEEEEEEEEEEEEEEEEEEEEEEEEEEEEEE
RRRRRRRRRRRRRRRRRRRRRRRRRRRRRRRRRRRRRRRRRRRRRRRRRRRRRRRRRRRRRRRRRRRRRRRRRRRRRRRRRRRRRRRRRRRRRRRRRRR
SSSSSSSSSSSSSSSSSSSSSSSSSSSSSSSSSSSSSSSSSSSSSSSSSSSSSSSSSSSSSSSSSSSSSSSSSSSSSSSSSSSSSSSSSSSSSSSSSSSS
MMMMMMMMMMMMMMMMMMMMMMMMMMMMMMMMMMMMMMMMMMMMMMMMMMMMMMMMMMMMMMMMMMMMMMMMMMMMMMMMMMMMMMMMMMMMMMMMMMMM
AAAAAAAAAAAAAAAAAAAAAAAAAAAAAAAAAAAAAAAAAAAAAAAAAAAAAAAAAAAAAAAAAAAAAAAAAAAAAAAAAAAAAAAAAAAAAAAAAAAA
RRRRRRRRRRRRRRRRRRRRRRRRRRRRRRRRRRRRRRRRRRRRRRRRRRRRRRRRRRRRRRRRRRRRRRRRRRRRRRRRRRRRRRRRRRRRRRRRRRR
TTTTTTTTTTTTTTTTTTTTTTTTTTTTTTTTTTTTTTTTTTTTTTTTTTTTTTTTTTTTTTTTTTTTTTTTTTTTTTTTTTTTTTTTTTTTTTTTTTT

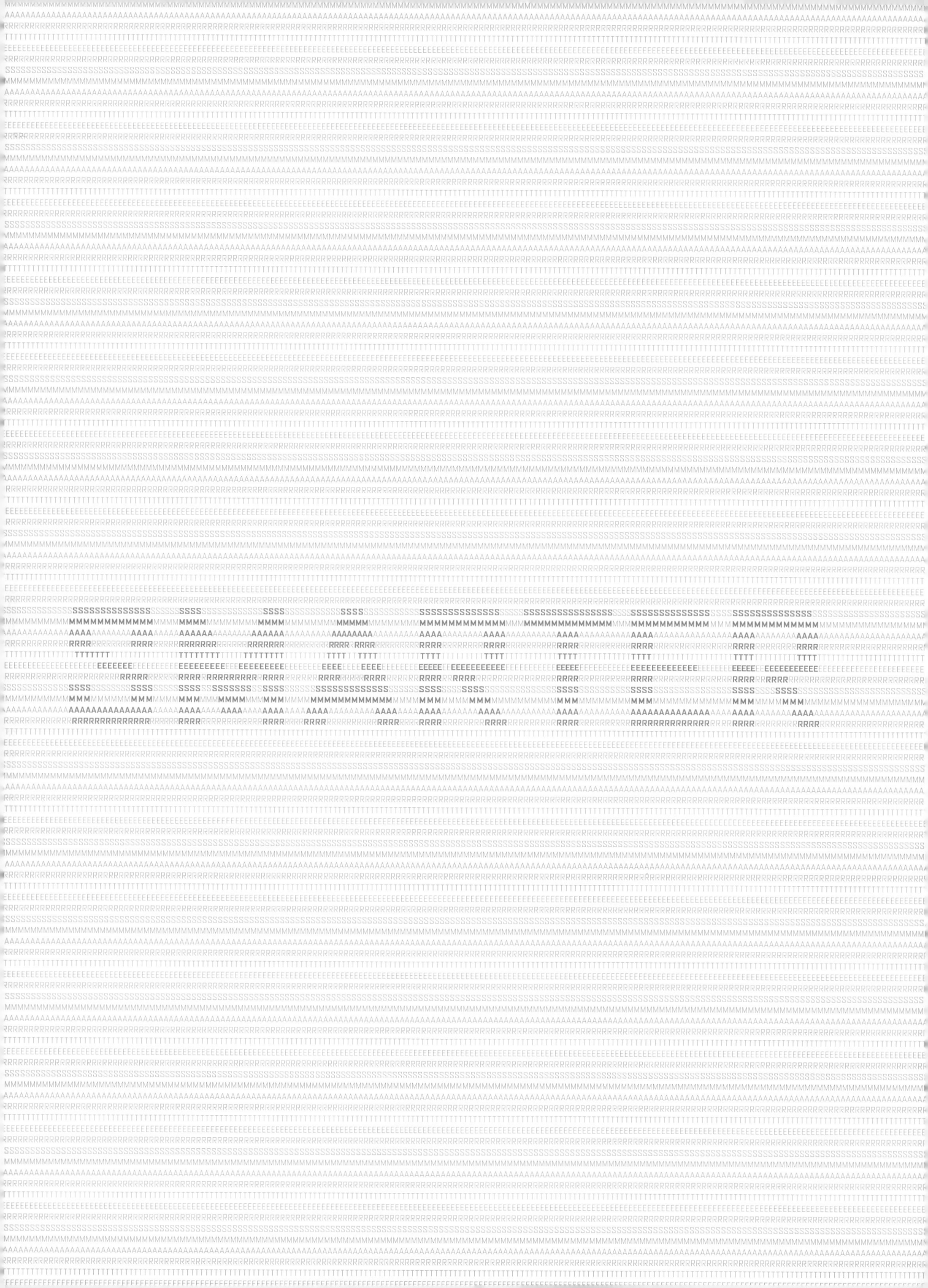

fig. 1
Electronic harness
Manufactured by Woven Electronics
U.S.A., 2003
Plain woven synthetic fibers

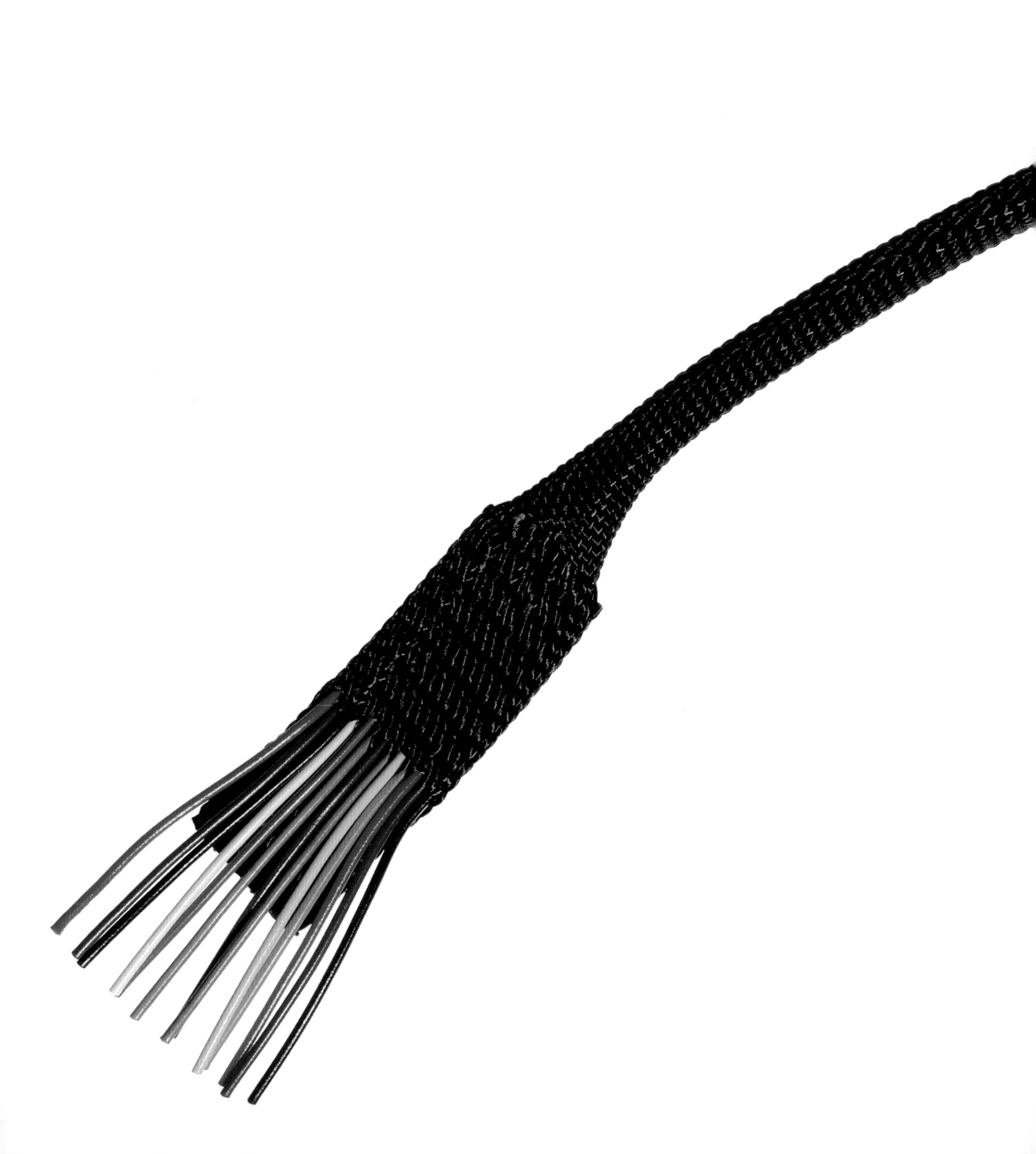

Patricia Wilson

# TEXTILES FROM NOVEL MEANS OF INNOVATION

Real innovations and leaps in technology and design are not the result of incremental improvements. Rather, they are the result of disruptive thinking by innovators, born of radical experience or challenges. To expand their thought processes, often the innovators must be put into contact with disparate ideas removed from their fields of expertise. This essay will examine advancements in smart textiles and systems and show how many develop from nontraditional professions joining together to make new discoveries. Many times it is the formation of communities of artists, designers, historians, scientists, and engineers, who challenge each other to dream of impossible possibilities for textiles, that have culminated in these products.

These impossible possibilities have led to new generations of textiles that begin to define what a smart textile could and will become. Often the adjective *smart* is applied to a material or system that is able to sense its surroundings and respond with an appropriate action. This could be a textile that recognizes that a child is sitting in a car and sends a signal to the computer to disable the airbag, or a fabric that senses a person's irregular heartbeat and helps the system respond with the needed medical alert. The most extreme version of this vision is a giant textile that could create its own power, sense the surroundings, and react through computation and sending of information. At first, these ideas seem the essence of science fiction, but over the last decade, researchers have been making steady progress experimenting with new fibers and yarns, and incorporating them into textile structures, to make these ideas a reality. In fact, a small number of products are already reaching the marketplace, enabling companies to introduce the marriage between fabrics and electronics to the consumer.

A number of terms have been coined to describe these types of smart textiles, including e-textiles, intelligent textiles, electronic textiles, and wearables. A textile is made up of individual fibers combined into yarns. These yarns are then woven, braided, knitted, embroidered, or felted to make a multi-dimensional object. With component yarns, it is possible to make fibers that have a range of conductivities via several methods, such as an all-metal fiber, a metal- or conductive-oxide coating, or a narrow metallic ribbon wrapping. The conductive fibers can be combined with either like or unlike fibers to make yarns with unique properties, producing the textile analog of a wire. These yarns can then be made into conductive fabrics or fabrics that have specific conductive traces—much like an electronic breadboard. Just as in high school electronics class, connecting certain conductive yarns together and adding electronic components such as processors, resistors, light emitting diodes (LEDs), and batteries can result in a multitude of devices—but with textiles these devices are mostly soft to the touch.

Companies such as ITN and Konarka are attempting to reach the next logical step and turn as many electronic components as possible into yarns themselves. ITN is working on a yarn coated with multiple solid conductive and electrolyte layers, which would act as a bendable battery. Konarka, a flexible solar-cell company, is developing solar-cell fibers that could charge battery fibers, enabling objects such as tents to collect the energy needed by the inhabitants when in remote locations, away from conventional power sources. With the advent of other efficient solid-state technologies such as blue and organic LEDs and electroluminescent pigments, many light-generation possibilities can be envisioned that use conductive fibers as a component in the manufacture. Therefore, the potential in the next decade of truly soft electronics is open to the designer's imagination.

The emergence of the third world as a powerhouse in the production of inexpensive and high-quality textiles and textile products has forced the West and Japan to rethink how they can add value to their textile industries. Smart textiles, which rely on the strengths of these countries to innovate in electronics and materials, have seemed like the perfect answer to stem the outflow of textile-manufacturing jobs. Combined with new generations of mobile, high-bandwidth technologies and the consumers' increasing acceptance of technology and its role in their lives, textiles are the newest frontier of the high-tech revolution. The current information age is driving devices to be smaller, more ubiquitous, and design savvy, complete with fashionable colors. Therefore, the connection between electronics and textiles starts to become a natural evolution. The following dreamers and their products are on the leading front of this revolution in textiles, which promises significant advancement in technology over the next decade. As usually happens with exceptional innovation, the spark that ignited this revolution was the posing of a truly simple but radical question: "Why do electronics have to be hard?"

### ARTIST AS MUSE

An early proponent of mixing artists, designers, scientists, and engineers to create a hotbed of innovation was Rich Gold of Xerox's Palo Alto Research Center (PARC). PARC has long been known for its free-wheeling progressive atmosphere, allowing scientists and engineers to explore their creativity with the hope that their efforts will yield products with commercial potential. Gold was himself an avant-garde music composer before he became a games designer and entered the computer industry. He felt that both the artist and scientist share a common belief in the impossible. Although these professions appear to be on the opposite side of the logic-creativity spectrum, Gold recognized that they often use a similar language to communicate their ideas: they are both highly visual, comfortable with the abstract, and focus on the unknown. Gold founded the PARC Artist-in-Residence Program (PAIR) in 1993 with the idea that if you put creative people together, no matter what their field, innovation will naturally occur. The PAIR program brings artists who use new media to PARC and joins them with researchers who use the same media, but with a different goal, to great creative result. The vision at PARC is not a solitary experiment, but an early foray as part of a larger movement to enable innovation through the cooperation of varied disciplines. Many other companies and universities have found enormous value in incorporating the artistic viewpoint as a catalyst to their traditional manufacturing process, inciting the evolution of the smart textile field.

Much of the beginnings of the smart textile field grew out of these diverse creative environments. One example is the MIT Media Lab. Another unique forum for creativity, the lab was formed in 1980 by Professor Nicholas Negroponte and former MIT President Jerome Wiesner. The lab began with a desire to blur the boundaries between traditional disciplines in academia and industry, with a focus on electronic content and digital technologies. Known around the world for its creative atmosphere and innovative development of new technologies, the lab quickly embraced students who were as interested in art as they were in engineering. Maggie Orth, a sculpture artist and graduate of Rhode Island School of Design, joined the lab as a student in 1997. Through the Opera for the Future group at the Media Lab, Orth explored interactive means for nonmusicians, such as children, to compose music. Using textiles to develop tactilely rich, nonthreatening input devices, Orth experimented with the embroidery of hair-thin stainless-steel fibers and yarns produced by Bekaert (fig. 2). Another student, Rehmi Post, interested

fig. 2
Musical Shapers
Designed and fabricated by Maggie Orth, Ph.D., music by Gili Weinberg, for Tod Machover's Toy Symphony, produced at MIT Media Lab
U.S.A., 2000–01
Natural and synthetic fibers, conductive yarns

in interfaces for wearable computers, joined Orth in fabricating a series of switches, keyboards, and other devices, using various techniques from embroidery to quilting. They realized early in their work that capacitive sensing, the principle behind most keyboards, could be used with textiles to register an electronic signal by simply touching an embroidered conductive patch. To produce more complicated devices, they used metallic fabrics woven for Indian saris as a means of connecting microprocessors to fabric. Through the Media Lab's industrial-academic partnerships and a seminal fashion show highlighting some of their prototypes, this early work in electronic textiles excited a host of companies to explore the intersection of fashion, design, textiles, and electronics (fig. 3).

After graduation, Orth founded International Fashion Machines (IFM) to produce flexible electronic art and develop new technology to bring these concepts to consumer products. Her work has expanded to new forms of

fig. 3 (above left)
Firefly Dress
Designed by Maggie Orth, Ph.D., Emily Cooper, and Derek Lockwood; produced at MIT Media Lab
U.S.A., 1998
Metallic silk organza, LEDs, conductive Velcro®, electronics

fig. 4 (above right)
Electronic tablecloth
Designed by Maggie Orth, Ph.D., created in collaboration with Rehmi Post; produced at MIT Media Lab
U.S.A., 1999
Linen, conductive yarns, electronics, RFID tags

information display through textiles. Orth describes her exploration of dynamically controlled textile printing as a natural result of her frustrations during graduate school at MIT: "All of the projects, such as the electronic tablecloth, were hampered by the need to display the output of the electronic textile (fig. 4). At the time, there wasn't any way to escape the need for a hard electronic display." Fueled by her desire to express change through the textile platform, she experimented with thermochromic pigments, whose color changes with heat. The existing palette of thermochromics was limited, but Orth knew that through color mixing a larger range could be produced. By layering the pigments, using textile-printing technologies, on top of an electronic textile with resistors woven into patterns, a dynamic change of the patterns could be generated using software. Orth dubbed this technology Electric Plaid. The latest incarnation of Electric Plaid, called Dynamic Double Weave, is inspired by early woven New England coverlets, and has applications for future home furnishings (fig. 5).

IFM has further expanded tactile textiles into the home-electronics market with the introduction of StitchSwitch, embroidered and woven light switches.

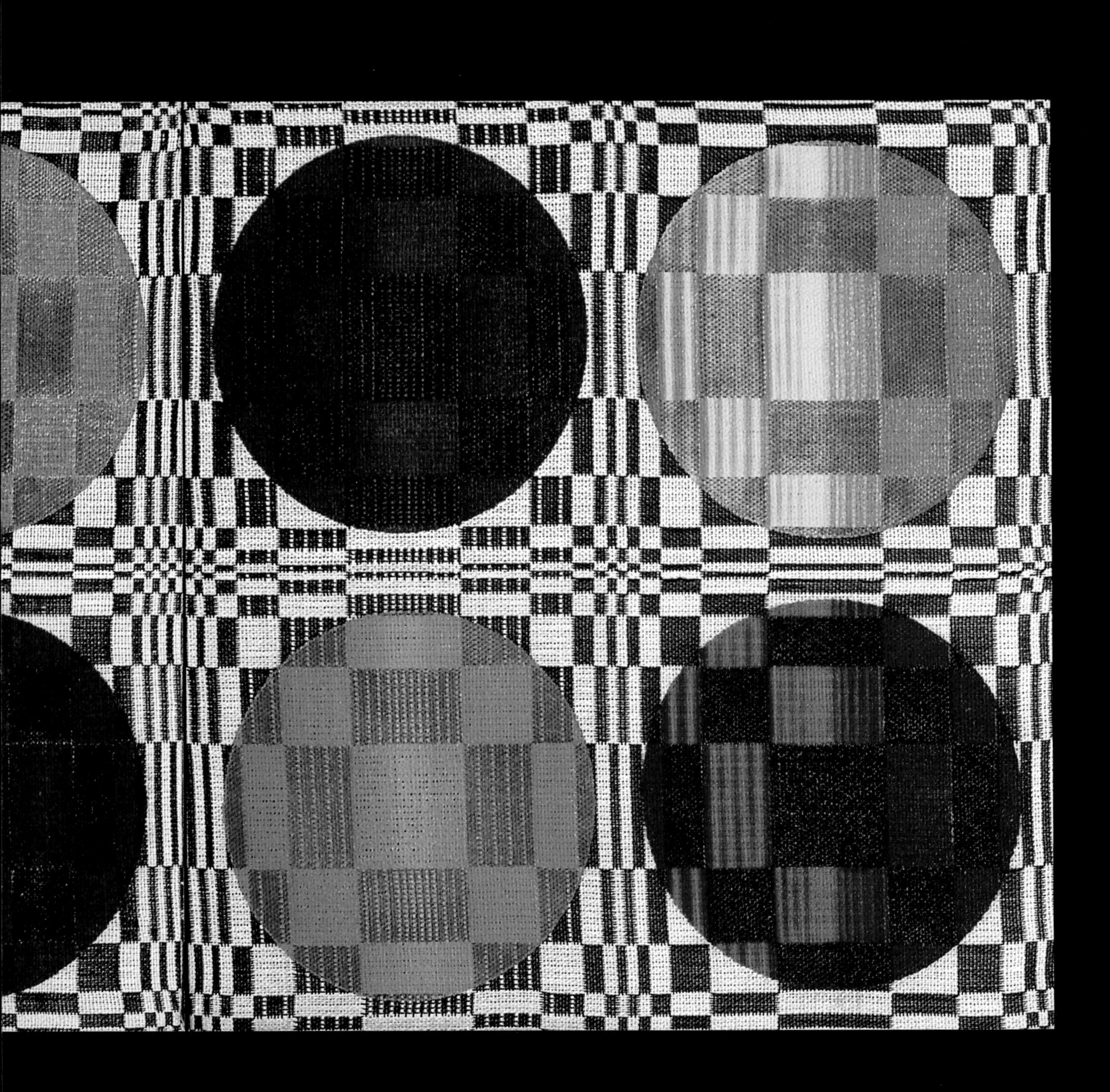

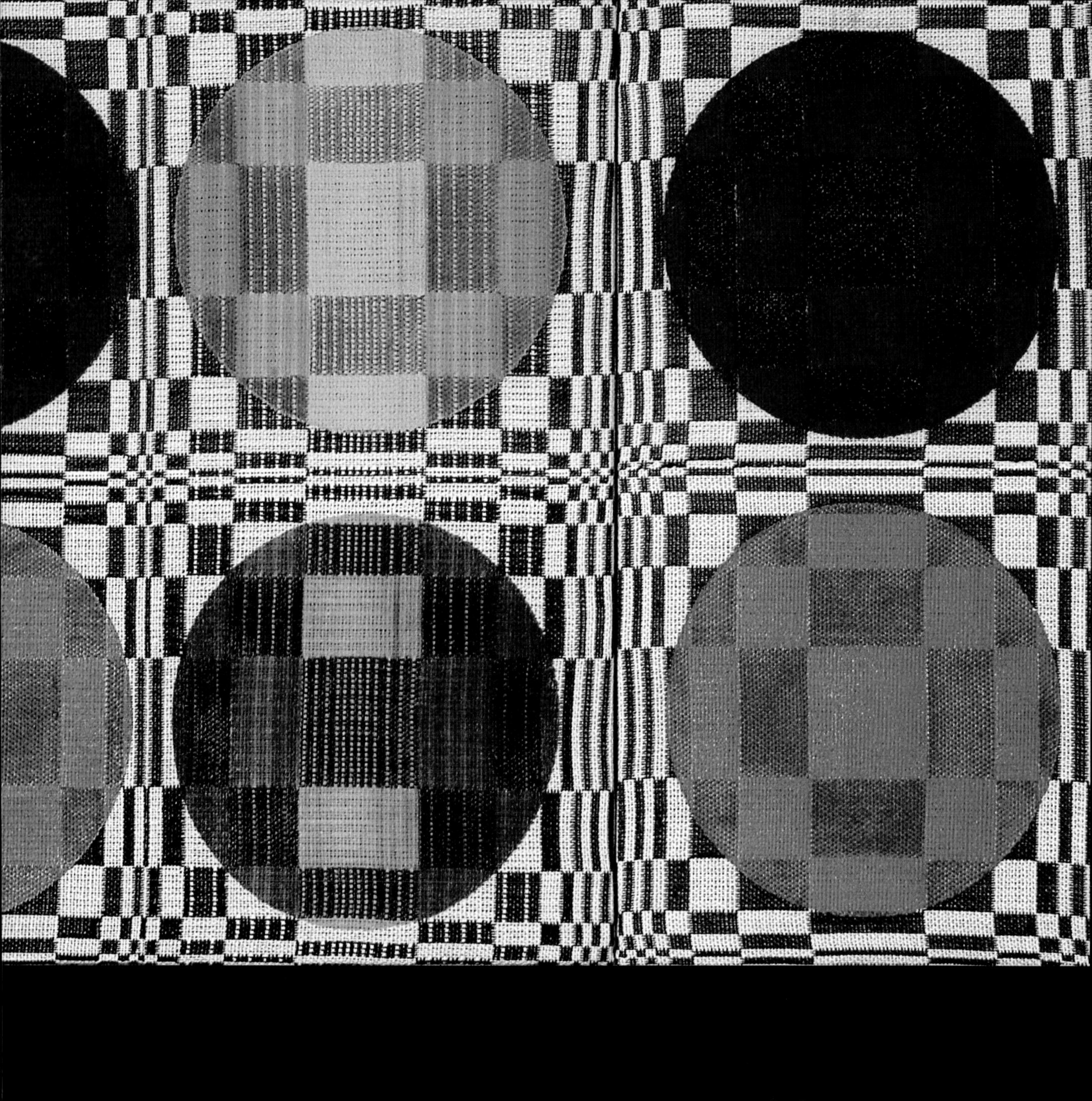

fig. 5.
Dynamic Double Weave
Designed by Maggie Orth, Ph.D., manufactured by International Fashion Machines Inc.
U.S.A., 2002
Cotton, rayon, conductive yarns, thermochromic inks, drive electronics

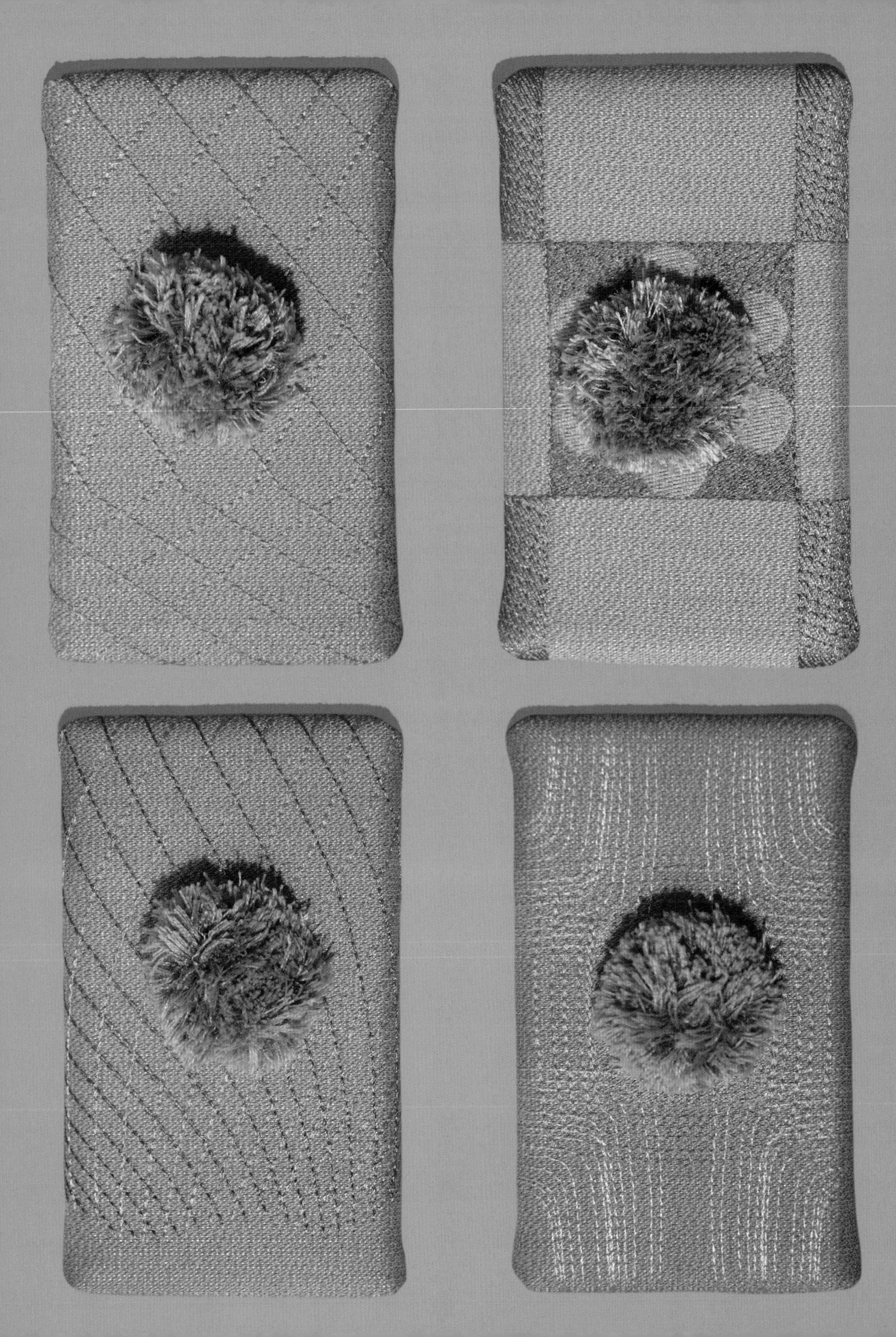

fig. 6 (facing page)
Textile switches
Designed by Maggie Orth, Ph.D., manufactured by International Fashion Machines Inc.
U.S.A., 2005
Sensors made from embroidered or knotted conductive yarns

These touch sensors allow control of lights or other home-electronic devices through a pleasing soft interface, such as the fuzzy pom-pom switch (fig. 6).

The incorporation of smart textiles into consumer electronics is also being explored at Philips. In the 1990s, Philips Design began envisioning, through its project Vision of the Future, the integration of technological functions into clothing, furniture, and environments. Later, Philips Design and Philips Research initiated a program, led by social scientist Clive van Heerden, to gather diverse groups of professionals together to inspire the future of consumer electronics. One of these projects, stimulated by early work at the MIT Media Lab on wearable computing, focused on personal electronics. Fashion designers, fabric printers, embroiderers, spinners, tailors, pattern makers, and industrial designers were placed with electrical engineers, computer scientists, and antenna experts to generate new, more dynamic products. The group proved to be the nexus of creativity that Rich Gold of PARC had predicted. At first, Philips's engineers reacted to challenging questions from the designers, such as "Why does an antenna have to be hard?," with incredulity. Redefining traditionally held ideas, however, the engineers provided solutions that used conductive textiles to form soft and conformal antennas. Respect developed between the disciplines, and a unique exploration in soft electronics resulted. In 2000, many of these explorations were published in the book *New Nomads*.

For the collaboration, it was important that the product concepts contained function while maintaining comfort and style. Among the many projects presented in *New Nomads*, the Feels Good concept is a perfect example of fashion and function (fig. 7). The conductive embroidered spine that runs along the back of the kimono dress is used to disperse an electrostatic charge via fibers close to the skin. This creates a tingling sensation, which induces relaxation. The level of stimulation is controlled by biometric sensors that monitor the relaxation of the wearer. This project has potential as a springboard to medical monitoring and drug delivery through the electrical stimulation of the skin using clothing.

fig. 7 (above)
Feels Good concept, *New Nomads* project
Developed and executed by Philips Design
The Netherlands, 2000

The design philosophy and the technological expertise that were developed around the New Nomads project led to commercial ventures that incorporated Philips's knowledge into other brands. An early product of Philips Design that combined electronics and consumer clothing was the Levi's-Philips jacket. This garment was one of the first to include in its design a means to discreetly hold a cell phone, MP3 player, and headphone network. While no electronic textiles were involved in this product, the electronics were customized to fit the jacket, an unusual step forward. The commercial success of this product excited many in the fashion and electronics world and prompted additional investigations.

One such company inspired by Philips's success is Infineon, a leader in microprocessor design. In trying to adapt their products into wearables, Infineon has tested the durability of their processors to resist damage during

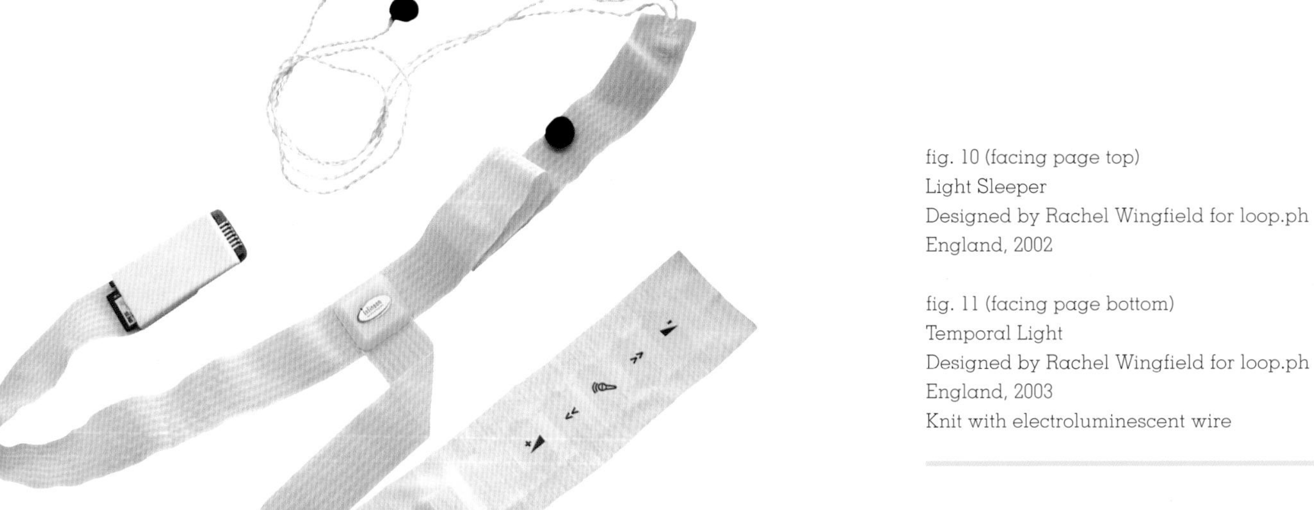

fig. 10 (facing page top)
Light Sleeper
Designed by Rachel Wingfield for loop.ph
England, 2002

fig. 11 (facing page bottom)
Temporal Light
Designed by Rachel Wingfield for loop.ph
England, 2003
Knit with electroluminescent wire

fig. 8 (above)
Components for a voice controlled MP3 player embedded into a jacket
Designed by Infineon Technologies with the German Master School of Fashion

fig. 9
Installation of fiber-optic textiles by Laurie Carlson
U.S.A., 1998
Foreground: *Passing Through...Hung Between Heaven and Earth* in white light (152 x 46 cm diameter [5 ft. x 18 in.]); background: *The River* shown in color light (183 x 61 x 15 cm [6 ft. x 24 in. x 6 in.])

washing, and modified their packaging to the needs of electronic textiles. Their garment-system demonstrations and initial product introductions of a more integrated MP3 jacket, developed with O'Neil, have given the entire smart-textile community a means to further incorporate consumer electronics and textiles (fig. 8).

Not every innovation in smart textiles has a direct link to the consumer. In some cases, artistic drive creates the impetus to explore the boundary between electronics and fabrics. Fiberartist Laurie Carlson philosophically investigates "total internal reflection" with her woven sculptures engineered with plastic optical lighting fibers (fig. 9). The placement of fiber optics and metallic yarns in hand-woven structures allows her to use materials that usually signify communications in a unique visual expression. Light streams from the textile in patterns through the controlled abrasion of the optical fibers. Carlson also collaborates with the Acoustic Laboratory in the Advanced Technology and Marketing Center at the University of Massachusetts at Dartmouth. Concept investigations include work with LED-sourced light connectors and Bragg-grated fibers used for monitoring stress and strain in woven composites. The Bragg grating acts like a mile marker etched within a single yarn. Theoretically, this characteristic permits location identification within a fabric grid structure, aiding in the detection of compromised areas. Her expertise in weaving dissimilar materials and coupling light into the fibers is invaluable to advancing research and future applications. Similarly, France Telecom weaves abraded fiber optics into simple displays that can be worn or used as flexible signage.

Rachel Wingfield of loop.ph is also an artist-turned-engineer, who creates reactive, luminous surfaces and objects by exploiting electroluminescence. Her designs are strongly influenced by the need for innovative solutions to relieve sufferers of Seasonal Affective Disorder (SAD). As an artist and designer, she acknowledges that light has a profound impact on our emotional and physiological state. The absence of light can cause

severe medical problems for those with extreme SAD, such as loss of energy, weight gain, migraines, and depression. Her design proposal to help remedy irregular circadian rhythms is to incorporate light generation into bedding. Her Light Sleeper is a set of cushions and a duvet embroidered with electroluminescent wire and silk (fig. 10). The bedding can simulate an artificial sunrise, and help the user reset his or her body clock. Several other projects explore the use of electroluminescent wire in unique textile structures, such as weaving, knitting, and embroidery, to enable familiar products to take on a new dimension of reactive light emission (fig. 11).

Another pioneer in the design of textiles as an integration media for light, information, and smart controls is Sheila Kennedy, a principal architect at Kennedy & Violich Architecture and a professor at Harvard University's Graduate School of Design. Kennedy works with a palette of nanoscaled materials to produce an architecture that transforms the ways in which

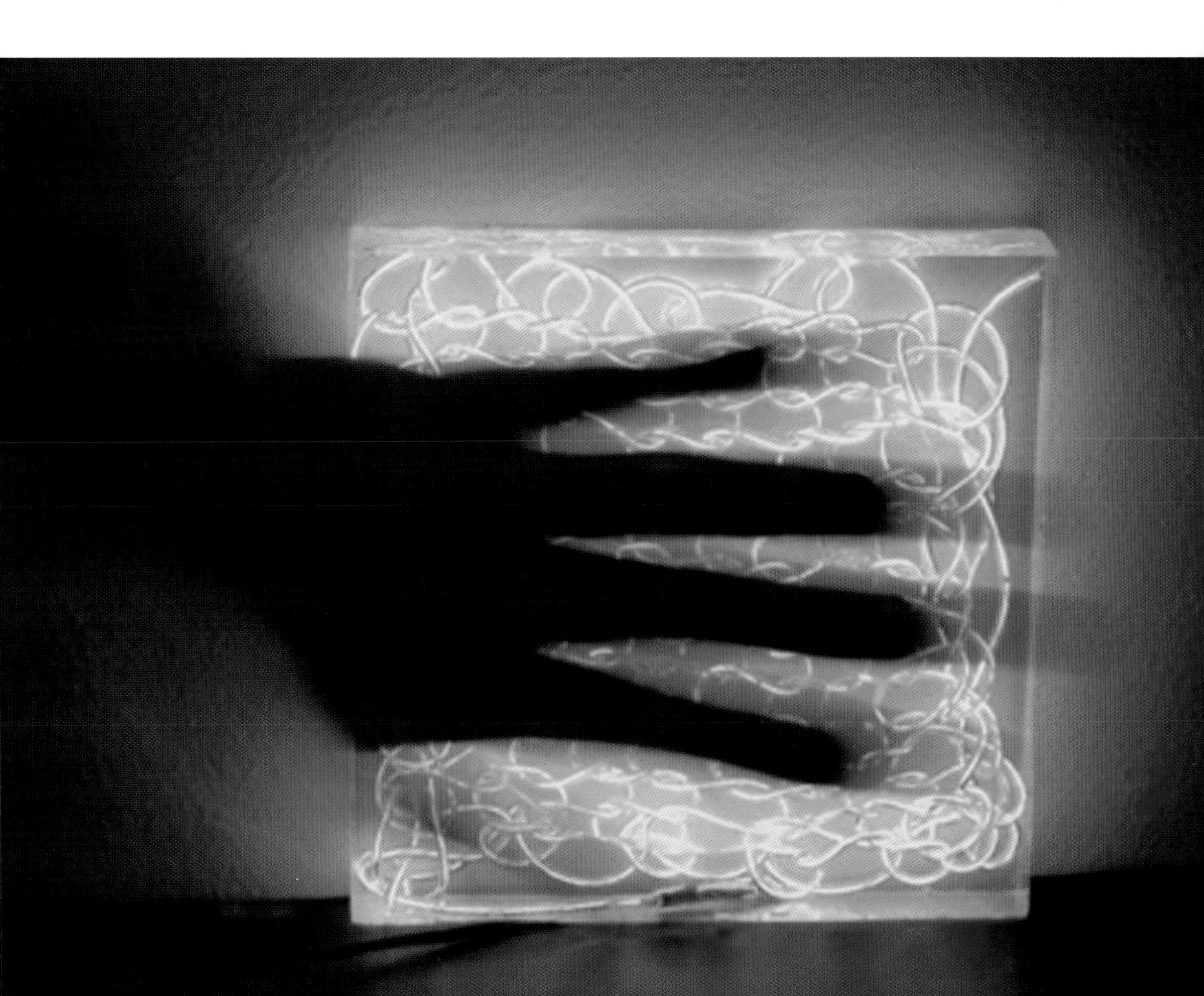

figs. 12, a–c
Zip Wall, a Nextwall application
Designed by KVA Design Team: Sheila Kennedy (Principal, Kennedy & Violich Architecture Ltd.), Senan Choe, Joseph Ho, Brian Price, Bill Yen; Purple Infrastructure technology package, system design concept by Sheila Kennedy, with Danny Hillis and Bran Ferren, of Applied Minds Inc.
U.S.A., 2003
Double needle bar 3-D raschel knit of 90% PET polyester, 10% elastomer with optical coatings

12a. (facing page) Nextwall prototype with DC power distribution system and plug in ports for digital tools such as cell phone, PDA, and solid-state lighting

people use and experience information infrastructure. The daughter of an eminent biochemist, Kennedy grew up playing among post-doctorate students in the laboratories of Woods Hole Oceanographic Institution. Today she is comfortable challenging widely held industry assumptions about the role and use of semiconductor technologies in contemporary global cultures. Kennedy's gift is to apply a form to architecture and material science that is both highly analytical and imaginative. She is able to engage an emerging domain of technology, rapidly assess the best thinking in the field, search for its contradictions, and discover new ideas and opportunities that redefine the uses, cultural impacts, and market value of the technologies she works with.

With MATx, the materials-research unit she founded in 2000, Kennedy has established a collaborative, interdisciplinary model for practice, which includes teaching, writing, and applied creative production across the fields of electronics, architecture, materials, fabrics, and product design. Kennedy and her colleagues have developed techniques to combine thin-film and semiconductor technologies into mass-customizable materials that function both as infrastructure and architecture.

Kennedy's Zip Room application, developed with Herman Miller, challenges the historically modern idea that digital technology is distinct from or opposes materiality (fig. 12). The fabric architecture of the Zip Room synthesizes design and technology to emphasize touch and adaptability. This expands the material properties of the fabric as an emissive, interactive surface, and creates a compelling set of new possibilities between the human body and the space of activities in architecture. The Zip Room is both materially specific and complex; the fabric surface is dynamic, becoming tactile, sheer, translucent, and light-reflective according to different conditions of use.

Unlike traditional woven fabrics, Kennedy's Zip Room design brings to a pliable fabric matrix the capacity to generate and conduct low-voltage DC power, store and access digital information, and emit digital light. Freed from distributing conduits and fixtures, the effects of color, light, and information become material, volumetric, and spatial. The impressive energy efficiency of semiconductors, and their ability to be effectively powered by photovoltaics, establishes a new paradigm for digital light and information delivery in architecture. "Think of it as a material diaspora," says Kennedy. "Light can now be harvested, stored, and carried in portable fabric surfaces. Digital light and information can be decentralized, dispersed, and distributed in cooperative networks."

Kennedy's work with textiles suggests a contemporary, nomadic vision for architecture. As textiles are a familiar form of expression in many cultures, the social and political implications of Kennedy's research extend to a new set of projects she is currently developing. Entitled 3-2-1, this initiative explores collaborative design practices that implement practical and humanitarian applications for portable light with nomadic indigenous peoples in

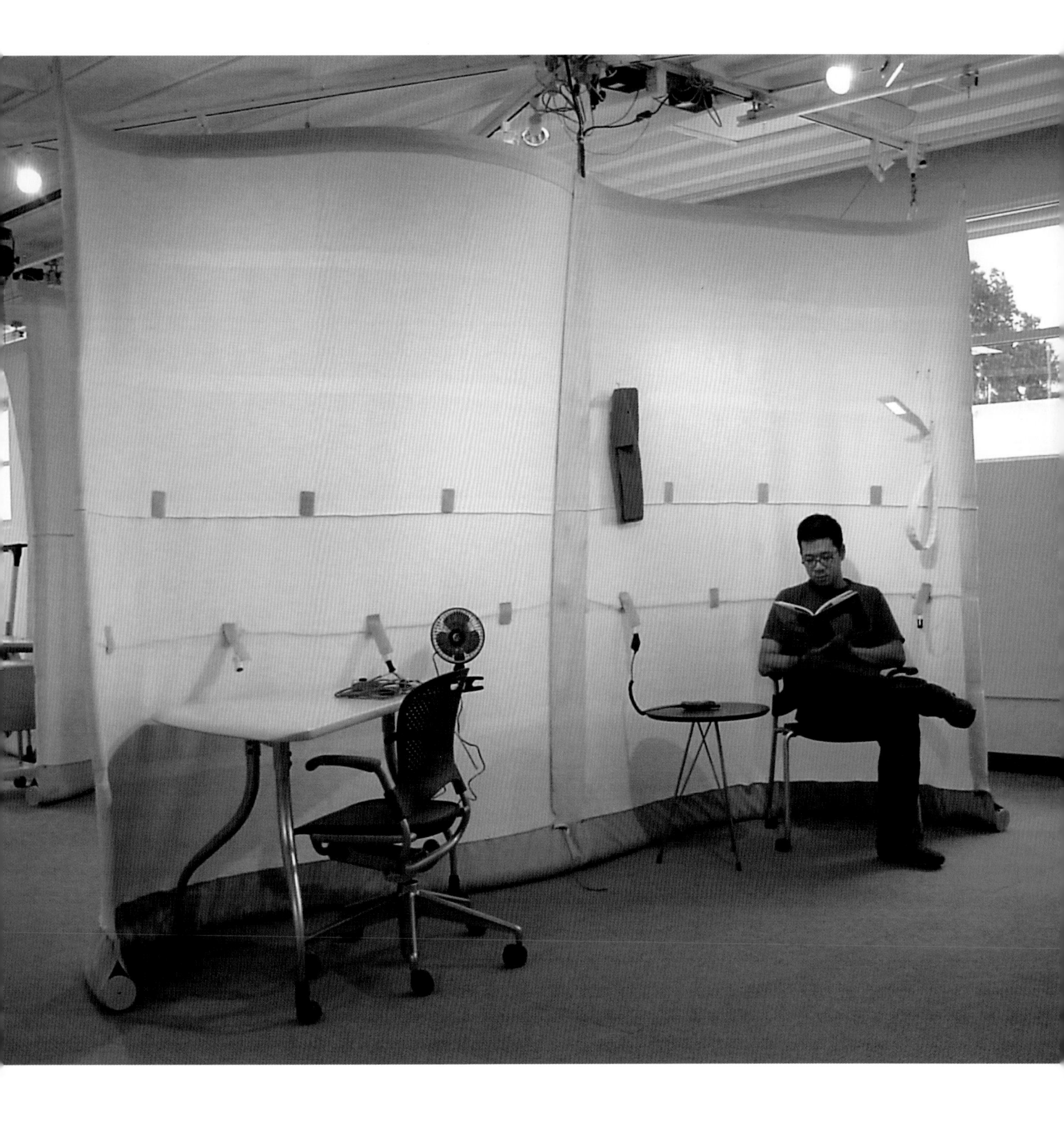

12b. Nextwall prototype of Zip Room "think tank" with integrated acoustical inserts and color-changing LED illumination

12c. Nextwall prototype of Zip Room interior with tactile switch and private sound- and lightscape

fig. 13
Electro-textile antenna vest
Developed by Foster-Miller Inc., BAE Systems Inc., Offray Specialty Narrow Fabrics Inc., Plastics One Inc., and the U.S. Army Research, Development, and Engineering Command, Natick Soldier Center
U.S.A., 2002
Merenda double-loop antenna integrated into a narrow-woven webbing, applied to vest

Mexico and South America. Kennedy's work invites us to consider a range of hybrid programs and forms that combine infrastructure and fabric, transforming the roles, responsibilities, and working relationships of user, manufacturer, engineer, scientist, and architect.

### HISTORY AS A GUIDE

Another avenue for smart-textile research has been driven by the needs of the superuser of technology versus the average consumer. Such individuals require mobile communications and information that allow them to work in extreme environments. These superusers include firemen, oil platform riggers, astronauts, and soldiers. In each of these situations, specialized protective clothing helps the individual do his or her job, and increasingly, technology has enabled a host of electronic devices to be carried to augment his or her abilities. During the 1990s, the researchers at the U.S. Army Natick Soldier Center became concerned that the number of electronics the soldiers were being asked to carry outweighed the soldiers' capacities. Carole Winterhalter and Jim Fairneny were two of the early visionaries who proposed the integration of as much of the function as possible into the soldier's clothing. The Army issued a call for innovative, efficient designs. Foster-Miller and a small number of scientists who had a love for historical textiles, including Justyna Teverovsky and myself, responded. We are materials scientists who have long shared a passion for embroidery and the textile arts. As we applied our team's knowledge of optics, electronics, and materials to the problem, we were equally inspired by textile solutions of the past.

Examining the situation, we recognized an immediate need to intranet the soldier so that their electronics could communicate and integrate information. Raw sensor data from physiological monitors could give information on the stress level, or need for water, whether the individual had been wounded, and how critical it was to evacuate the soldier. At this early stage of wearable electronics, the focus in the textile community was on weaving conductors into the fabric of the clothing. The team at Foster-Miller quickly realized that a fabric with woven conductors would be a nightmare to cut, sew, and connect through conventional clothing fabrication processes—a seamstress would need to be an expert in electronics as well as pattern-making! We took inspiration from the art of Victorian millinery and manipulation of ribbons, which the team leaders were teaching outside of work at the time. Wire-edged ribbon was easily formed into flowers for hat decoration, and could obviously be applied flat onto the soldier's clothing to simplify routing of digital signals and power. Thus began a long collaboration between Foster-Miller, the Army, and the Offray Specialty Narrow Woven Fabrics division of C. M. Offray and Sons, the largest producer of wire-edged decorative and technical ribbons in the world.

Our team developed many versions of ribbons, dubbed narrow woven busses, which contained textile-based wires (fig. 14). Soon the need to

fig. 14
Electro-textile bus for high amperage applications and antenna prototypes
Developed by Foster-Miller Inc. and Offray Specialty Narrow Fabrics Inc. for the U.S. Army Research, Development and Engineering Command, Natick Soldier Center
U.S.A., 2002

fig. 15
Polartec® Heat™ blanket
Engineered by Foster-Miller Inc. and Altitude Inc.; manufactured by Malden Mills® Industries Inc.
U.S.A., designed 1998–2001, manufactured 2005
Circular knit double velour fleece of 96% polyester and 4% metallic fiber

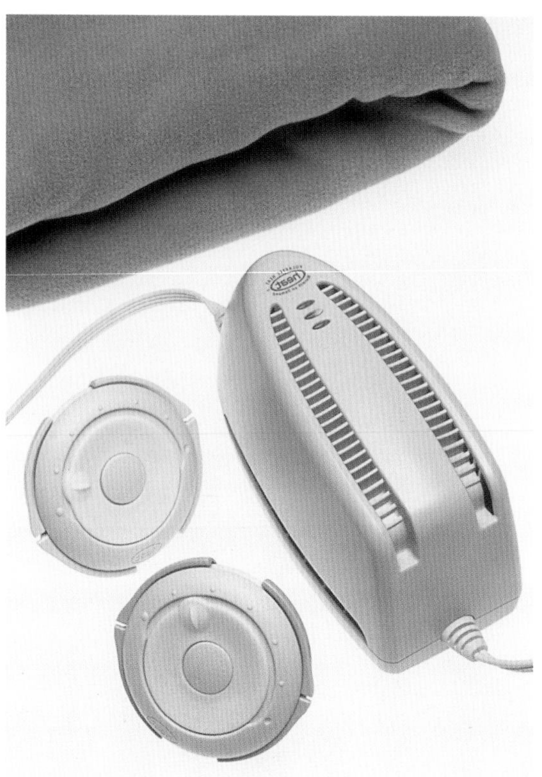

fabricate high-data-rate cables for video required another look into the annals of history. To produce the complex cables, methods for ensuring high and low electromagnetic frequency shielding were required. Our team recalled the wonderful variety of braided cords made by the Japanese to tie obi around the kimono, and purchased a traditional kumihimo stand to make electronic versions of these ancient cords. The variety of braid structures could equally be used to provide differing levels of electromagnetic shielding around a wire when conductive fibers of different metals were substituted for the colored silks of the historical patterns. Through this research, soft and pliable textile cables capable of transmitting high-speed data, such as the USB protocol, were manufactured.

The next extension of our work paralleled the developments at Philips Design. Much like the questions being asked at Philips, the Army-sponsored team started to investigate how antennas could be made soft. A particular problem for antennas is that they usually scale in size with the wavelength of the communications band, and, therefore, can easily become larger than a human. A novel solution, developed by BAE Systems, consisted of loops that could wrap around a person's torso, but it was made of rigid circuit sections. The Army asked the two teams to collaborate and fabricate a soft version of the antenna system that could actually be worn and tested (fig. 13). The antenna technologies and personal-area network textiles are being applied to the Future Force Warrior project, under development by the Army as the next generation of equipment enabling the soldier to have vital information wherever he or she goes (fig. 24).

Not only applicable to the soldier, this narrow woven electronic bus technology was soon transitioned to the consumer in a project with Malden Mills®, a high-tech textiles company known through their high-performance Polartec® fleece products. Aaron Feuerstein, the owner of Malden Mills famous for refusing to move his plant overseas, has pushed his products to new frontiers incorporating hair-thin stainless steel fibers into fleece knit to make a number of heated textile products. The first of this family of products was to be a comfortable, safe, low-voltage, heated blanket that would revolutionize the consumers' impression of heated textiles.

When Malden Mills looked to Foster-Miller to apply the Army-funded bus technology to the blanket, they had already proven the concept, but needed a flexible and easily manufactured power bus to feed the tiny resistors in the blanket. Again the team turned to their knowledge of historical embroidery to find solutions. At the time I was actively studying and teaching Elizabethan gold embroidery, and had been looking at the structure of gold threads and their metallurgy. These lessons inspired ways to modify flexible wires to enable a low-cost, washable electronic bus that could be applied to the blanket. The resulting product, the Polartec Heat blanket, was an instant success with the consumer as it provided even, controllable heating with a wonderful tactile feel and drape (fig. 15).

figs. 16, 17
Space suit glove with embedded robotic controls
Developed by NASA,
ILC Dover Inc., and Softswitch Ltd.
U.S.A., 2001
Embroidered keypad with quantum tunneling composites embedded in spacesuit extravehicular activity glove (opposite); as shown on astronaut with rover (above)

While the Army team was concerned with the soldier, another team, focused on the astronaut as a superuser, was turning to electronic textiles for solutions. The design team at ILC Dover used smart textiles to solve problems posed by traditional spacesuits. The result is the I-Suit, a higher mobility suit that contains enhancements such as textile switches and cabling (fig. 37, p. 179). These switches are embedded into the material layers of the Rover Control glove that allows the astronaut to operate a robotic rover assistant and such items as the helmet-mounted lamps (figs. 16, 17).

ILC Dover paired up with Softswitch to create pressure-sensitive textile switches with their proprietary quantum tunneling composites (fig. 18). These composites use multiple layers of conductive textiles and an elastomer loaded with fine metallic particles that allow electron tunneling between them when brought closer together by the pressure of a person's touch. Astronauts have trouble with gripping external controllers because of the need to work against the glove's internal pressure. By incorporating the textile switches, the simple touch of the glove finger is enough for activation. This Softswitch technology is rapidly being introduced to the non-space-traveling public through a series of jackets that are designed to hold digital music players, such as the iPod by Apple. These jackets, manufactured by Burton Snowboards, allow the snowboarder to listen to music and control the player by touching the sleeve of the jacket while the player stays in the pocket (fig. 19).

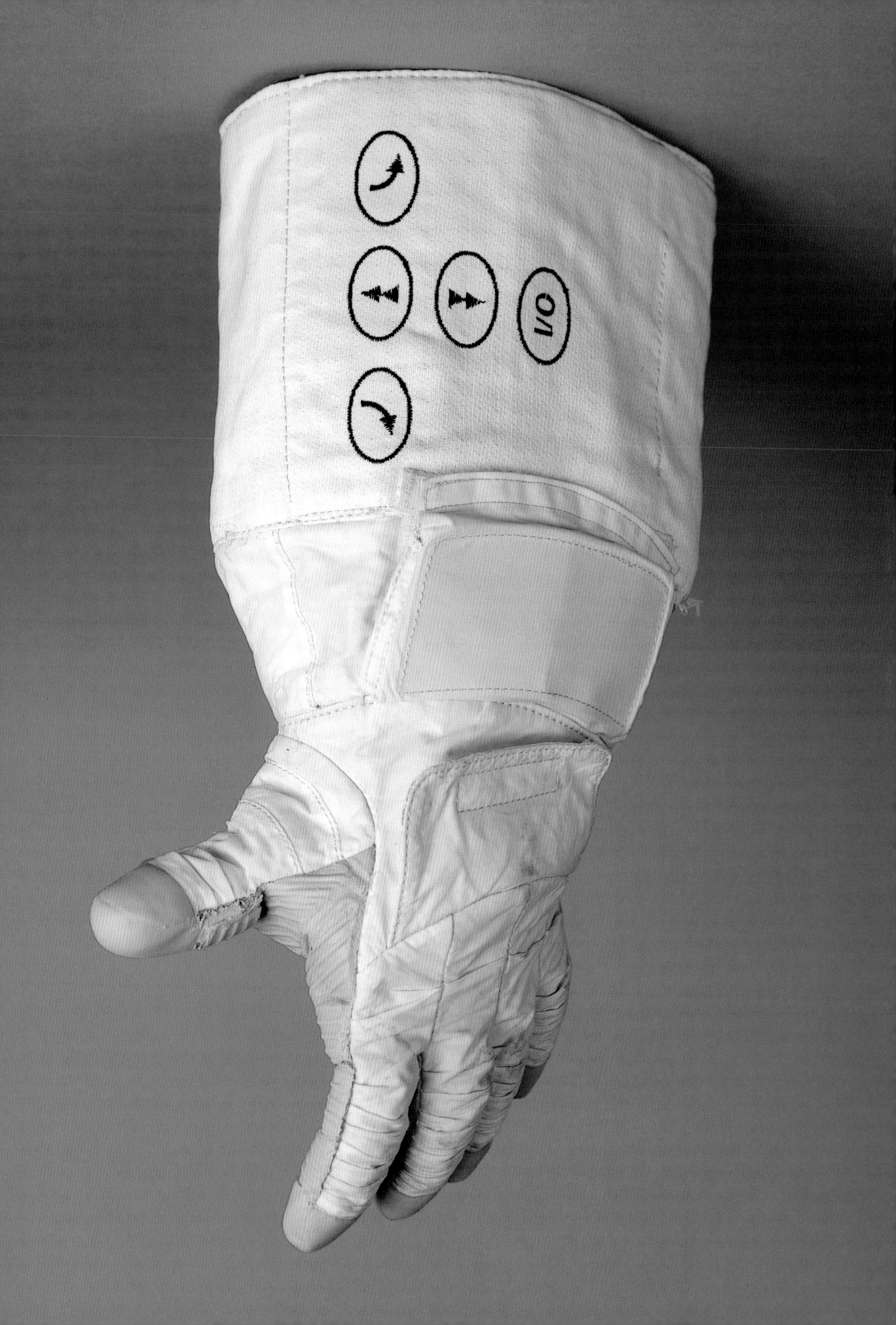
I/O

fig. 18
Softswitch keypad
Developed by Softswitch Ltd.
England, 2003
Flexible, embroidered keypad on cotton, with quantum tunneling composite

Efforts by the companies profiled above have inspired the Defense Advanced Research Projects Agency (DARPA) to look at ways that very large electronic systems can be incorporated into giant area fabrics, like carpets, wall coverings, awnings, and truck covers. But, surprisingly, new ideas are sometimes old ideas that occurred before their time. The product developments of Woven Electronics, formerly Southern Weaving Company, fall under this category. Their research and development in electronic textiles occurred in the early 1960s, when the printed circuit board was in its early stages of development. Edgar Ross, an engineer at the company, felt that the large and fragile circuit boards were not going to be accepted in all applications, and began experimenting with weaving wire to make a flexible, woven circuit. His early trials looked promising, and a new division, Woven Electronics, was spawned to fabricate circuits and cable systems. Unfortunately, the miniaturization of silicon and improvements in the printed circuit board soon overtook their developments, and the woven circuit board was abandoned (fig. 20). Nearly forty years later, other developments in miniaturization of technology would inspire thoughts of large-scale systems that could again rely on the woven circuit to distribute highly efficient computational devices over vast areas. Ross had proved to be, like many visionaries, decades before his time. Spinoffs of his early work manufactured by Woven Electronics are used today as electronic harnesses for the military, and digital signal transmission line cables for commercial computers and test equipment (fig. 1).

Hopefully, unlike Ross, the following individuals will be recognized for their contributions to the field of wearables during their time. Chris Kasabach and Ivo Stivoric did not realize their college friendship would someday spark a revolution in thinking about wearable devices. Already

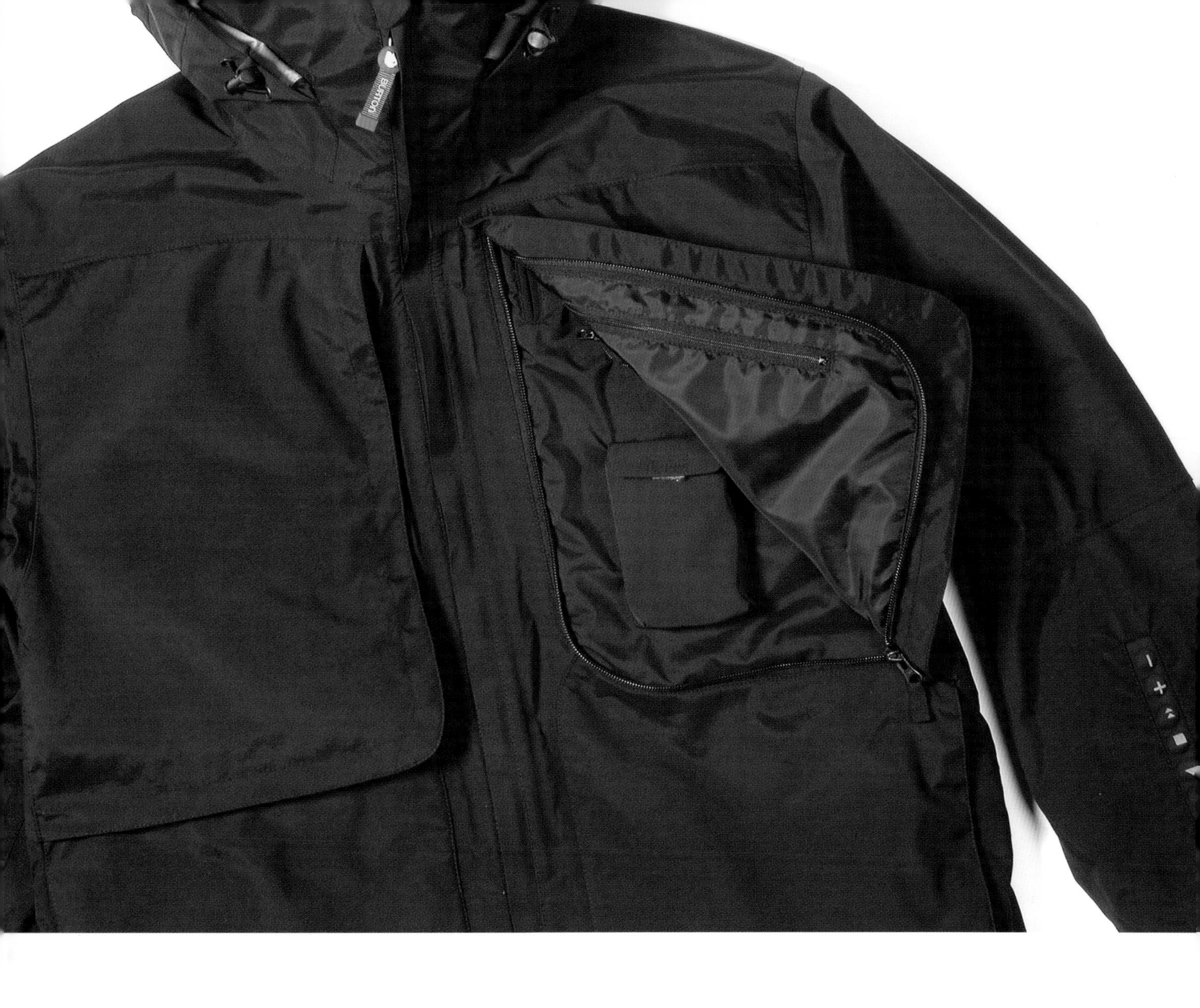

fig. 19
Analog Clone MD jacket
Designed by Burton Snowboards
and Softswitch Ltd.
U.S.A., 2003

industrial designers, they became researchers at Carnegie Mellon University in 1991 to develop and codirect a unique lab devoted to the intersection of design, technology, and mobility. Throughout the 1990s, the Mobile and Wearable Lab at the Engineering Design Research Center (now the Institute for Complex Engineered Systems) challenged and redefined what computers are by reinventing how computers are interacted with.

Collaborating with companies including Boeing and Daimler-Benz, the group discovered significant advantages to conforming technology to the bodies of dynamic workers (such as aircraft inspectors) who required thousands of complex drawings and instructions during their shifts but did not have the space, time, or ability to maneuver traditional computers into confined spaces and motion-sensitive tasks. The lab's designs broke away from the box-as-computer paradigms, replacing them with highly body-contoured wearable solutions evolved from anatomical research and empirical movement studies. A few of their designs became almost muscles in themselves, floating and flexing on top of the skin, respecting the circulatory and thermal conditions that are continuously in flux in the body.

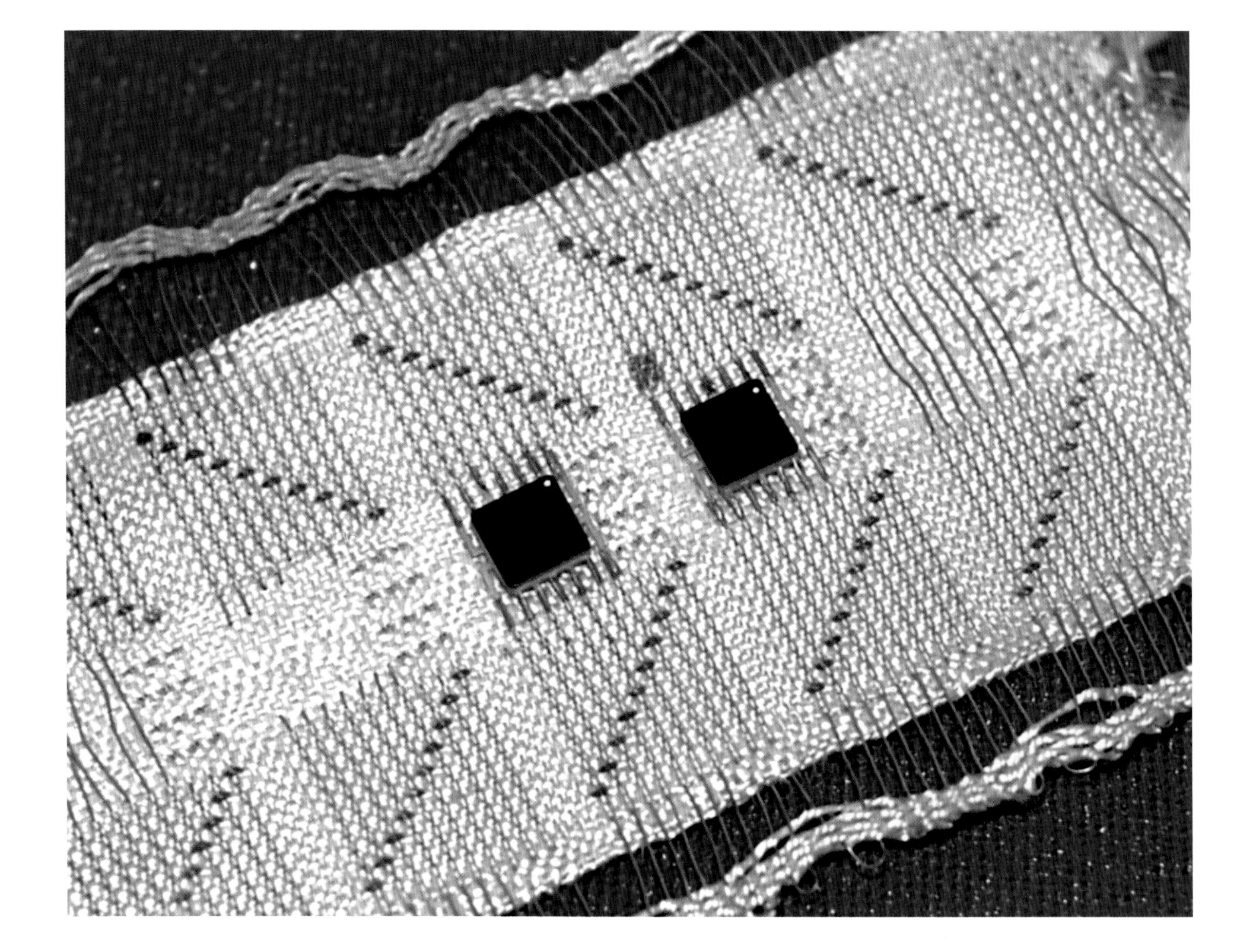

fig. 20
Woven circuit board
Designed by Edgar Ross,
manufactured by Woven Electronics
U.S.A., ca. 1960
Plain woven synthetic and
metallic fibers

Along the way, their team systematized their work, defining the best areas for locating technology on the body without encumbering the wearer. Then they looked at how humankind has historically ornamented its clothing or body with devices such as watches, hats, earrings, buttons, and pockets. These objects are naturally placed in areas where they are comfortable and accessible. They codified their findings into a detailed Wearability Map, which consisted of more than twenty-five physical forms that, once applied to the body, visualized the maximum sizes and shapes that could fit comfortably on men and women of vastly different sizes (fig. 21). The publication of their seminal wearability study caused a sensation in the mobile-computing world and played an early role in the founding of BodyMedia®, a leader in continuous body-monitoring systems, which they formed with Astro Teller and Chris Pacione in 1999. BodyMedia's monitors are small, wearable multi-sensor products that can collect, store, and transmit their wearer's physiological and lifestyle information to an array of devices such as computers, cell phones, or watches, providing the wearer, as well as care providers or researchers, with actionable information about their health and well-being.

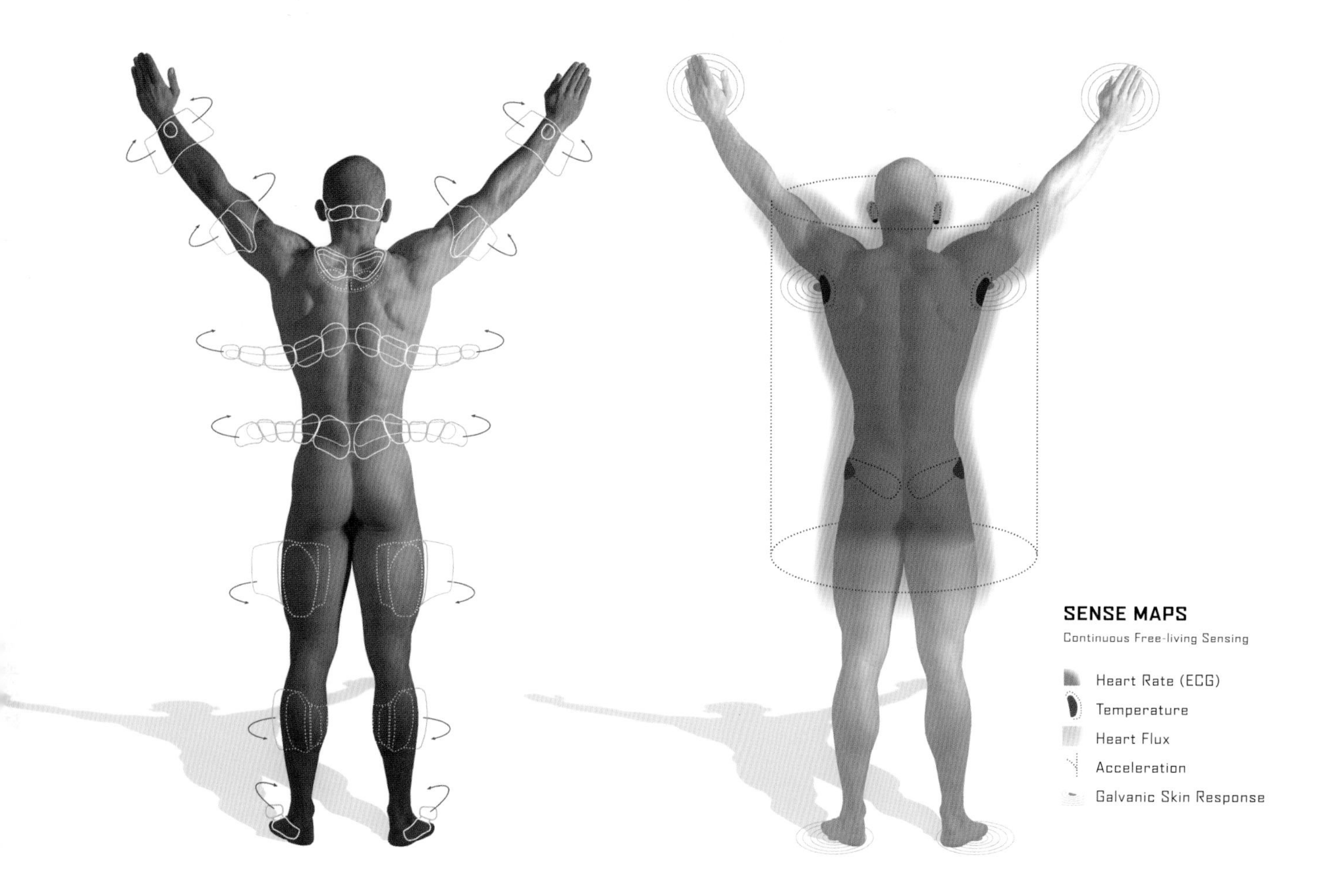


fig. 21
Wearability map
Designed by Chris Kasabach and John (Ivo) Stivoric, BodyMedia® Inc., and Francine Gemperle, Institute for Complex Engineered Systems, Carnegie Mellon University
U.S.A., designed 1995–98

fig. 22
Sensing map
Designed by BodyMedia® Development Team.
U.S.A., designed 1999–2004

Their multidisciplinary team consists of hardware, informatics, interaction, mechanical, and software designers, as well as medical experts, who have collaborated over the last five years to extend the wearability maps to include sense maps that define the best locations for collecting an array of physiological parameters including heart rate, temperature, and acceleration from the body in motion (fig. 22). The overlapping of their sense and wearability maps informed the development of BodyMedia's SenseWear® armband, and its more recent relative, HealthWear®, a personalized weight-management system marketed by the Swiss giant Roche Diagnostics. In its latest prototype, BodyMedia has reshaped technology and materials to create a system that incorporates a multi-sensor conductive textile for collecting heart rate and contextual data about its wearer. The textile, composed of woven conductive threads, provides a flexible and breathable interface to the skin ensuring optimum contact, accuracy, and comfort over prolonged wearing (fig. 23).

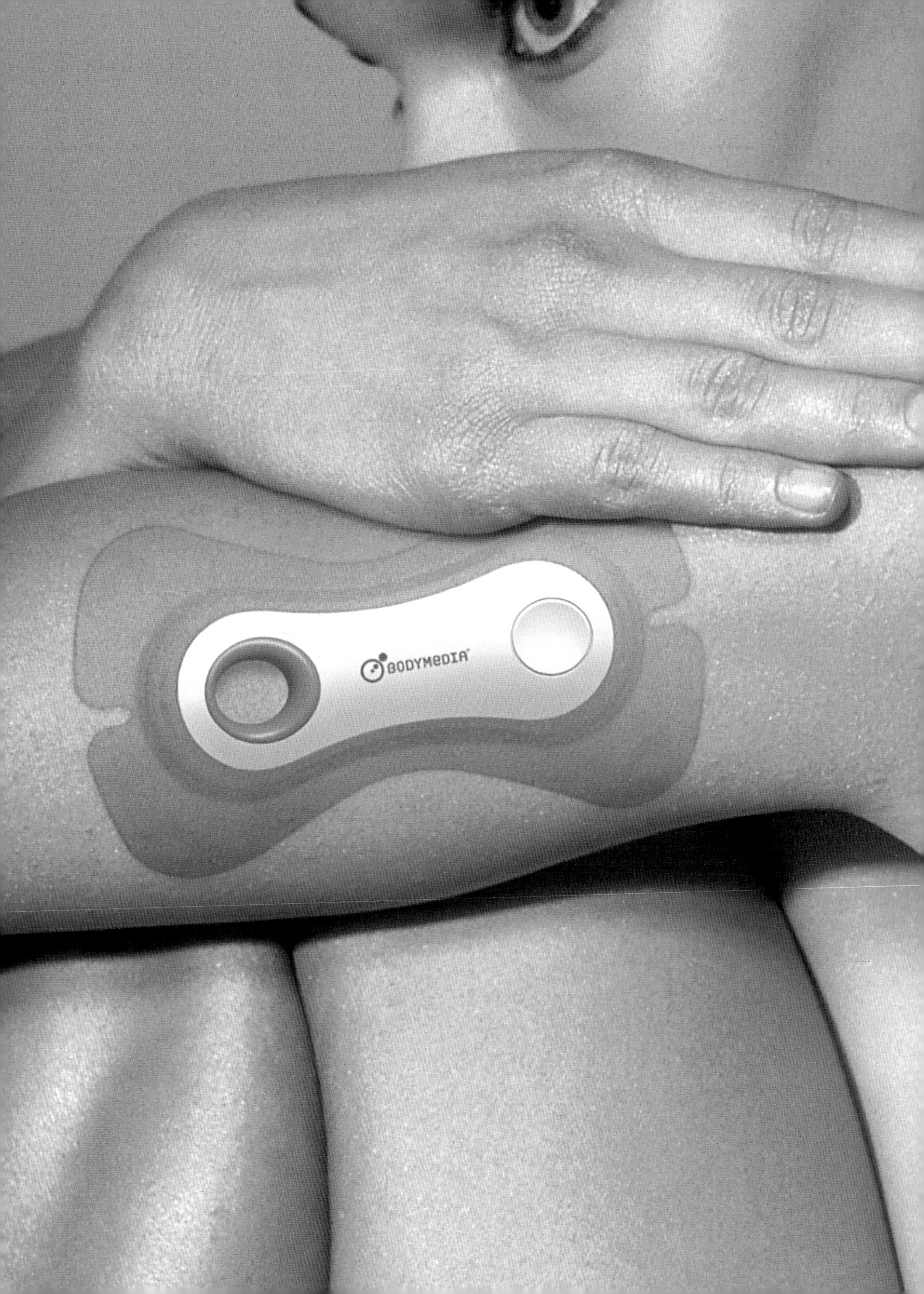
BODYMEDIA

fig. 23
SenseWear® Patch body monitor
Designed and engineered by BodyMedia® Development Team, engineering support by K Development Inc., manufactured by BodyMedia® Inc.
U.S.A., designed 2004, manufactured 2005
Multi-sensor conductive textile
92.2 x 50.8 x 8.4 mm (3⅝ x 2 x 5/16 in.)

## THE FUTURE

While significant progress has been made in the last decade to manufacture the first generation of smart-textile prototypes and products, there are certainly many challenges ahead. Two important obstacles to the growth of smart textiles must be overcome before the consumer will have a plethora of product options to choose from. The first challenge is industrial standardization. In the early history of solid-state electronics, it was realized that customized solutions would result in expensive products and a stagnation of growth. To solve this problem, regulations and standards were developed to govern the size of a resistor, spacing between prongs on a processor, and other such examples. That way a designer could be assured that a low-cost, high-volume component manufactured by another company would fit their final design. Standardization led to short turnaround times between product conceptualization and market launch. This same process must occur in the smart-textile field to enable fast growth and lower the cost of products and their development.

A significant impediment to the standardization process is embodied in the second challenge. In an industry that traverses so many fields, from fashion design and textile fabrication to antenna engineering and electronic component manufacture, to name a few, often the "language" barrier is the greatest stumbling block. I have found myself at times deep into discussions about new products with fabulous fashion designers when we both suddenly realize that an information gulf lies between us, and it must be crossed to complete the project. As an example, I might have to explain that a circuit must connect in a closed continuous loop to work, and they have to explain to me that if the light-up ribbon cannot cost twenty cents a yard and be put on ten different garments for this season, it is useless to them. It was in the unique collaborations profiled above that a common language was hammered out of hard work and time, as the disparate fields tenaciously worked together to form the common vision of the group. The interactions of artists, designers, engineers, and scientists must continue to build on this new language and understanding in order for the industry to grow.

Even though challenges exist, certainly the exciting innovations in the smart textiles demonstrated here tell of a bright future for the area. No longer can we look at a piece of cloth and think of it as a technology as old as time. We must now realize that it holds infinite possibilities when put into the hands of creative teams who strive to answer the question, "What if?"

fig. 24
Future Force Warrior
Developed by U.S. Army Natick Soldier Center, Individual Protection Directorate
U.S.A., 2004

Future Force Warrior, the soldier of the future, is the result of years of research and development by U.S. Army Natick Soldier Center's Individual Protection Directorate, along with a surprising number and variety of industry partners. With the ultimate goal of improving soldiers' mobility, effectiveness, and survivability, the team set out to reduce the average soldier's load from about 120 pounds to less than 40. The emphasis on lightness spurred intensive research in textile-based solutions for a broad range of problems. Many of these are protective in nature: protecting against ballistics, cuts and abrasion, biochemical threats, and extreme weather conditions. Others improve effectiveness by adding functionality in the form of networked team communication systems, enhanced situational awareness, and visual or electromagnetic signature concealment. Still others provide a system of sensors for physiological monitoring and integrated emergency care. Much of this increased functionality results from developments in the area of electronic textiles, making a vision for a lightweight, low-bulk, multifunctional system a reality.

fig. 26 (left)
MET 5 jacket
Textile designed and manufactured by Malden Mills® Industries Inc., jacket designed by The North Face Design Team, engineered by Thomas Laakso
U.S.A., 2002
Polyester fleece with conductive heating fibers, four-way stretch nylon shell, Powershield® semi-air-permeable membrane

Advances in electronic textiles have affected not only communications but comfort, and The North Face's MET 5 jacket, produced in collaboration with Malden Mills®, which makes Polartec® fleeces, demonstrates how extreme cold can be endured over extended periods. Powered by two lightweight lithium ion batteries, this jacket conducts heat through a network of small elements made up of microscopic conductive fibers, each finer than a human hair. The heating element is laminated between the shell and liner, and is invisible, flexible, and waterproof, producing a comfortable, durable, and machine-washable jacket. The conductive fiber is concentrated around the chest area where major blood circulation occurs and gives warmth on an as-needed basis to the wearer. The garment is controlled by a unique textile switch sensor integrated to the body of the jacket that is completely flexible, washable, and durable. By activating the control unit (located in the chest pocket of the jacket) to medium (42°C or 107°F) or high (45°C or 114°F), one can maintain body temperature for up to five hours. The jacket is designed to provide flexible added warmth during activities like skiing or multipitch climbing, in which the user is alternating frequently between periods of rest and high-output. The jacket eliminates the need for the user to take layers on and off as activity levels fluctuate.

fig. 25 (left)
Nike ACG Commvest
Garment designed by Scott Hutsenpiller, industrial design by Shane Kohatsu, textile cable switching system designed by Softswitch Ltd., manufactured by Nike Inc.
U.S.A., 2003

The Nike Commvest was designed in collaboration with Softswitch Ltd., Portland Mountain Rescue, and Mount Hood Meadows Ski Patrol, who had an urgent need for a communications system to use in extreme weather conditions during rescue missions. Utilizing innovative materials and technologies integrated in a vest, such as the push-to-talk (PTT) button and removable speaker and microphone, the Nike Commvest solved three major problems for rescuers: poor audio, inaccessibility to the button with gloved hands, and ergonomic fit of the radio. In addition, the design of the vest allows it to be washed. The Nike Commvest ultimately met the needs of the rescue professionals while providing a unique collaborative design effort involving various companies and individuals across the globe.

fig. 27
Warp-knitted stainless steel
Designed by Ekkehard Preuss,
engineered by Bekaert Fibre Technologies,
manufactured by Preuss Industrial Fabrics
Germany, designed 1996,
manufactured 2004
Warp-knitted 100% stainless steel spun yarn

This knitted steel band is used as a damping and transportation device in the production of car windshields. The glass becomes very hot, and this fabric can withstand temperatures up to 700°C (1,292°F).

# NOTES

## INTRODUCTION McQuaid

1 Ian Lidell, "Very Large Flexible Barges," in *Patterns 14* (newsletter, Buro Happold Consulting Engineers for Aquamarine Transportation, forthcoming).
2 Brian P. O'Rourke (chief composites engineer, Williams Grand Prix Engineering, Ltd.), in correspondence with the author, June 2004.
3 For a complete history of the Wright brothers, see Tom D. Crouch and Peter L. Jakab, *The Wright Brothers and the Invention of the Aerial Age* (Washington, D.C.: Smithsonian National Air and Space Museum, 2003).
4 Rick Young, in discussion with the author, June 2004.
5 David Cadogan, Tim Smith, Ryan Lee, Stephen Scarborough, David Graziosi, "Inflatable and Rigidizable Wings for Unmanned Aerial Vehicles" (paper, 44th AIAA/ASME/ASCE/AHS Structures, Structural Dynamics, and Materials Conference, April 7–10, 2003), (American Institute of Aeronautics and Astronautics, 2003), 6.
6 Ibid., 1.
7 Evelyne Orndoff, "Fine Gems: The Rare Fabrics of NASA," *Industrial Fabric Products Review* (July 2001): 61.

## TEXTILES: FIBER, STRUCTURE, AND FUNCTION Brown

1 "Six Minutes of Terror," *Mars Exploration Rovers: The Challenge of Getting to Mars* (Pasadena, CA: NASA, Jet Propulsion Laboratory, California Institute of Technology, 2004).
2 Brian Doyle, "Aramid, Carbon, and PBO Fibers/Yarns in Engineered Fabrics or Membranes," *Journal of Industrial Textiles* 30, no.1 (July 2000): 43.
3 E. J. W. Barber, *Pre-Historic Textiles: The Development of Cloth in the Neolithic and Bronze Ages* (Princeton, N.J.: Princeton University Press, 1991), 127. Non-loom techniques, such as twining, were in use much earlier.
4 Ibid., 11.
5 Jet Propulsion Laboratory, "Airbag Tracks on Mars," *Planetary Photojournal*, NASA, http://photojournal.jpl.nasa.gov/catalog.
6 This and other information on the construction of the airbags from D. Cadogan, C. Sandy, and M. Grahne, "Development and Evaluation of the Mars Pathfinder Inflatable Airbag Landing System" in 49th International Astronautical Congress, September 28–October 2, 1998 (Melbourne, Australia, and Paris: International Astronautical Federation, 1998) and J. Stein and C. Sandy, "Recent Developments in Inflatable Airbag Impact Attenuation Systems for Mars Exploration" (Frederica, DE: ILC Dover Inc., 2003).
7 Carbon fiber was first patented by Thomas Edison in 1879 for use as filament in his electric lightbulbs. Through controlled pyrolysis, or burning of acrylic precursors, fibers that are at least 90% carbon are obtained. Carbon fibers are ideally suited to applications where extreme strength and stiffness are desired. Katie Harholdt, "Carbon Fiber, Past and Future," *Industrial Fabric Products Review* 88, no. 4 (April 2003): 14–16.
8 This and other information on the Advantage racing shell from conversation with Dr. Edward S. Van Dusen, Van Dusen Racing Boats, June 2003–June 2004.
9 Owen Edwards, "The Shell of Champions: Van Dusen Racing Shells," *Departures*, January/February 2003.

10 This and other information on the manufacture of composite sail masts from conversation with Van Dusen, June 2003–June 2004.
11 Foster-Miller Inc., "Rocket Engine Turbopump Blisk Fabrication," http://www.fostermiller.com/pages.
12 Industrial News Room, "Pneumatic Actuator Works like Human Muscle," http://www.industrialnewsroom.com/fullstory/6158 (accessed January 17, 2002).
13 Linda Marsa, "Heartening News: Doctors Are Excited about a Fiber Brace that Helps Ailing Organs Keep the Beat," *Los Angeles Times*, February 17, 2003.
14 Anndrea Vorobej, ed., "Textile Technology Used in Infection-Resistant Polyurethane Biomaterials," News & Analysis@TextileWeb via TextileWeb, http://www.textileweb.com/content/news/article.asp?docid={e3637b18-f69b-11d3-82c25-00927d20829}.
15 "Embroidery Technology for Medical Textiles and Tissue Engineering." *Technical Textiles International* (July/August 2000): 9–12. Also personal correspondence with Julian Ellis, Ellis Developments Ltd., September 2003–June 2004.
16 This and other information on the manufacture of 3DL sails from conversation and correspondence with Bill Pearson, North Sails Nevada, September 2003–June 2004.
17 Janine M. Benyus, *Biomimicry: Innovation Inspired by Nature* (New York: William Morrow and Company Inc., 1997).
18 Jay-Anne Casuga, "Electrospinners: Technique Could Produce Natural Blood Vessels, Cartilage and Bandages." Richmond.com, http://www.richmond.com/printer/cfm?article=2462232 (accessed April 18, 2003).

## ROPES Hearle

1 For further information, see John W. S. Hearle, Henry McKenna, and Nick O'Hear, *Handbook of Fibre Rope Technology* (Cambridge, England: Woodhead Publishing, 2004).

## HIGH-PERFORMANCE FIBERS Becker

1 F. R. Jones, "Glass Fibers," in *High-Performance Fibres*, ed. John W. S. Hearle, 191–238 (Cambridge, England: Woodhead Publishing, 2000). A. Sircar, "Introduction to Glass and Glass Fiber Manufacturing Technology with an Application to Nonwoven Process," *TAPPI Journal* 76, no. 4 (April 1993): 167–76.
2 N. Gokarneshan, "Recent Trends in Synthetic Fibers: An Overview," *Textile Magazine*, April 2002, 30–32.
3 Jones, 191–238.
4 Pushpa Bajaj, "Heat and Flame Protection," in *Handbook of Technical Textiles*, ed. A. R. Horrocks and S. C. Anand, 224–63 (Cambridge, England: Woodhead Publishing, 2000).
5 Anthony R. Bunsell and Marie-Hélène Berger, "Ceramic Fibres," in Hearle, 239–58.
6 3M, "3M Nextel Ceramic Textiles: Aerospace Solutions," http://www.3m.com/market/industrial/ceramics/solutions/aerospace_solutions.jhtml (accessed June 2004).
7 Bajaj, 224–63. Bunsell and Berger, 239–58.
8 Katie Harholdt, "Carbon Fiber, Past and Future," *Industrial Fabric Products Review* 88, no. 4 (April 2003): 14–16. J. Gerard Lavin, "Carbon Fibres," in Hearle, 157–90.
9 Harholdt, 14–16. Neil Saville, "Thermally Resistant Fibres," in Hearle, 301–10.
10 Lavin, 157–90.
11 Brian Doyle, "Aramid, Carbon and PBO Fibers/Yarns in Engineered Fabrics or Membranes," *Journal of Industrial Textiles* 30, no. 1 (July 2000): 43–49.
12 Bajaj, 224–63.
13 Chen X and R. H. Gong, "Technical Yarns," in Horrocks and Anand, 42–61.

14 Serge Rebouillat, "Aramids," in Hearle, 23–61.

15 David Beers, "Other High Modulus-High Tenacity (HM-HT) Fibres from Linear Polymers," in Hearle, 93–101. Vectran, "Vectran Fiber: A Unique Combination of Properties for the Most Demanding Applications," http://www.vectranfiber.com/index1.html (accessed June 2004).

16 Beers, 93–101.

17 Jan L. J. Van Dingenen, "Gel-Spun High-Performance Polyethylene Fibres," in Hearle, 62–91.

18 Spectra, "Spectra Fiber Product Information," http://www.spectrafiber.com/products (accessed June 2004).

19 Toyobo, "What is Zylon?" http://www.toyobo.co.jp/e/seihin/kc/pbo/menu/fra_menu_en.htm (accessed June 2004).

20 Robert J. Young and C. L. So, "Other High Modulus-High Tenacity (HM-HT) Fibres from Linear Polymers," in Hearle, 101–108.

21 Magellan Systems International, "M5 Product Applications," http://www.m5fiber.com/magellan/m5_product_applications.htm (accessed June 2004). Doetze J. Sikkema, "Other High Modulus-High Tenacity (HM-HT) Fibres from Linear Polymers," in Hearle, 108–15.

## A TRANSFORMED ARCHITECTURE Beesley and Hanna

1 This relationship between fabric and building can be traced to the very roots of architecture. *Textile, technology*, and *connection* all derive from the same Proto-Indo-European root *tek*, from which we get "architecture." The word "technology" comes from the Latin *texere*, meaning to weave or construct. *Oxford English Dictionary* (London: Oxford University Press, 1971).

2 A classical view of architecture would claim that permanence and solid mass are at the core of architecture. When the ancient Roman writer Vitruvius wrote his famed treatise he set permanence and eternal forms above everything else. The ephemeral qualities of new textile-based buildings are antithetical to this view. Vitruvius, *The Ten Books of Architecture*.

3 This kind of rigid system is relatively easy to analyze, so engineers quite naturally favor it. Hierarchical structures may be the easiest to understand, but they are not necessarily the most efficient systems. The total energy consumed by a system accumulates at each step. This means that systems organized into multiple "steps" often waste large amounts of energy.

4 R. Buckminster Fuller, with E. J. Applewhite, *Synergetics: Explorations in the Geometry of Thinking* (England: Macmillan Publishing, 1975, 1979), proposition 101.01. Fuller's work 1947 and 1948 was self-titled "Energetic-Synergetic Geometry." Fuller borrowed *synergy* from medicine, where the term is used to mean the combined action of a group of bodies, organs, or drugs. The origin is *synergos*, or "working together." Robert K. Barnhart, ed., *Chambers Dictionary of Etymology* (Edinburgh: Chambers, 1988).

5 In the middle of the twentieth century, after the visions of Corbett and Fuller, a new image of the city emerged based upon complexes of interconnected relationships. The American urbanist Jane Jacobs ended her 1961 book, *The Death and Life of Great American Cities*, with the concept that architecture at the largest scale is a complex, highly organized system of many small parts that behave in unpredictable ways. To illustrate this idea, she cited Dr. Warren Weaver's account of the progress of scientific thought, in which he paints a picture of progress leading from simple problems to problems of disorganized complexity using statistical analysis of massive data, and finally to the emerging science of organized complexity. In contrast to previous methods that

relied on hierarchies, this new approach studied emergent behavior from the "bottom" up. This approach changes the way architects design. New analytical technologies such as Finite Element Analysis finally allow designers to understand the complex dynamics of textiles sufficiently to use them in general building systems. Jane Jacobs, *The Death and Life of Great American Cities* (New York: Random House, 1961).

## NASA: ADVANCING ULTRA-PERFORMANCE McCarty

1 Roger D. Launius, *NASA: A History of the U.S. Civil Space Program* (Melbourne, Fla.: Krieger Publishing Company, 2001), 65.
2 Lynn Ermann, "Technology Transfer Remains a Nascent Movement, but More Architects Take Up the Challenge," special issue, *Architectural Record*, October 2003, 47.
3 Debra Rosenberg, "Plastics," special issue, *Newsweek*, Winter 1997–98, 45.
4 U.S. Senator Ron Wyden, interview, *Small Times*, January/February 2004, 10.
5 National Aeronautics and Space Administration, *Spinoff 2000* (Washington, D.C.: U.S. Government Printing Office, 2000), 13.

The author would like to give special thanks to Pamela Stewart, Julian Beinart, and Gay Lynn Montgomery.

## THE SPACESUIT Young

1 They were Scott Carpenter, L. Gordon Cooper Jr., John H. Glenn Jr., Virgil I "Gus" Grissom, Walter Schirra Jr., Alan B. Shepard Jr., and Donald K. "Deke" Slayton.

# SELECTED REFERENCES


## GENERAL

Edwards, J. Vincent, and Tyrone Vigo, eds. *Bioactive Fibers and Polymers*. Washington, D.C.: American Chemical Society, distributed by Oxford University Press, 2001.

Hearle, John W. S., ed. *High-performance Fibres*. London: CRC Press, 2001.

Hongu, Tatsuya, and Philips, Glyn O., eds. *New Fibers*, second edition. Cambridge, England: Woodhead Publishing Limited, in association with The Textile Institute, 1997.

Horrocks, A. R., and S. C. Anand, eds. *Handbook of Technical Textiles*. Cambridge, England: Woodhead Publishing Limited, in association with The Textile Institute, 2000.

McKenna, H. A., N. O'Hear, and J. W. S. Hearle, eds., *Handbook of Fibre Rope Technology*. Cambridge, England: Woodhead Publishing Limited, in association with The Textile Institute, 2004.

O'Mahony, Marie, and Sarah Braddock, eds. *Textiles and New Technology*. London: Artemis, 1994.

Tao, Xiaoming, ed. *Smart Fibres, Fabrics and Clothing*. Cambridge, England: Woodhead Publishing Limited, in association with The Textile Institute, 2001.

Vigo, Tyrone, and Albin F. Turback, eds. *Hi-Tech Fibrous Materials: Composites, Biomedical Materials, Protective Clothing, and Geotextiles*. Washington, D.C.: American Chemical Society, 1991.

## COMPOSITES

Bucquoye, Moniek E., ed. *From Bakelite to Composite: Design in New Materials*. Stichting Kunstboek, in association with The Design Museum Ghent, 2002.

*Composites: An Insider's Technical Guide to Corporate America's Activities, 3rd edition, a Turner Moss Company Materials Handbook*. New York: Turner Moss Company, 1998.

## AEROSPACE

Bradley, Edwards C., Ph.D., and Eric A. Westling. *The Space Elevator: A Revolutionary Earth-to-Space Transportation System*. Houston, Tex.: B. C. Edwards, 2003.

Crouch, Tom D. *Aiming for the Stars: the Dreamers and Doers of the Space Age*. Washington, D.C.: Smithsonian Institution Press, 1999.

Crouch, Tom D., and Peter L. Jakab. *The Wright Brothers and the Invention of the Aerial Age*. Washington, D.C.: Smithsonian National Air and Space Museum: National Geographic, 2003.

Kozloski, Lillian D. *U.S. Space Gear: Outfitting the Astronaut*. Washington, D.C.: Smithsonian Institution Press, 1994.

Launius, Roger D. *NASA: A History of the U.S. Civil Space Program*. Malabar, Fla.: Krieger Pub. Co., 1994.

## ARCHITECTURE

Benkers, Adriaan, and Ed Van Hinte. *Lightness: Inevitable Renaissance of Minimum Energy Structure*. Commissioned by the Netherlands Design Institute. Rotterdam: 010 Publishers, 2001.

Frei, Otto, and Bodo Rasch. *Finding Form: Toward an Architecture of the Minimal*. Fellbach, Germany: Axel Menges, 1995.

*FTL Architects: Innovations in Tensile Structures*. Chichester, England: Academy Editions, 1997.

Kronenburg, Robert, ed. *Ephemeral/ Portable Architecture*. London: Academy Editions: Architectural Design, 1998.

Kronenburg, Robert. *Houses in Motion: The Genesis, History and Development of the Portable Building, 2nd edition*. Chichester, West Sussex: Wiley-Academy, 2002.

Koerner, R. M. *Designing with Geosynthetics*, 4th edition. Upper Saddle River, N.J.: Prentice Hall, 1997.

Mollaert, Marijke, Sven Hebbelinck, and Jurga Haase, eds. *The Design of Membrane and Lightweight Structures: From Concept to Execution*. Belgium: VUB Brussels University Press, 2002.

Robbin, Tony. *Engineering a New Architecture*. New Haven, Conn.: Yale University Press, 1996.


## PERIODICALS AND WEB SITES


*American Fabrics and Fashions*. New York: Doric Publishing Co.

*American Institute of Aeronautics and Astronautics*. A publication of the American Institute of Aeronautics and Astronautics Inc. http://www.aiaa.org.

*Fabric Architecture: Design Solutions for the Future*. Bimonthly. Official publication of Professional Awning Manufacturers Association, the Banner, Flag and Graphics Association, and the Lightweight Structures Association.

Fisher, Geoff, and Peter Lennox-Kerr, eds., *Medical Textiles* [newsletter]. Monthly. UK: International Newsletters.

*IFAI: Industrial Fabrics Association International*. http: //www.ifai.com.

*Industrial Fabric Product Review*. Monthly. Industrial Fabrics Association International.

*International Fiber Journal*. Bimonthly. Charlotte, N.C.: Invista Inc., USA. http://www.fiberjournal.com.

*MRS Bulletin*. Pennsylvania: The Materials Research Society. http://www.mrs.org.

*Nonwovens Industry*. Monthly. N.J.: Rodman Publications Inc. http://www.nonwovens industry.com.

*T3 Technical Textile Technology: The Forum for Performance Textiles and Nonwovens*. Quarterly. Charlotte, N.C.: International Media Group Inc.

*Textiles*. Quarterly. Manchester: The Textile Institute.

*Technical Textiles International*. UK: International Newsletter, 2000–2004. http://www.technical-textiles.net.

*Techtextil: International trade fair for technical textile and nonwovens*. USA: Messe Frankfurt Inc. Germany: Messe Frankfurt GmbH, Techtextil Team. http://www.techtextil.com.


## SPORTS


Fung, Walter, and Mike Hardcastle Lancaster. *Textiles in Automotive Engineering*. Pa.: Technomic; Cambridge, England: Woodhead Publishing Limited, 2001.

Harvey, Derek. *Sails: The Way They Work and How to Make Them*. New York: Sheridan House Inc., 1997.

Jenkins, Mike, ed. *Materials Used in Sports Equipment*. Cambridge, England: Woodhead Publishing Limited, 2003.

Olympic Museum Lausanne in collaboration with Ecole polytechnique fédérale de Lausanne, Institut des materiaux. *New Materials for Success!* Lausanne: The Museum, 2002.

O'Mahoney, Marie, and Sarah E. Braddock. *SportsTech: Revolutionary Fabrics, Fashion and Design*. New York: Thames and Hudson, 2002.

# BIOGRAPHIES

Alyssa Becker is a Mellon fellow in textile conservation at the National Museum of the American Indian, and a Smithsonian Center for Materials Research and Education post-graduate fellow at Cooper-Hewitt, National Design Museum. She received a B.F.A. (1997) from the Alberta College of Art and Design, and an M.Sc. (2002) in Textiles and Clothing at the University of Alberta in Edmonton, Alberta, Canada.

Philip Beesley practices architecture and art in Waterloo and Toronto, Canada. His work within the practice of Philip Beesley Architect concentrates on public buildings, textile lattices in architecture, and stage and exhibition design.

Susan Brown is a Curatorial Assistant in the Textiles department at Cooper-Hewitt, National Design Museum. Before joining the Museum four years ago, she worked in textile conservation at the Metropolitan Museum of Art and was a costume designer for theater, opera, and television.

Dr. John W. S. Hearle is a senior consultant to Tension Technology International and a leading authority on the mechanics of fiber assemblies and ropes. He has written, edited, or contributed to over twenty books on science and technology, and has served as a consultant to DuPont, Monsanto, Burlington Industries, Goodrich, Albany International Research Co., WRONZ, and other companies.

Cara McCarty is the Grace L. Brumbaugh and Richard E. Brumbaugh Curator of Decorative Arts and Design at the Saint Louis Art Museum, where she has assembled one of the country's leading collections of modern and contemporary design. She has written catalogs and article, and has organized seminal exhibitions on contemporary topics ranging from architecture to the design of microchips. Transfer technology has been a consistent theme throughout her research.

Matilda McQuaid is Exhibitions Curator and Head of Textiles at Cooper-Hewitt, National Design Museum. She has curated architecture and design exhibitions at both Cooper-Hewitt and The Museum of Modern Art, where she worked until 2002. Currentl, she oversees one of the most important textile collections in the United States, which includes more than thirty thousand textiles produced over twenty-three centuries.

Dr. Patricia Wilson is the principal scientist on the Wearable Electronics Team at Foster-Miller, Inc. Her team has been investigating how to use common conductive textile materials and techniques to build integrated networks for wearable electronic systems. She has won numerous awards for her silk-and-metal-thread embroidery.

Amanda Young is a Museum Specialist in the Space History division of the Smithsonian's National Air and Space Museum. As the official "keeper" of the spacesuits for almost thirty years, her knowledge of the evolution and composition of the spacesuit is unsurpassed.

Sean Hanna is a research engineer at the University College London. He obtained his professional architecture degree from the University of Waterloo and received the American Institute of Architects' gold medal in 1998. He has also studied intelligent systems at the University College London and virtual environments at the Bartlett School of Architecture, and has published in the fields of artificial intelligence, robotics, and optimization of structures and materials. Since 1999, he has worked with Foster and Partners in London, most recently on new techniques of parametric design.

# PHOTO CREDITS

## INTRODUCTION

fig. 1. Courtesy of Invista. Photo Matt Flynn; fig. 2. Courtesy of Invista; fig. 3. Courtesy of B.A.G. Corp.®; fig. 4. Courtesy of Buro Happold Consulting Engineers; figs. 5, 6. Courtesy of BMW WilliamsF1 Team; fig. 7. Photo courtesy of Laurent Spore; figs. 8, 9. Courtesy of Beat Engel Speed Design. Photo Matt Flynn; fig. 10. Courtesy of Alban Geissler; figs. 11, 12. Courtesy of Atair Aerospace Inc.; fig. 13. Courtesy of Smithsonian Institution, National Air and Space Museum; fig. 14. Courtesy of ILC Dover Inc. Photo Matt Flynn; fig. 15. Courtesy of Smithsonian Institution, National Air and Space Museum; figs. 16, 17. Courtesy of SuperFabric® brand material, made by HDM Inc. Photo Matt Flynn; figs. 18, 19. Courtesy of U.S. Army Natick Soldier Center. Photo Matt Flynn; figs. 20, 21. Courtesy of Intersport Fashions West Inc.; fig. 22. Courtesy of Squid:Labs LLC

## STRONGER

fig. 1. Courtesy of North Sails Nevada.; fig. 2. © 2002 Cornell University. All rights reserved. This work was performed by the Jet Propulsion Laboratory, California Institute of Technology, sponsored by the United States Government under Prime Contract #NAS7-1407 between the California Institute of Technology and NASA. Copyright and other rights in the design drawings of the Mars Exploration Rover are held by the California Institute of Technology (Caltech)/Jet Propulsion Laboratory (JPL). Use of the MER design has been provided to Cornell courtesy of NASA, JPL, and Caltech; fig. 3. Courtesy of NASA, Jet Propulsion Laboratory, and ILC Dover Inc.; fig. 5. Courtesy of NASA, Jet Propulsion Laboratory, and Cornell University; fig. 7. Courtesy of Van Dusen Racing Boats; fig. 8. Courtesy of Fothergill Engineered Fabrics. Photo Matt Flynn; figs. 9, 10. Courtesy of CarbonSports GmbH; fig. 11. Courtesy of Foster-Miller Inc.; fig. 12. Photo Matt Flynn; figs. 13, 14. Courtesy of Festo AG & Company KG; figs. 16–18. Courtesy of Acorn Cardiovascular Inc.™; figs. 19, 20. Courtesy of Boston Scientific Company Inc.; fig. 21. Photo Matt Flynn; figs. 22, 23. Courtesy of Ellis Developments Ltd. and Pearsalls Ltd. Photo Matt Flynn; fig. 24. Courtesy of North Sails Nevada. Photo Sharon Green; figs. 25, 26. Courtesy of North Sails Nevada; fig. 27. Courtesy of U.S. Army Natick Soldier Center. Photo Cary Wolinsky/Aurora; fig. 29. Courtesy of Medtronic Heart Valves; fig. 30. Courtesy of Sefar Filtration Fabrics Inc.; fig. 31. Courtesy of Vereinigte Filzfabriken AG. Photo Matt Flynn; fig. 32. Courtesy of Buck Enterprises LLC. Photo Matt Flynn; fig. 33. Courtesy of Marathon Belting Ltd. Photo Matt Flynn; fig. 34. Courtesy of Mammut tec AG; fig. 35. Courtesy of Samson Rope Technologies; fig. 36. Courtesy of Edelrid. Photo Matt Flynn. figs. 37, 38. Courtesy of Marlow Ropes Ltd., U.K.; figs. 39–41. Courtesy of Hills Inc.

## FASTER

fig. 1. Photo Billy Black; figs. 2–11. Courtesy of Goetz Custom Boats; fig. 12. Courtesy of Toshiko Mori Architect; figs. 13, 14. Photos Billy Black; figs. 15, 16. Courtesy of Vanguard Sailboats; figs. 17, 18. Courtesy of Ossur North America; fig. 19. Courtesy of Cor Vos; figs. 20–22. Courtesy of ADA Carbon Wheels; fig. 23. Courtesy of Hoyt U.S.A.; fig. 24. Courtesy of SaddleCo., BoomBang, Dahti Technology, and Quantum Group

## LIGHTER

fig. 1. Courtesy of Foster and Partners; fig. 2. Courtesy of *Scientific American*; fig. 3. Collection of the Estate of Buckminster Fuller, Sebastopol, California; figs. 4, 5. Photo Philip Beesley; fig. 6. Courtesy of Foster and Partners; fig. 7. Courtesy of Front Inc.; fig. 8. Photo Philip Beesley; figs. 9–11. Courtesy of Foster and Partners; figs. 12–15. Courtesy of Testa Architecture and Design; fig. 16. Courtesy of Michael Maltzan Architecture; fig. 17. Courtesy of Vertigo Inc.; figs. 18–20. Courtesy of ILC Dover Inc. Photo Matt Flynn; fig. 21. Courtesy of Maccaferri Canada Ltd.; fig. 22. Courtesy of S. I. Corporation, Performance Fabrics Division/Geosolutions. Photo Matt Flynn; figs. 23–24. Courtesy of Colbond Inc.; fig. 25. Courtesy of Colbond Inc. Photo Matt Flynn; fig. 26. Collection Peter Cook; fig. 27. Photo Philip Beesley; fig. 28. Courtesy of Greenstreak Inc. Photo Matt Flynn; fig. 29. Photo Matt Flynn; figs. 30, 31. Courtesy of Foster-Miller Inc. Photo Robert Bossart; fig. 32. Courtesy of Karl Mayer Textilmaschinenfabrik Obertshausen. Photo Matt Flynn; fig. 33. Courtesy of Foster-Miller Inc.; fig. 34. Courtesy of Foster-Miller Inc. Photo Robert Bossart; figs. 35–37. Courtesy of Institut für Textiltechnik (ITA), University of Aachen, and Institute of Materials Science and Engineering (WKK), Universität Kaiserslautern

## SAFER

fig. 1. Courtesy of NASA, Jet Propulsion Laboratory; fig. 2. Photo Mark Avino; fig. 3. Courtesy of Lion Apparel Inc.; fig. 4. Courtesy of NASA *Spinoff*; fig. 5. Courtesy of Mine Safety Appliances; fig. 6. Courtesy of NASA *Spinoff*; fig. 7. Courtesy © 1994 Robert Reck; figs. 8–10. Courtesy of W. L. Gore & Associates Inc.; fig. 11. Courtesy of 3M, Boeing, and NASA; fig. 12. Courtesy of NASA, Johns Hopkins University Applied Physics Laboratory, and Carnegie Institution of Washington; figs. 13, 14. Courtesy of NASA; fig. 15. Courtesy of NASA, Jet Propulsion Laboratory; fig. 16. Courtesy of NASA, Marshall Space Flight Center; fig. 17. Courtesy of Sefar America Inc. Photo Matt Flynn; figs. 18, 19. Courtesy of Slickbar Products Corporation; fig. 20. Courtesy of Riri USA Inc. Photo Matt Flynn; fig. 21. Courtesy of Palmhive® Technical Textiles Ltd. Photo Matt Flynn; fig. 22. Republished with permission of Globe Newspaper Company, Inc., from a November edition of *The Boston Globe*, © 1995; fig. 23. Courtesy of Shoei Co. Ltd.; fig. 24. Photo Eric Long; figs. 25–35. Photo Mark Avino; fig. 36. Courtesy of NASA; fig. 37. Courtesy of ILC Dover Inc.

## SMARTER

fig. 1. Courtesy of Woven Electronics. Photo Matt Flynn; figs. 2–6. Courtesy of International Fashion Machines Inc.; fig. 7. Courtesy of Philips Design. Photo by Korff & van Mierlo, Eindhoven; fig. 8. Courtesy of Infineon Technologies AG; fig. 9. Courtesy of Laurie Carlson; figs. 10, 11. Courtesy of loop.ph; figs. 12 a–c. Infrastructure prototype courtesy of Herman Miller Creative Office, Milliken Inc., and Osram Opto Semi-Conductors; fig. 13. Courtesy of U.S. Army Natick Soldier Center and Foster-Miller Inc.; fig. 14. Courtesy of Foster-Miller Inc. Photo Robert Bossart; fig. 15. Courtesy of Malden Mills® Industries Inc.; fig. 16. Courtesy of ILC Dover Inc.; fig. 17. Courtesy of ILC Dover Inc. Photo Matt Flynn; fig. 18. Courtesy of Softswitch Ltd.; fig. 19. Courtesy of Burton Snowboards Inc.; fig. 20. Courtesy of Woven Electronics; figs. 21–23. Courtesy of BodyMedia® Inc.; fig. 24. Courtesy of U.S. Army Natick Soldier Center; fig. 25. Courtesy of Nike Inc.; fig. 26. Courtesy of The North Face Inc.; fig. 27. Courtesy of Theodor Preuss GmbH & Co KG. Photo Matt Flynn

# INDEX

Published on the occasion of the exhibition *Extreme Textiles: Designing for High Performance*, organized by Cooper-Hewitt, National Design Museum, Smithsonian Institution, April 8–October 23, 2005.

*Extreme Textiles: Designing for High Performance* is made possible by

TARGET

Generous support is provided by

maharam

Additional funding is provided by The Coby Foundation, Ltd., Stephen McKay, Inc., Furthermore: a program of the J. M. Kaplan Fund, Elise Jaffe + Jeffrey Brown, and Foster-Miller, Inc.

Museum Editor: Chul R. Kim
Editor: Megan Carey
Designer: Tsang Seymour Design, NYC

First published in the United Kingdom in 2005 by
Thames & Hudson Ltd, 181A High Holborn,
London WC1V 7QX
www.thamesandhudson.com

First published in the United States in 2005 by
Princeton Architectural Press
37 East Seventh Street
New York, New York 10003
www.papress.com

In association with
Cooper-Hewitt, National Design Museum,
Smithsonian Institution
2 East 91st Street
New York, New York 10128
www.cooperhewitt.org



British Library Cataloguing-in-Publication Data
A catalogue record for this book is available from the British Library

ISBN-13: 978-0-500-51225-8

ISBN-10: 0-500-51225-6

Printed and bound in China